# SOIL AND WATER QUALITY

## An Agenda for Agriculture

Committee on Long-Range Soil and Water Conservation

Board on Agriculture

National Research Council

NATIONAL ACADEMY PRESS
Washington, D.C. 1993

**NATIONAL ACADEMY PRESS** • 2101 Constitution Avenue • Washington, D.C. 20418

NOTICE: The project that is the subject of this report was approved by the Governing Board of the National Research Council, whose members are drawn from the councils of the National Academy of Sciences, the National Academy of Engineering, and the Institute of Medicine. The members of the committee responsible for the report were chosen for their special competencies and with regard for appropriate balance.

This report has been reviewed by a group other than the authors according to procedures approved by a Report Review Committee consisting of members of the National Academy of Sciences, the National Academy of Engineering, and the Institute of Medicine.

This report has been prepared with funds provided by the U.S. Department of Agriculture, Soil Conservation Service, under agreement number 68-3A75-9-56; the U.S. Environmental Protection Agency, Office of Policy, Planning, and Evaluation under agreement number C X 818573-01-1; and The Joyce Foundation. Dissemination was supported in part by Pioneer Hi-Bred International, Inc., The Joyce Foundation, The W. K. Kellogg Foundation, the U.S. Department of Agriculture, Soil Conservation Service, and the Environmental Protection Agency, Office of Policy, Planning and Evaluation.

**Library of Congress Cataloging-in-Publication Data**
Soil and water quality: an agenda for agriculture/Committee on Long-Range Soil and
    Water Conservation, Board on Agriculture, National Research Council.
        p.    cm.
    Includes bibliographical references and index.
    ISBN 0-309-04933-4
    1. Soil management—United States.   2. Soils—United States—Quality.   3. Water
quality management—United States.   4. Sediment control—United States.   5. Agricultural ecology—United States.   I. National Research Council (U.S.). Committee on
Long-Range Soil and Water Conservation.
S599.A1S62    1993
333.76′0973—dc20                                                             93-35470
                                                                                 CIP

Any opinions, findings, conclusions, or recommendations expressed in this publication are those of the author(s) and do not necessarily reflect the view of the organizations or agencies that provided support for this project.

Printed in the United States of America

# Committee on Long-Range Soil and Water Conservation

SANDRA S. BATIE, *Chair*, Michigan State University[*]
J. WENDELL GILLIAM, North Carolina State University
PETER M. GROFFMAN, Institute of Ecosystem Studies, Millbrook, New York
GEORGE R. HALLBERG, Iowa Department of Natural Resources
NEIL D. HAMILTON, Drake University Law School
WILLIAM E. LARSON, University of Minnesota (Retired)
LINDA K. LEE, University of Connecticut
PETER J. NOWAK, University of Wisconsin
KENNETH G. RENARD, Agricultural Research Service, U.S. Department of Agriculture, Tucson, Arizona
RICHARD E. ROMINGER, A. H. Rominger and Sons, Winters, California[†]
B. A. STEWART, Agricultural Research Service, U.S. Department of Agriculture
KENNETH K. TANJI, University of California
JAN VAN SCHILFGAARDE, Agricultural Research Service, U.S. Department of Agriculture
R. J. WAGENET, Cornell University
DOUGLAS L. YOUNG, Washington State University

Staff

CRAIG COX, *Project Director*
JOSEPH GAGNIER, *Project Associate*
JANET OVERTON, *Editor*
CRISTELLYN BANKS, *Senior Secretary and Project Assistant*

---

[*]Virginia Polytechnic Institute & State University until September 1993.
[†]Sworn in as Deputy Secretary, U.S. Department of Agriculture, May 12, 1993.

*iii*

*v*

The National Academy of Sciences is a private, nonprofit, self-perpetuating society of distinguished scholars engaged in scientific and engineering research, dedicated to the furtherance of science and technology and to their use for the general welfare. Upon the authority of the charter granted to it by the Congress in 1863, the Academy has a mandate that requires it to advise the federal government on scientific and technical matters. Dr. Bruce M. Alberts is president of the National Academy of Sciences.

The National Academy of Engineering was established in 1964, under the charter of the National Academy of Sciences, as a parallel organization of outstanding engineers. It is autonomous in its administration and in the selection of its members, sharing with the National Academy of Sciences the responsibility for advising the federal government. The National Academy of Engineering also sponsors engineering programs aimed at meeting national needs, encourages education and research, and recognizes the superior achievements of engineers. Dr. Robert M. White is president of the National Academy of Engineering.

The Institute of Medicine was established in 1970 by the National Academy of Sciences to secure the services of eminent members of appropriate professions in the examination of policy matters pertaining to the health of the public. The Institute acts under the responsibility given to the National Academy of Sciences by its congressional charter to be an adviser to the federal government and, upon its own initiative, to identify issues of medical care, research, and education. Dr. Kenneth I. Shine is president of the Institute of Medicine.

The National Research Council was organized by the National Academy of Sciences in 1916 to associate the broad community of science and technology with the Academy's purposes of furthering knowledge and advising the federal government. Functioning in accordance with general policies determined by the Academy, the Council has become the principal operating agency of both the National Academy of Sciences and the National Academy of Engineering in providing services to the government, the public, and the scientific and engineering communities. The Council is administered jointly by both Academies and the Institute of Medicine. Dr. Bruce M. Alberts and Dr. Robert M. White are chairman and vice-chairman, respectively, of the National Research Council.

# Preface

The list of environmental problems on the agricultural agenda has grown in the past 15 years. The long-standing concerns about soil erosion and sedimentation have been supplemented with new concerns about soil compaction, salinization, and loss of soil organic matter. The transfer of nitrates, phosphorus, pesticides, and salts from farming systems to surface water and groundwater has also become more important.

Efforts to address the larger complex of environmental problems has been hampered by concerns about trade-offs. For example, best-management practices designed to reduce soil loss are now scrutinized for their role in increasing the leaching of nitrates and pesticides to groundwater. Other trade-offs arise between efforts to improve agriculture's environmental performance and efforts to reduce costs of production and maintain U.S. agriculture's share of world markets.

In 1989 the Board on Agriculture of the National Research Council was asked to convene a committee to assess the science, technical tools, and policies needed to protect soil and water quality while providing for the production of food and fiber from U.S. croplands. More specifically, the committee was asked to

- investigate the threats to soil resources and recommend criteria to guide soil management;
- analyze fate and transport of agricultural chemicals to identify changes in farming systems required to improve water quality;
- identify remedial approaches that minimize trade-offs between

improving soil or water quality, surface water or groundwater quality, or between different pollutants; and

• recommend policy and program options to improve long-term conservation of soil and water quality.

The committee focused primarily on water quality, rather than water quantity, problems and on croplands rather than on forestlands or rangelands. The committee presents its work in two parts. Part One contains Chapters 1 through 4 and presents the committee's synthesis of the technical, economic and policy issues relating to soil and water quality. In Part Two, Chapters 5 through 12 describe in greater detail the scientific and technical knowledge on which the chapters in Part One are based.

During its deliberative process, the committee first analyzed the physical, chemical, and biological processes that determine farming systems' impact on soil and water quality. The committee analyzed the effects of farming practices on soil, the role of soil in mediating the effect of farming systems on water quality, as well as the processes leading to the loss of nitrogen, phosphorus, pesticides, sediment, salts, and trace elements from farming systems. The committee studied the special problems posed by managing animal wastes and examined the important influence of the landscape in shaping the effects of farming systems on soil and water quality. The results of these analyses are presented in Chapters 5 through 12 of this report. The Appendix describes the methods used by the committee to estimate national, regional, and state nutrient budgets.

Chapter 1 reviews the status of soil and water quality and discusses how and why emphasis has changed over the years from simply soil erosion and sedimentation to include soil degradation and water pollution. The presence of nutrients, pesticides, salts, and trace elements in crops, soil, and drinking water has created new problems that require new solutions. The search for solutions includes recognizing the importance of state and local policies as well as the needs and characteristics of the agricultural sector in efforts to improve soil and water quality.

Based on the understanding gained by analyzing the processes that govern the interaction of farming systems and the environment, the committee identified promising opportunities for managing those processes in ways that protect soil and water quality and are profitable for the producer. The committee's analysis identified four major objectives for the management of soil and water resources:

• conserve and enhance soil quality as a fundamental first step to environmental improvement;

• increase nutrient, pesticide, and irrigation use efficiencies in farming systems;

- increase the resistance of farming systems to erosion and runoff; and
- make greater use of field and landscape buffer zones.

These objectives and the technologies available to implement them in agricultural production are presented in Chapter 2.

The task then became to develop strategies to implement those objectives and to identify the changes in concepts, technologies, and policies that might be needed. The farming system concept was central to the development of this report, and the need for a farming system approach at the farm enterprise, regional, and national levels underlies all of the recommendations that the committee developed. The advantages of using a farming systems approach to direct and target soil and water quality programs are presented in Chapter 3.

Ultimately, to achieve long-term improvements in soil and water quality, the behaviors of some producers must be changed. A constant challenge in preparing this report was the attempt to link the social and economic factors that determine producer behavior with the physical, chemical, and biological factors that determine the effects of that behavior on soil and water quality. The committee used the understanding gained from studying these links to recommend a combination of policy and program reforms that will be needed to achieve long-term improvements in soil and water quality. The policy and program reforms recommended by the committee are discussed in Chapter 4.

The debate over national policy to protect soil and water quality has intensified during the course of the committee's deliberation. The 1990 Food, Agriculture, Conservation and Trade Act and the 1990 Coastal Zone Act Reauthorization Amendments created new programs and new authorities that can be used to implement many of the committee's recommendations. Reauthorization of the Clean Water Act and the prospect of a new farm bill in 1995 provide more opportunities to move ahead with an agenda to protect soil and water quality.

A great deal of progress could be made—even in the absence of new legislation—by integrating the multitude of federal, state, and local programs that are already addressing pieces of the soil and water quality problem. The opportunities to make current programs more effective are great and, in many cases, the authorities needed are already provided by legislation. It is the committee's hope that this report will help provide a framework to facilitate the integration of existing and new programs.

SANDRA S. BATIE, *Chair*
Committee on Long-Range Soil
and Water Conservation Policy

# Acknowledgments

A report of this magnitude represents the combined efforts of many individuals from a variety of backgrounds. The committee thanks all those who contributed their ideas and experiences in technical and policy areas. During the course of its deliberations, the committee sought advice and special assistance. Among those who gave generously of their time were Raymond R. Allmaras, University of Minnesota; Ramon Aragues, Agricultural Research Service, Zaragosa, Spain; Peter E. Avers, U.S. Department of Agriculture, Forest Service at Washington, D.C.; Russell R. Bruce, U.S. Department of Agriculture, Agricultural Research Service at Watkinsville, Georgia; H. H. Cheng, University of Minnesota; C. V. Cole, U.S. Department of Agriculture, Agricultural Research Service at Colorado State University; Cornelia Butler Flora, Virginia Polytechnic Institute and State University; George R. Foster, U.S. Department of Agriculture, Agricultural Research Service at Oxford, Mississippi; Robert Grossman, U.S. Department of Agriculture, Soil Conservation Service at Lincoln, Nebraska; Benjamin F. Hajek, Auburn University; Roger Hanson, North Carolina State University; Fawzi Karajeh, University of California at Davis; Jean A. Molina, University of Minnesota; Gary B. Muckel, U.S. Department of Agriculture, Soil Conservation Service at Lincoln, Nebraska; Mathias J. Romkens, U.S. Department of Agriculture, Agricultural Research Service at Oxford, Mississippi; C. Ford Runge, University of Minnesota; David L. Schertz, U.S. Department of Agriculture, Soil Conservation Service at Washington, D.C.; Steven J. Taff, University of Minnesota; and Ward B. Voorhees, U.S.

Department of Agriculture, Agricultural Research Service at Morris, Minnesota.

The committee is particularly grateful to Christopher D. Koss, President of the J. N. "Ding" Darling Foundation, Key Biscayne, Florida, for his generous assistance in providing the four Ding Darling cartoons that help illustrate this report.

The committee also acknowledges the special efforts of Amy Gorena, who served as a senior project assistant during the early stages of the study; Michael Hayes, who provided editorial expertise during development of the manuscript; and Rolla Chuang, who assisted as a student intern sponsored by the Midwest Universities Consortium for International Activities, Inc.

# Contents

# Tables and Figures

## TABLES

## FIGURES

# SOIL AND WATER QUALITY

## An Agenda for Agriculture

# Executive Summary

The U.S. economy and the livelihood of citizens depend on soil, water, air, plants, and animals, and natural and managed ecosystems as fundamental resources. Agricultural production, by its very nature, has pervasive effects on all these resources. Agricultural production takes place within farming systems. Those systems are defined by the pattern and sequence of crops; the management decisions regarding the inputs and production practices used; the management skills, education, and objectives of the producer; the quality of the soil and water; and the nature of the landscape and the ecosystems within which production takes place. This report focuses on the opportunities to manage farming systems in ways that protect two of these fundamental resources—soil and water.

## BASIC CONCEPTS

The committee's deliberations were based on three basic concepts of soil and water resource management: (1) the fundamental importance of the soil and of the links between soil quality and water pollution, (2) the importance of preventing rather than mitigating water pollution, and (3) the need to sustain profitable and productive farming systems to provide the food and fiber society demands.

### Soil Quality

*Protecting soil quality, like protecting air and water quality, should be a fundamental goal of national environmental policy.*

The quality of a soil depends on attributes such as the soil's texture,

*1*

The Manhantango Creek watershed near Klingerstown, Pennsylvania. The combination of farm management, land use, soil properties, and hydrogeology largely determine the vulnerability of surface water and groundwater to contamination by agricultural waste. Credit: Agricultural Research Service, USDA.

depth, permeability, biological activity, capacity to store water and nutrients, and the amount of organic matter contained in the soil. Soils are living, dynamic systems that are the interface between agriculture and the environment. High-quality soils promote the growth of crops and make farming systems more productive. High-quality soils also prevent water pollution by resisting erosion, absorbing and partitioning rainfall, and degrading or immobilizing agricultural chemicals, wastes, or other potential pollutants. The quality of some U.S. soils, however, is degenerating because of erosion, compaction, salinization, loss of biological activity, and other factors. The full extent of soil degradation in the United States is unknown, but current estimates of damage from erosion understate the true extent of soil degradation.

The 1990 Clean Air Act (PL 101-549) and the Clean Water Act (PL 100-104) give national recognition to the fundamental importance of air and water resources. Soil resources are equally important components of environmental quality, and national policies to protect soil resources should be based on the fundamental functions that soils perform in natural and agroecosystems.

## Pollution Prevention

*Preventing surface water and groundwater pollution by reducing the sources of contamination by nutrients, pesticides, sediments, salts, and trace elements should be the goal of national policies.*

Treatment of drinking water to remove nitrates and pesticides is expensive and in some cases ineffective. The disruption of aquatic ecosystems caused by excessive nutrients, pesticides, sediments, salts, and trace elements may be difficult to reverse at a reasonable cost or in a reasonable length of time. Preventing pollution by changing farming practices, rather than treating problems after they have occurred, should be the primary approach to solving water pollution problems caused by farming practices.

The goal of pollution prevention should be to reduce the total mass of nutrients, pesticides, salts, and trace elements that are lost to the environment. Solutions that reduce loadings of one pollutant by increasing the loadings of a different pollutant or that reduce loadings to surface water by increasing loadings to groundwater are not likely to be acceptable or effective in the long term.

## Profitability and Productivity

*National policies should take advantage of opportunities to protect soil and water quality while sustaining profitable production of food and fiber.*

Policymakers face a dilemma. Society needs and wants the food and fiber that agriculture produces. Producing that food and fiber inescapably alters the environment, and some effect on soil and water quality is inevitable. Unfortunately, comprehensive national data on soil degradation and water pollution caused by farming practices are often lacking. Available data are sufficient to cause concern but often not sufficient to confidently determine priorities.

Given this dilemma, national policy should, in the short term, take advantage of opportunities to refine the management of farming systems in ways that protect soil and water with minimal or even positive effects on profitability. The committee found a diverse set of technologies and management methods that promise to improve soil and water quality and, at the same time, maintain or even enhance profit. The magnitude and nature of these opportunities, however, vary from region to region, crop to crop, and farm enterprise to farm enterprise.

Preventing soil degradation and water pollution in the present may deter forcing solutions that impose serious costs on producers in the future. Time, however, may run out. In some regions, soil degradation and water pollution may already be serious enough that solutions will entail economic losses to the agricultural sector. Concerted action now is needed to prevent the list of such regions from getting longer.

## THE AGENDA

The committee defined four broad opportunities that hold the most promise of preventing soil degradation and water pollution while sustaining a profitable agricultural sector.

*National policy should seek to (1) conserve and enhance soil quality as a fundamental first step to environmental improvement; (2) increase nutrient, pesticide, and irrigation use efficiencies in farming systems; (3) increase the resistance of farming systems to erosion and runoff; and (4) make greater use of field and landscape buffer zones.*

These four approaches are interrelated. Emphasis on one objective to the exclusion of the others may exacerbate one environmental problem while solving another. Reducing runoff, for example, without improving nutrient management may reduce the amount of nutrients reaching surface water but increase the amount leaching to groundwater. The balance of emphasis between objectives may necessarily change from one region to another to best address local conditions. For example, in some cases, shifting emphasis to creating buffer zones, as the cost of refining input management increases, may be the least expensive way for producers and taxpayers to prevent pollution. Ultimately, the decision to emphasize one approach over another is, at least implicitly, a political and social judgement on the importance of protecting particular soils or water bodies (see Chapter 2).

### Enhancing Soil Quality

National policies to protect soil resources are too narrowly focused on (1) controlling erosion and (2) conserving soil productivity. Erosion is not the only, and in some cases not the most important, threat to soil quality. Salinization and compaction are important and often irreversible processes of soil degradation. More important, erosion, salinization, compaction, acidification, and loss of biological activity interact to accelerate soil degradation. Comprehensive policies that address all processes of soil degradation are needed.

"Sometime those boys should get together" (July 1, 1947). Credit: Courtesy of the J. N. "Ding" Darling Foundation.

Soil productivity is not the only, and in some regions may not be the most important, reason to protect soil resources. Soil and water quality are inherently linked. Preventing water pollution by nutrients, pesticides, salts, sediments, or other pollutants will be difficult and more expensive if soil degradation is not controlled. Protecting soil quality alone, however, will not prevent water pollution unless other elements of the farming system are addressed (see Chapter 5).

## Efficient Use of Inputs

Agricultural production inevitably generates a certain mass of residual products including nutrients, sediments, pesticides, salts, and trace elements that can become pollutants. The emphasis of traditional conservation programs has been to prevent pollutants from leaving the farming system by reducing erosion and runoff. New programs are needed that reduce the amount of potential pollutants produced as a by-product of farming by improving the way nutrients, pesticides, and irrigation water are used.

Increasing the efficiency of nutrient, pesticide, and irrigation water use reduces the total residual mass of nitrogen, phosphorus, pesticides, salts, and trace elements that can become pollutants. In some cases, efficiency can be achieved by using fewer nutrients or pesticides or both or less irrigation water to produce the same yield; in other cases, efficiency can be achieved by increasing the yield while using the same mass of inputs. Many technologies and management methods are already available that promise to dramatically increase the efficiency of nutrient, pesticide, and irrigation water use; but they need to be more widely implemented. In many cases, the cost of achieving greater efficiency in input use is offset by reduced costs of production. In those regions and farming systems where these economic incentives are significant, substantial and rapid progress toward preventing water quality problems may be possible. (See Chapters 6, 7, 8, 10, and 11 for discussions of nitrogen, phosphorus, pesticide, irrigation, and manure management.)

## Resisting Erosion and Runoff

Conservation tillage and residue management systems are well-understood and effective means of reducing erosion and runoff. A great diversity of tillage and residue management systems is available to producers. Many of these systems result in dramatic decreases in erosion and runoff from farming systems and from agricultural watersheds. The major opportunity to improve the effectiveness of these systems is to increase their use on lands that are most vulnerable to soil quality degradation or that most contribute to water pollution. In some regions the applicability of these systems may be limited, however, because of unfavorable physical or economic factors.

Water runoff from cropped fields carries with it soil, nutrients, and pesticides that may pollute surface water. Protecting and improving soil quality helps reduce the amount and erosive force of runoff water by increasing the amount of rainfall that percolates into the soil. Better management of nutrients and pesticides can reduce the amount lost in runoff water. Credit: U.S. Department of Agriculture.

Much of the damage from erosion and runoff can occur during storms that occur infrequently. Incorporating the probability of storm events into the design of farming systems should help identify approaches that combine residue management with changes in cropping systems to provide more protection to the soil during periods when storms are likely. Current computer simulation capacities coupled with available climatic data should be used to identify opportunities to design farming systems that can resist damage from storm events of various duration and intensities.

### Field and Landscape Buffer Zones

Field-by-field efforts to conserve soil quality, improve input use efficiency, and increase resistance to erosion and runoff will not be adequate to protect soil and water quality in regions where overland and subsurface movements of nutrients, pesticides, salts, and sediment are pervasive. Buffer zones to intercept or immobilize pollutants and reduce the amount and energy of runoff need to be created. Existing buffer zones need to be protected in such regions to prevent soil degradation and water pollution. New and existing buffer zones need to be connected across fields and farm boundaries. Buffer zones can include natural riparian corridor vegetation (vegetation along waterways); simple, but strategically placed, grass strips; or sophisticated artificial wetlands. Federal, state, and local government programs to protect existing riparian vegetation, whether bordering major streams or small tributaries, lakes, or wetlands, should be promoted. The creation or protection of field or landscape buffer zones, however, should augment efforts to improve farming systems. They should not be substitutes for such efforts. (See Chapter 12 for a discussion of buffer zones.)

## IMPLEMENTING THE AGENDA

A range of technical opportunities to improve the management of farming systems exist. In many cases, better use of available technologies, understanding, and information would result in immediate gains in preventing soil degradation and water pollution; many producers have already made substantial improvements in their farming systems. There are, however, important obstacles to achieving more widespread use of the new technologies and management methods needed to prevent soil degradation and water pollution. Substantial changes must occur in the way current programs are implemented before it will be possible to take advantage of the technical opportunities to improve farming systems (see Chapter 3).

### Problem Areas, Problem Farms

The committee strongly emphasizes the importance of targeting—that is, attempting to direct technical assistance, educational effort, financial resources, or regulations at those regions where soil degradation and water pollution are most severe. It is also important to target those farm enterprises that cause a disproportionate amount of soil and water quality problems. The inability or unwillingness to target policies, whether voluntary or nonvoluntary, at problem areas and problem farms is a major obstacle to preventing soil degradation and water pollution.

#### Problem Areas

*The Secretary of the U.S. Department of Agriculture (USDA), the Administrator of the Environmental Protection Agency (EPA), and the U.S. Congress should undertake a coordinated effort to identify regions or watersheds that should be highest priority for federal, state, and local programs to improve soil and water quality.*

Federal, state, and local governments have, at least implicitly, identified priority areas for various soil and water quality problems and for various programs to improve soil and water quality. These priority areas, however, have been established by different agencies, for different purposes, and on different measures of soil or water quality. The problem areas that have already been identified in current programs or legislation need to be categorized to create a clear set of national priorities that can be used by USDA, EPA, and Congress to direct programs and resources to areas consistently defined as problem areas.

### Problem Farms

*Soil and water quality programs should be targeted at problem farms that, because of their location, production practices, or management, have greater potential to cause soil degradation or water pollution.*

Although systematic data on production practices, input use, and management systems are scarce, available data indicate that some farm enterprises cause more soil and water problems than others. Targeting programs, whether voluntary or nonvoluntary, at problem farms is an opportunity to reduce the cost and increase the effectiveness of soil and water quality programs. The Secretary of USDA and the Administrator of the EPA should initiate a multiagency effort to assemble available data on production practices and enterprise characteristics to identify problem farms within problem areas.

### Farming Systems

Encouraging or requiring the adoption of single-objective best-management practices is not a sufficient basis for soil and water quality programs at the farm level. Inherent links exist among the components of a farming system and the larger landscape. Adoption of a tillage system that increases soil cover to reduce erosion, for example, may require changes in the methods, timing, and amounts of nutrients and pesticides applied. Failure to recognize and manage these links increases the cost, slows the rate of adoption, and decreases the effectiveness of new technologies or management methods.

*The development and implementation of approved integrated farming system plans should be the basis for delivery of education and technical assistance, should be the condition under which producers become eligible for financial assistance, and should be the basis for determining whether producers are complying with soil and water quality programs.*

Integrated farming system plans should become the basis of federal, state, and local soil and water quality programs. Receipt of cost-sharing or other financial assistance should depend on developing and implementing integrated farming system plans, rather than on implementing single-objective best-management practices. In the long term, implementation of an integrated farming system plan should be required of producers, regardless of their participation in federal farm programs, in regions where soil degradation and water pollution caused by farming practices are severe.

## Better Tools and Information

Substantial progress can be made toward improving the management of farming systems using available technology and information. Much greater progress could be made if producers had better tools and information to refine the management of their farming systems.

### Better Management Tools

*Developing and implementing cost-effective diagnostic and monitoring methods to refine the management of soils, nutrients, pesticides, and irrigation water should be a high priority of USDA and EPA research and technology transfer programs.*

Progress has been made in developing technologies to match farming practices to variations in soil quality, to monitor and assess the nutrient and water status of crop plants, and to monitor and determine economic levels of pest problems. It is important to accelerate the development of the diagnostic and monitoring tools producers need to refine their management of soil, nutrients, pest control, and irrigation water. The degree to which the management of farming systems can be improved will be determined, in large part, by the management tools available to producers and how well they are used.

### Better Information

*Keeping and using records of production practices, crop and livestock yields, and other elements of the farm management system should be a fundamental component of programs to improve the management of farming systems.*

The systems established to manage the flow and analysis of information are as important as the specific production practices specified in the plan. Policies that encourage or mandate the collection and use of information by the producer may prove more effective than encouraging or mandating the use of specific farming practices. The information needed to manage a farm operation to maximize profit, if properly organized, complements the information needed to improve soil and water quality. The collection and synthesis of this information can point out ways to improve both profitability and soil and water quality. Record keeping should be mandatory when integrated farming system plans are the basis for granting financial assistance. It should also be mandatory when integrated plans are the basis for ensuring compliance with soil or water quality programs.

### New Cropping Systems

*Research and development of economically viable cropping systems that incorporate cover crops, multiple crops, and other innovations should be accelerated to meet long-term soil and water quality goals.*

Innovative cropping systems use cover crops, companion crops, strip-cropping, reduced reliance on fallow, or other changes in the timing or sequence of crops. Such systems can be designed to increase soil cover; reduce insect, disease, and weed problems; utilize excess nutrients; and control runoff and leaching from farming systems. These innovations in cropping systems may prove to be the most effective way to protect soil and water quality while sustaining profitable food and fiber production. Guiding the research to develop new cropping systems requires a long-term perspective and a vigorous imagination. Existing cropping systems have little resemblance to the systems common 75 years ago. It is reasonable to expect that future systems will be equally different from current systems.

### Criteria and Standards

*USDA and EPA should initiate an integrated research effort to develop quantifiable standards that can be used to evaluate the management of farming systems.*

Current understanding of the effect of farming systems on soil and water quality is generally sufficient to identify the best available production practices or management systems; it is not, however, sufficient for making quantitative estimates of how much soil and water quality will improve as a result of the use of alternative practices or management methods. In the short term, the Secretary of USDA and the Administrator of EPA should convene an interagency task force to develop

To prepare the soil for planting, this mulch tiller is designed to penetrate mulch cover. The mulch protects the soil from erosion and supplies organic matter to the soil. Credit: Deere & Company.

standards that can be used to implement and evaluate integrated farming system plans. Clear standards will increase the confidence that soil and water quality will be improved and provide a basis for determining whether plans are being adequately implemented. In the long term, however, the inability to provide more quantitative predictions of the effect of changes in farming systems on soil and water quality will be a serious constraint to efforts to meet soil and water quality goals. There is an urgent need to develop the scientific capacity to provide producers, policymakers, and program managers with more rigorous methods to determine how much improvement in the management of farming systems is needed to meet specific soil and water quality goals.

## INFLUENCING PRODUCERS' DECISIONS

Targeting programs at problem areas and farms, basing programs on a farming system rather than a best-management practice approach, and filling gaps in technology and information are all important steps toward preventing soil degradation and water pollution. Federal, state, and local programs could be made much more effective if these three steps were taken. Ultimately it is the millions of management decisions producer's make each year that determine the effect of farming systems on soil and water quality. The role of national policy should be to create incentives that influence the information and technologies that producers use to manage their farming systems.

Many factors influence those choices including market prices for inputs and products, the cost of new technologies, the labor and capital available to the producer, agricultural policy, environmental regulations, and the goals of the individual producer or enterprise. The inadequacy of empirical data and predictive models of producer behavior and the diversity of enterprises that make up the agricultural sector make it difficult to pinpoint the precise effect of alternative policies on the behavior of producers. General understanding of the factors that influence producers' decisions can guide the development of national policies to induce producers to change their farming systems.

### Barriers Imposed by Price and Supply Control

*Federal agricultural price support and supply control programs should be reformed to increase the flexibility participants have to diversify their cropping systems.*

The incentives created to grow only program crops and protect base acreage are barriers to the adoption of more diverse cropping systems to prevent soil degradation and water pollution. The 1985 Food Security Act (PL 99-198) and the 1990 Food, Agriculture, Conservation and Trade Act (PL 101-624) have reduced these barriers by freezing established yields at 1986 levels, applying severe constraints on the expansion of base acres, and increasing the flexibility to plant a variety of crops on base acres. Continued reform along these lines will help to remove barriers to adoption of farming systems that protect soil and water quality and permit time for farmers to adjust to new incentives.

## Soil and Water Quality as Policy Objectives

*Long-term protection of soil and water quality should be based on policies and programs that are independent of price support, supply control, or income support mechanisms; policies are needed that target problem areas and problem farms, regardless of participation in federal commodity support programs.*

Research suggests that price support and supply control programs exacerbate soil degradation and water pollution. These programs, however, are not the cause of those problems—soil degradation and water pollution problems would remain even if these programs were eliminated. Incremental changes in conventional agricultural policies will most likely not result in major changes in farming practices and will likely result in only modest gains in soil and water quality.

In agricultural policy, environmental objectives have traditionally been closely linked with income support and supply control objectives. Since 1985, the traditional priority given to income support and supply control has been reversed on highly erodible cropland through programs such as Conservation Compliance and Sodbuster that make receipt of federal farm program benefits conditional on adoption of soil conservation measures. These programs should be fully implemented as an important step toward preventing soil degradation and water pollution.

Long-term protection of soil and water quality should not be an adjunct of income or price support policies. Price and income stability have been and may remain important objectives of national agricultural policy. Protecting soil and water quality will require gearing incentives and penalties toward problem farms in problem areas. Producers in problem areas may be different than those requiring income support and may be producing commodities not subject to price or supply control.

## Policy Instruments

Producers decide to use new information and technologies for different reasons. There is no single solution to preventing soil degradation and water pollution. A range of policy instruments—from purely voluntary to compulsory—will be required. Integration of policy instruments to create consistent and lasting incentives for producers to use new information and technologies is essential.

### Research Applications

*Two types of research should be high priorities for USDA and EPA research programs: (1) research directed at identifying the nature and magnitude of factors influencing producers' management of cropping and livestock production systems and (2) research leading to the development and implementation of new technologies, cropping systems, and methods to manage farming systems that are profitable and protect soil and water quality.*

The application of science and technology to agricultural production has had a revolutionary effect on productivity. Scientific understanding coupled with improvements in production and information technology present agriculture a second opportunity to revolutionize production to meet the twin goals of profitability and environmental compatibility. In many localities, this second revolution is already under way as producers, researchers, and educators develop farming systems to solve local soil and water quality problems.

New technologies and management methods, however, are only successful if they can be efficiently used by producers. New programs are only effective if they are based on an understanding of the factors that influence producers' decisions to change their farming systems. Economic and social research should be a fundamental component of the development of new technology and new policies.

### Technical Assistance

*Aggressive public sector programs, based on modern marketing methods, are needed.*

The voluntary approach to change based on the provision of technical assistance has achieved substantial improvements in farming practices, particularly when there have been opportunities to improve environmental and financial performances simultaneously. New and more targeted approaches are needed, however, rather than wider use of the current approaches. Modern marketing methods should be used to tailor technical assistance and educational programs to target audiences.

*Mechanisms should be developed to augment public sector efforts to deliver technical assistance with nonpublic sector channels and to certify the quality of technical assistance provided through these channels.*

Crop-soil consultants, dealers who sell agricultural inputs, soil testing laboratories, farmer-to-farmer networks, and nonprofit organizations are increasingly important sources of information for producers. In many cases, these private sources of information have become more important direct sources of advice and recommendations than public sources. Soil and water quality programs need to take advantage of the capacity of the private and nonprofit sectors to deliver information and education to producers. The potential to accelerate the delivery of technical assistance and information is great if methods can be developed to certify the quality of the technical assistance provided through these channels.

*Research should be directed at the design of market-based incentives to protect soil and water quality.*

Market-based incentives are being increasingly explored as alternative to strict command and control approaches to environmental regulation. Economic incentives such as tradable pollutant permits, taxes, or fees promise, at least conceptually, to reduce the cost and increase the efficiency of protecting soil and water resources. Marketable permits are already used in the Clean Air Act and mechanisms to trade water rights are being used to allocate water between agricultural and urban users.

Many questions remain to be answered before economic incentives to address agricultural soil and water quality problems could be considered feasible alternatives to current approaches. The flexibility and potential efficiency of market-based incentives suggest that analysis of the feasibility of using market-based incentives should be accelerated.

### Long-Term Easements

*A program to purchase selective use rights from producers through long-term easements should be developed to protect environmentally sensitive lands.*

Some croplands, because of their soils, landscape position, or hydrogeological setting, cannot be profitably farmed without causing soil degradation or water pollution. Other lands, if managed as buffer zones or wetlands rather than as croplands, could help improve soil and water quality. Long-term easements are an effective way to encourage producers to change the land use on these environmentally sensitive lands. The specific set of rights purchased as part of long-term easements should depend on the environmental problem that needs to be solved and the farming system in use.

### Nonvoluntary Change

*Nonvoluntary approaches may be needed in problem areas where soil and water quality degradation is severe and where there are problem farms unacceptably slow in implementing improved farming systems.*

Although the opportunities to accelerate voluntary adoption of improved farming systems are great when such approaches also lead to increased profits, reliance on voluntary change alone may not achieve the improvements in soil and water quality increasingly demanded by the public. Financially optimal improvements in farming systems also may not be sufficient to solve soil and water quality problems. In some watersheds, refinements that impose real costs on producers may be required to meet soil and water quality goals. Voluntary approaches that achieve general improvements in farming systems may not be enough if problem farms fail to respond. Nonvoluntary approaches will be needed to provide more permanent protection when commodity prices are high, damage to soil and water quality is severe, and problem farm owners or managers resist voluntary change.

## Rights and Responsibilities

*The legal responsibilities of landowners and land users to manage land in ways that do not degrade soil and water quality should be clarified in state and federal laws.*

The philosophy that it is the responsibility of landowners and land users, as stewards of the land, to protect soil and water quality is a powerful ideal that is reflected in many of the traditional approaches to soil and water protection in U.S. policy. The ideal has been promoted through education, financial incentives, ethical imperatives, or legal mandates, and many landowners and land users manage their lands in ways that prevent soil degradation and water pollution. The lack of clarity and consistency in the legal definition of the responsibilities as well as rights of landowners and land users has impeded long-term comprehensive efforts in which publicly funded soil and water quality gains are made permanent. A policy that clearly establishes the responsibilities of landowners and land users—to manage their lands in ways that protect soil and water quality—would provide a consistent and uniform basis for implementing soil and water quality protection efforts on a permanent basis.

# PART ONE

# 1

# Soil and Water Quality: New Problems, New Solutions

Since 1970, agricultural policymakers have been confronted with a new and vexing set of problems. Water quality problems resulting from the presence of nutrients, pesticides, salts, and trace elements have been added to an historical concern for soil erosion and sedimentation. Economic problems in the 1980s intensified concern about the loss of family farms and rural development issues. Maintaining the ability of U.S. agriculture to compete in international markets became a central tenet of agricultural policy, and agriculture became a central issue in international trade talks (e.g., General Agreement on Tariffs and Trade). At the same time profound structural changes were occurring in the agricultural sector and new technologies were changing the face of agricultural production. The search for solutions to these different but related problems has dominated debate over agricultural policy.

## SOIL AND WATER QUALITY PROBLEMS

Soil and water quality problems caused by agricultural production practices are receiving increased national attention and are now perceived by society as environmental problems comparable to other national environmental problems such as air quality and the release of toxic pollutants from industrial sources.

Severe soil degradation from erosion, compaction, or salinization can destroy the productive capacity of the soil and exacerbate water pollution from sediment and agricultural chemicals. Sediments from eroded croplands interfere with the use of waterbodies for transportation; threaten investments made in dams, locks, reservoirs, and other developments; and degrade aquatic ecosystems. Nutrients accelerate the rate of eutrophication of lakes, streams, and estuaries; and nitrogen in the form of nitrates can cause health problems if ingested by humans in drinking water. Pesticides in drinking water can become a human health concern and have been suggested to disrupt aquatic ecosystems. Salts can be toxic at high enough levels and can seriously reduce the uses to which water can be put. In some areas, toxic trace elements in irrigation drainage water have caused serious damage to fish, wildlife, and aquatic ecosystems.

## Soil Quality

Renewed concern about soil erosion led to major new initiatives in the 1985 Food Security Act (PL 99-198; also known as the 1985 farm bill) (Table 1-1). For the first time, to be eligible for farm program benefits, agricultural producers were required to implement a soil conservation plan for their highly erodible croplands. A conservation plan was required for highly erodible land converted to cropland, and Congress also established the Conservation Reserve Program to pay producers to take highly erodible land out of production.

Sheet and rill erosion remains an important problem, causing soil degradation on about 25 percent of U.S. croplands (Figure 1-1). Other forms of erosion—such as wind, gully, and ephemeral gully erosion—are also important and, if quantified, would expand the reported area of cropland on which erosion causes soil degradation. Conservation Compliance and Sodbuster, which are provisions of the 1985 Food Security Act, should result in substantial reductions in erosion caused by both wind and water. If these provisions are fully implemented and if the conservation practices remain in place, the United States will have taken a large step toward solving a soil erosion problem that has plagued U.S. agriculture since settlement by Europeans began.

Even as major strides toward erosion control are being taken, however, new concerns about the soil resource are emerging. Compaction is increasingly noted as a factor that degrades soils and reduces crop yields, but no comprehensive data on the extent or severity of compaction are available. Salinization of soils, particularly in the western part of the United States, is causing serious and often irreversible damage where it is occurring (Table 1-2 and Figure 1-2).

Investigators are also concerned about more subtle forms of soil degradation, such as declining levels of organic matter in the soil and

TABLE 1-1   U.S. Department of Agriculture and U.S. Environmental Protection Agency Soil and Water Quality Programs

| Program | Description |
|---|---|
| *U.S. Department of Agriculture Programs* | |
| Conservation Reserve Program | Provides annual rental payments to landowners and operators who voluntarily retire highly erodible and other environmentally critical lands from crop production for 10 years. |
| Conservation Compliance Program | Requires that producers who produce agricultural commodities on highly erodible cropland implement approved erosion control plans by January 1, 1995, or lose eligibility for USDA agricultural program benefits. |
| Sodbuster Program | Requires that producers who convert highly erodible land to cropland for the production of agricultural commodities do so under an approved erosion control plan or forfeit eligibility for USDA agricultural program benefits. |
| Swampbuster Program | Bars producers who convert wetlands to agricultural commodity production from eligibility for USDA agricultural program benefits, unless USDA determines that conversion would have only a minimal effect on wetland hydrology and biology. |
| Agricultural Conservation Program | Provides financial assistance to farmers for implementing approved soil and water conservation and pollution abatement practices. |
| Conservation Technical Assistance | Provides technical assistance by the Soil Conservation Service through county Conservation Districts to producers for planning and implementing soil and water conservation and water quality improvement practices. |
| Great Plains Conservation Program | Provides technical and financial assistance in Great Plains states to producers who implement total conservation treatment of their entire farm or ranch operation. |
| Small Watershed Program | Provides technical and financial assistance to local organizations for flood prevention, watershed protection, or water management. |
| Resource Conservation and Development Program | Assists multicounty areas in enhancing conservation and water quality, wildlife habitat and recreation, and rural development. |
| Rural Clean Water Program | An experimental program that ends in 1995 that provides cost-sharing and technical assistance to producers who voluntarily implement best management practices to improve water quality. |
| Extension | Provides information and recommendations on soil and water quality practices to landowners and operators, in cooperation with the Soil Conservation Service and county Conservation Districts. |

*(continued)*

TABLE 1-1 (*Continued*)

| Program | Description |
| --- | --- |
| Water Bank Program | Provides annual rental payments for preserving wetlands in important migratory waterfowl nesting, breeding, or feeding areas. |

*U.S. Environmental Protection Agency Programs*

| Program | Description |
| --- | --- |
| Nonpoint Source Pollution Control Program | Requires states and territories to file assessment reports with EPA identifying navigable waters where water quality standards cannot be attained or maintained without reducing nonpoint source pollution. States must also file management plans with EPA identifying steps that will be taken to reduce nonpoint source pollution in those waters identified in the state assessment reports. Grants are available to states with approved management plans to help implement nonpoint source pollution control programs. |
| National Estuary Program | Provides for identification of nationally significant estuaries threatened by pollution, preparation of conservation and management plans, and federal grants to prepare the plans. |
| Clean Lakes Program | Requires states to submit assessment reports on the status and trends of lake water quality, including the nature and extent of pollution loading from point and nonpoint sources, and methods of pollution control to restore lake water quality. Financial assistance is provided to states to prepare assessment reports and to implement watershed improvements and lake restoration activities. |
| Regional Water Quality Programs | Provides for cooperation between EPA and other federal agencies to reduce nonpoint source pollution in specified regional areas such as the Chesapeake Bay Program, the Colorado River Salinity Control Program, the Gulf of Mexico Program, and the Land and Water 201 Program in the Tennessee Valley region. |
| Wellhead Protection Program | Requires each state to prepare and submit to EPA a plan to protect from pollution, including from agricultural sources, the water recharge areas (areas where water leaching below the land surface replenishes the groundwater supplies tapped by wells) of wells that supply public drinking water. |
| Coastal Zone Program | Requires the implementation of enforceable management measures to protect coastal zones from nonpoint source pollution. |

SOURCE: Adapted from U.S. Department of Agriculture, Economic Research Service. 1989. Conservation and water quality. Pp. 21-35 in Agricultural Resources: Cropland, Water, and Conservation Situation and Outlook. Report No. AR-16. Washington, D.C.: U.S. Department of Agriculture.

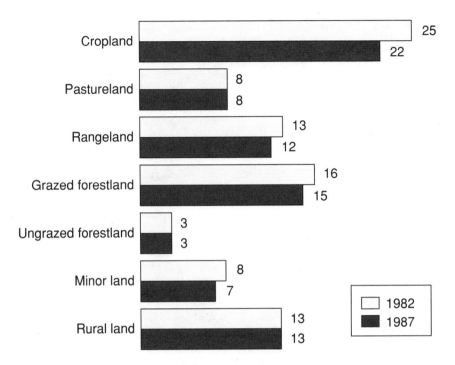

FIGURE 1-1  Percentage of land eroding by sheet and rill erosion at greater than the soil loss tolerance level. Minor land includes farmsteads, strip mines, quarries, gravel pits, borrow pits, permanent snow and ice, small built up areas, and all other land uses that do not fit into any other category. Source: Derived from U.S. Department of Agriculture, Soil Conservation Service. 1989. Summary Report: 1987 National Resources Inventory. Statistical Bulletin No. 790. Washington, D.C.: U.S. Department of Agriculture.

the attendant degradation of soil structure, the soil's water-holding and nutrient-holding capacities, and biological activity (Larson and Pierce, 1991). The effect of soil degradation on carbon dioxide emissions is also receiving greater attention (Lal and Pierce, 1991).

## Water Quality

Even as the 1985 Food Security Act was being debated, policymakers began to recognize that the intensification of agricultural production that gained speed in the 1970s was leading to a new set of environmental problems. Clark and colleagues (1985), for example, reported that sediments in U.S. waterways caused $2.2 billion in damage every year.

TABLE 1-2   Cropland and Pastureland Soils Affected by Saline or Sodic Conditions

| | Thousands of Hectares | | |
| --- | --- | --- | --- |
| Region | Total Affected Soils | Total Cropland or Pastureland | Percent Affected |
| Northeast | 0 | 10,562 | <1 |
| Appalachia | 1 | 16,681 | <1 |
| Southeast | 12 | 12,348 | <1 |
| Lake States | 1,118 | 21,797 | 5 |
| Corn Belt | 158 | 47,623 | <1 |
| Delta | 328 | 13,796 | 2 |
| Northern Plains | 8,110 | 41,198 | 20 |
| Southern Plains | 2,000 | 27,973 | 7 |
| Mountain | 6,075 | 20,516 | 30 |
| Pacific | 1,821 | 11,085 | 16 |
| Other | 6 | 1,082 | <1 |
| Total | 19,630 | 224,659 | 9 |

NOTE: "Other" refers to Hawaii and the Caribbean region.

SOURCE: Adapted from U.S. Department of Agriculture, Soil Conservation Service. 1989. The Second RCA Appraisal: Soil, Water, and Related Resources on Nonfederal Land in the United States. Washington, D.C.: U.S. Department of Agriculture.

Nitrates, pesticides, salts, and trace elements were increasingly reported in the nation's lakes, rivers, and groundwater bodies.

These new concerns for the broader environmental effects of agricultural production led to increased attention to agriculture as a source of nonpoint source pollution problems in the 1987 amendments to the Federal Water Pollution Control Act (PL 100-4) and the 1990 Coastal Zone Act Reauthorization Amendments (PL 101-508), as well as to new initiatives in the 1990 Food, Agriculture, Conservation and Trade Act (PL 101-624; also known as the 1990 farm bill) (Tables 1-1 and 1-3).

### Surface Water Quality

Agricultural production has been identified as a major source of nonpoint source pollution in U.S. lakes and rivers that do not meet water quality goals (Figure 1-3). Nutrients (nitrogen and phosphorus) and sediments, major pollutants closely associated with agricultural production, affect surface water quality in the United States (Figure 1-3) and loadings of these pollutants have increased in agricultural watersheds (R.A. Smith et al., 1987). Pesticides have also been reported in surface waters, often at high concentrations in the spring following pesticide application to

FIGURE 1-2   Farm production regions used in this report. Alaska and Hawaii are included in the Pacific region.

crops (Baker, 1985; Thurman et al., 1991), although the mean annual concentrations were low. The total loadings of nutrients and pesticides into estuaries such as the Chesapeake Bay have become serious problems (U.S. Environmental Protection Agency, 1990a). In the western United States, the pollution of surface waters with salts in waters drained from irrigated agricultural lands has become both a national and an international problem (National Research Council, 1989a). The long-standing concern about salt damage from irrigated agriculture has now been augmented with concerns about the delivery of toxic trace elements such as selenium (National Research Council, 1989a).

*Groundwater Quality*

Agricultural chemicals are also being detected in groundwater bodies. Nitrates have been widely reported in both shallow and deep aquifers, although rarely at levels exceeding health standards (Holden et al., 1992; Power and Schepers, 1989; U.S. Environmental Protection Agency, 1988, 1990b). Pesticides have been found less frequently and at much lower levels than nitrates, usually at concentrations below human health standards (Holden et al., 1992; U.S. Environmental Protection Agency, 1990b), although pesticides have been found at greater concentrations in surficial aquifers (Hallberg, 1989a).

TABLE 1-3  New Initiatives in the 1990 Food, Agriculture, Conservation and Trade Act

| Initiative | Description |
| --- | --- |
| Conservation Compliance, Sodbuster, and Swampbuster Programs | Potential penalties for violating provisions of these programs increased to include loss of eligibility for Agricultural Conservation Program, Emergency Conservation Program, Conservation Reserve Program, Agricultural Water Quality Protection Program, Environmental Easement Program, and assistance under the Small Watersheds Program. USDA is given more flexibility in assessing penalties. |
| Conservation Reserve Program | Provides for the extension of enrollment of land into the Conservation Reserve Program until 1995 and establishes priority areas for the enrollment of lands in Chesapeake Bay, Great Lakes, and Long Island Sound regions. |
| Wetland Reserve Program | Creates a new Wetland Reserve Program to offer long-term easements to producers who restore wetlands or who protect riparian corridors and critical wildlife habitats. |
| Agricultural Water Quality Protection Program | Provides for annual incentive payments to producers who implement a USDA-approved water quality protection plan. Incentive payments are for 3 to 5 years in duration and require the producer to keep records of the inputs used, yields achieved, and results of well water tests, soil tests, or other tests for each year in which incentive payments are received. |
| Environmental Easement Program | Provides for long-term protection of environmentally sensitive lands or to reduce water pollution by offering long-term or permanent easements to producers who retire lands already enrolled in the Conservation Reserve Program, in the Water Bank Program, or lands in riparian areas, critical wildlife habitats, or other environmentally sensitive areas that, if cropped, would prevent a producer from complying with state of federal environmental goals. |

SOURCE: U.S. Department of Agriculture, Economic Research Service, Resources and Technology Division. 1991. Conservation and water quality. Pp. 23-41 in Agricultural Resources: Cropland, Water, and Conservation Situation and Outlook. Report No. AR-23. Washington, D.C.: U.S. Department of Agriculture.

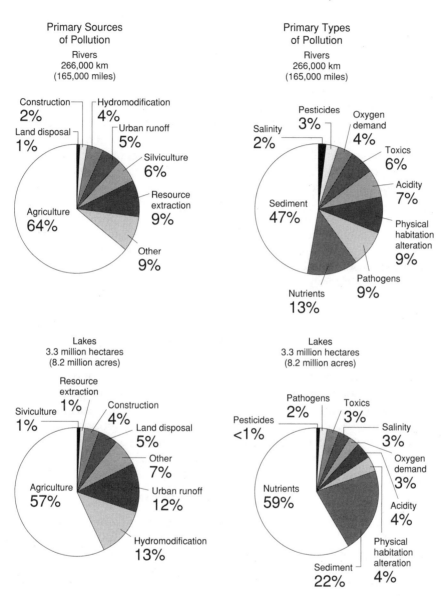

FIGURE 1-3 Sources and types of nonpoint source pollution in affected U.S. rivers and lakes. Source: A. E. Carey. 1991. Agriculture, agricultural chemicals, and water quality. Pp. 78–85 in Agriculture and the Environment: The 1991 Yearbook of Agriculture. Washington, D.C.: U.S. Government Printing Office.

## Environmental Risks

The damage to agricultural productivity caused by soil degradation and the effects of drinking water contaminated with nitrates, pesticides, salts, or trace elements on human health have, to date, been the driving forces behind the increased concerns over soil and water quality. More recently, however, the pervasive effects of human activities, particularly agricultural activities, on ecosystems and the ecological risks of these activities have received more attention. The effects of sediment, pesticide, nutrient, salt, and trace element loads on aquatic ecosystems may, in the long term, prove to be more important than their potential effects on human health. In surface water and groundwater, levels of these pollutants that are below human health standards may still be high enough to damage ecosystems. Assessment of the ecological risks of soil and water quality degradation may increasingly become the yardstick used to measure the damage caused by soil and water quality degradation (U.S. Environmental Protection Agency, Science Advisory Board, 1990).

## SEARCH FOR SOLUTIONS

The expansion of environmental issues on the agricultural agenda has led to calls for a reassessment of agricultural production practices and for the development of sustainable production systems that are environmentally sound as well as profitable (Harwood, 1990; Madigan, 1991; National Research Council, 1989b). Development of policies and programs that can be used to change agricultural production practices, however, has not proved easy.

### Factors Influencing Solutions

Each year U.S. food and fiber producers make millions of individual decisions that ultimately affect soil and water quality. Producers do not, however, make these decisions in a vacuum. They are influenced by their personal situations, the quantity and quality of the resources and technologies to which they have access, market prices, agricultural policies, environmental regulations, the use rights producers hold for the resources on their property, and the recommendations producers receive from public- and private-sector experts.

Figure 1-4 shows how the market environment, agricultural policies, environmental regulations, and private- and public-sector recommendations influence producers' decisions (Creason and Runge, 1990).

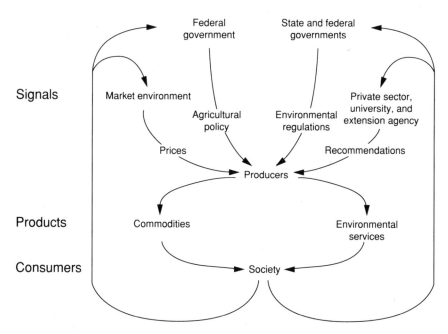

FIGURE 1-4    Interactions of factors that influence producers' decisions. Source: J. R. Creason and C. F. Runge. 1990. Agricultural Competitiveness and Environmental Quality: What Mix of Policies Will Accomplish Both Goals? St. Paul: University of Minnesota, Center for International Food and Agricultural Policy.

Each of these influences signals to producers the commodities they should produce and the technologies they should use. These choices, in turn, influence the farm's impact on the environment. This interaction of signals makes a policymaker's job difficult. It is not often clear how a change in policy will ultimately affect the decisions that producers make.

Worries about the potential for trade-offs between protecting soil versus water, protecting surface water versus groundwater, and reducing loadings of nitrates versus loadings of pesticides have also confounded the policy making process. The multiplicities of potential objectives and best management practices suggested to address those objectives have also made the choice of policies seem complicated. Simple recommendations that call for increased residue cover to reduce erosion or that suggest the installation of grassed waterways are no longer adequate to deal with the broader environmental problems facing agricultural producers.

## State and Local Government Policies

The policies made by local and state governments are increasingly important factors. Local and state governments have often taken the lead in developing new programs and approaches for dealing with soil and water quality problems. Integrating the activities of various levels of government with various federal agencies has become an increasingly important element of environmental policy for agriculture.

## Characteristics of the Agricultural Sector

The agricultural sector is too often discussed as if it were a homogeneous collection of uniform farms managed by similar producers. Many policies and programs are also based on the assumption of a "typical" producer.

---

## SOIL CONSERVATION IN COON CREEK, WISCONSIN

In the early 1930s the Coon Creek Basin in southwestern Wisconsin was designated the first Soil Erosion Control Demonstration Area in the United States. The area is marked by steep slopes and narrow valleys, with relief of about 135 m (430 feet). The productive soils were formed from loess (an unstratified calcareous silt that overlies various sedimentary rock units on the steep slopes) and alluvial deposits. Settlers arrived in about 1850, and land clearing and cultivation continued until about 1900.

In the early 1930s, when the soil conservation program began, the area showed the effects of 80 years of poor land management. At that time the soils were both degraded and eroded. The levels of sediments from sheet, rill, gully, and channel erosion were more than the streams could transport: more than 2 m (6 feet) of sediment was deposited in one 10-year period (McKelvey, 1939).

The area was characterized by rectangular fields on steep slopes, up-and-down plowing of slopes, poor crop rotations, lack of cover crops, and overgrazed and eroding pastures and woodlands. Erosion, compaction, and depletion of organic matter and nutrients had degraded soil quality. Active rills and gullies were widespread, and the channels of the small upland tributary streams were entrenched and eroding. Studies here and in the general region showed that the conversion to agricultural land use was accompanied by increased flooding as well as erosion and sedimentation. The hydrologic changes, in turn, also caused major changes in the physical or geomorphic characteristics of the stream channels (Knox, 1977).

The conservation demonstration project instituted widespread land treatment measures. The project increased the use of contour tillage and contour strip-cropping, instituted longer rotations with various cover crops, and

In reality, farms differ in the commodities they produce, their soil quality, and their topography. Ownership patterns differ, too. Beef cattle farms are often small-scale, part-time farm operations with only a few head of cattle, whereas poultry enterprises tend to resemble vertically integrated industries ("vertically integrated" refers to an industry in which a single company provides the control) (Reimund and Gale, 1992). Cash grain farms most closely match the popular perception of agriculture: family farms run by owner-operators (Reimund and Gale, 1992).

Just as farms are diverse, producers are also a diverse set of people who have a variety of goals: profit maximization, minimization of management time, maintenance of a certain life-style, protection of personal independence, desire to obtain a certain social status, and observation of a particular environmental or religious ethic. In addition, producers have different levels of skills, different levels of access to resources, and different sources of information. Such differences—

---

incorporated manure and crop residues into the soil. By the 1970s, when the area was reinvestigated by researchers from the U.S. Geological Survey, the conversion was complete and conservation tillage was being introduced as well (Trimble and Lund, 1982).

Even with these changes, aggregate land use had changed little since 1930; the proportion of land in row crops, cover crops, and pastureland had changed little. Land management, however, had improved dramatically. The calculated erosion rates decreased by more than 75 percent, from more than 3,400 metric tons/km² (15 tons/acre) in 1934 to about 720 metric tons/km² (3 tons/acre) in 1975. The linear extent of gullies was reduced by 76 percent, with medium and large gullies nearly eliminated by 1978 (Fraczek, 1988). Trimble and Lund (1982) also systematically studied sedimentation rates in sediment basins and along the bottomlands of streams. Sediment deposition rates decreased by 98 percent or more from 1936 to 1945.

Although erosion and sedimentation rates were still greater in the 1970s than before settlement and cultivation, the soil conservation programs greatly reduced erosion and sedimentation. Rills and gullies had mostly disappeared. The improved land management and soil quality increased water infiltration, decreasing runoff, reducing peak runoff and flood flows, and decreasing the erosion potential of streams as well.

The area is a good example of the changing concerns for water quality and the need for improvements in input efficiency and input management approaches. In the 1990s, some watersheds in this area have been established as demonstration areas for Wisconsin's Nutrient and Pest Management Program and some areas have been designated atrazine (a pesticide) management areas, to focus on more recent concerns for nitrogen, phosphorus, and pesticide impacts on groundwater and surface water quality.

particularly when coupled with the differences in farm characteristics—can mean considerable differences in the reasons why producers choose to adopt or reject new farming systems (Nowak, 1992).

## TIME TO MOVE AHEAD

Much has changed since 1970. The 1985 Food Security Act, the 1987 amendments to the Federal Water Pollution Control Act, the 1990 Coastal Zone Act Reauthorization Amendments, and the 1990 Food, Agriculture, Conservation and Trade Act produced a combination of new programs and mandates that can be used to address soil and water quality problems. Accelerated research has improved understanding of the physical, chemical, and biological processes that determine how agricultural systems affect soil and water quality. Experimentation with new production technologies and farming systems by researchers and producers has produced a wealth of information on the applicabilities and efficacies of innovative farming systems. Many of the mandates, programs, and knowledge needed to move ahead with an expanded agenda of soil and water quality improvement are available. What is needed is consensus on the broad objectives that the mandates, programs, and knowledge should achieve. This report is intended to help achieve that consensus.

# 2

# Opportunities to Improve Soil and Water Quality

The list of soil and water resource problems on the agricultural agenda has increased enormously over the past 15 years. Long-standing concerns about soil erosion and sedimentation have been supplemented with renewed concerns about soil compaction, salinization, acidification, and loss of soil organic matter. The loss of nitrates, phosphorus, pesticides, and salts from farming systems to surface water and groundwater has, in some ways, supplanted traditional concerns about soil degradation.

Efforts to address this larger complex of resource problems have been hampered by concerns about trade-offs. Management practices that have been designed to reduce soil erosion are now scrutinized for their role in increasing leaching of nitrates or pesticides to groundwater. Practices designed to reduce the amounts of sediment-borne pollutants delivered to waterways are sometimes thought to increase the amounts of the soluble forms of those pollutants delivered in runoff water. Such findings have raised doubts about society's ability to manage what appears to be unavoidable environmental trade-offs. Efforts to improve agriculture's environmental performance must be weighed against efforts to reduce costs of production, increase production, and maintain U.S. agriculture's share of world markets.

Uncertainty about trade-offs makes policy development difficult because it is hard to determine the best approaches for improving the effects of farming systems on soil and water resources. This difficulty is further complicated by the inherent regional and local variabilities in farm enterprises and soil and water resources. Soil and water quality problems would be easier to solve if the most promising opportunities for improving farming systems were more clearly defined. Policies could then be developed to take advantage of those opportunities.

The committee analyzed the processes that cause soil degradation and that result in the delivery of sediments, salts, and agricultural chemicals from croplands to surface water and groundwater. The committee also analyzed the ways that farming systems affect these processes. Through its analysis, the committee attempted to develop general solutions to soil and water resource degradation that can lead to productive and environmentally sound farming systems and to provide practical guidance for implementing such solutions.

The committee defined four broad opportunities that hold the most promise for improving the environmental performance of farming systems while maintaining profitability. Current soil and water resource policies should

1. conserve and enhance soil quality as a fundamental first step to environmental improvement;

2. increase nutrient, pesticide, and irrigation use efficiencies in farming systems;

3. increase the resistance of farming systems to erosion and runoff; and

4. make greater use of field and landscape buffer zones.

These four opportunities are related to the fundamental processes that determine how farming systems affect the environment. The soil is the mediator between farming practices, agricultural chemicals, and the environment. Soil degradation directly and indirectly affects agricultural productivity and water quality. Increasing nutrient, pesticide, and irrigation use efficiencies addresses the input side of the equation. The goal of increased efficiency is to reduce the total mass of residuals from inputs, thus making less mass available for loss to the environment. Increasing the soil's resistance to erosion and runoff addresses the output side. Erosion and runoff are the major pathways by which sediment, nutrients, pesticides, and other pollutants reach surface water, and erosion remains the greatest threat to soil quality. Finally, farming systems exist in a landscape, and landscape processes determine the ultimate effects of farming systems on soil and water quality. The creation of field and landscape buffer zones is a way to manipulate those landscape processes to gain further improvements in soil and water quality by intercepting pollutants and reducing the erosive force of runoff water.

Since agriculture and its associated soil and water resources vary

Whitman County, Washington, is one of the nation's primary producers of winter wheat. It also has some of the region's most erodible soil. The Food Security Act of 1985 provides incentives to protect soil on the cropland most likely to erode. Credit: U.S. Department of Agriculture.

dramatically across the United States, it is difficult to make specific recommendations for preventing soil degradation and water pollution that apply equally well to all farming systems. The four opportunities recommended here as goals for national policy, however, can be applied to policies and farming systems generally. The opportunity emphasized may change from one farming system or region to another, but the realization of only one of the four opportunities will not address the full complement of soil and water quality issues confronting the U.S. agricultural system. A combination of policies and policy instruments will be needed to pursue all four opportunities. This chapter examines each of the four opportunities in some depth to reveal their implications for soil and water quality policy.

## CONSERVING AND ENHANCING SOIL QUALITY

The 1990 Clean Air Act (PL 101-549) and the 1987 Federal Water Pollution Control Act (PL 100-4) give national recognition to the fundamental importance of air and water resources. The fundamental importance of soil resources, however, is usually overlooked, even though the soil is the interface between human activities and the environment. The quality of the soil and its management in large part determine whether

---

## THREE FUNCTIONS OF SOIL

Soils are living systems that are vital for producing the food and fiber humans need and for maintaining the ecosystems on which all life ultimately depends. Soil directly and indirectly affects agricultural productivity, water quality, and the global climate through its function as a medium for plant growth, a regulator and partitioner of water flow, and an environmental buffer.

**Soils make it possible for plants to grow.** Soils mediate the biological, chemical, and physical processes that supply nutrients, water, and other elements to growing plants. The microorganisms in soils transform nutrients into forms that can be used by growing plants. Soils are the water and nutrient storehouses on which plants draw when they need nutrients to produce roots, stems, and leaves. Eventually, these become food and fiber for human consumption. Soils—and the biological, chemical, and physical processes that they make possible—are a fundamental resource on which the productivities of agricultural and natural ecosystems depend.

**Soils regulate and partition water flow through the environment.** Rainfall in terrestrial ecosystems falls on the soil surface where it either infiltrates the soil or moves across the soil surface into streams or lakes.

agriculture or other land uses will cause or prevent water pollution. The increasingly urban U.S. population is quick to recognize when air quality is degraded. Indeed, newspapers and radio and television weather forecasts regularly report air quality indicators. People are aware of water quality every time they turn on a water tap, and the presence of contaminants in water is widely reported. Soil quality degradation, however, often goes unnoticed because most people rarely encounter soil in their daily lives and because soil quality degradation is often difficult to see and measure. Soil quality is, nevertheless, as fundamental as air and water quality to future environmental quality and ecological integrity. The threat posed by soil degradation needs to receive the same kind of attention given to air and water quality degradation by national, state, and local policymakers.

Society needs to change the way it thinks about soils. Society generally views soils simply as the rooting medium for plants. Society often fails to recognize that soils also regulate and partition water flow and buffer environmental changes. The way society thinks about soils affects the kinds of policies, programs, and research that investigators devise to manage soil resources. This narrow conception of soil is no longer adequate to address the linked problems of soil degradation and water pollution that agriculture faces.

---

The condition of the soil surface determines whether rainfall infiltrates or runs off. If it infiltrates the soil, it may be stored and later taken up by plants, move into groundwaters, or move laterally through the earth, appearing later in springs or seeps. This partitioning of rainfall between infiltration and runoff determines whether a storm results in a replenishing rain or a damaging flood. The movement of water through soils to streams, lakes, and groundwater is an essential component of recharge and base flow in the hydrological cycle.

**Soils buffer environmental change**. The biological, chemical, and physical processes that occur in soils buffer environmental changes in air quality, water quality, and global climate. The soil matrix is the major incubation chamber for the decomposition of organic wastes including pesticides, sewage, solid wastes, and a variety of other wastes. The accumulation of pesticide residues, heavy metals, pathogens, or other potentially toxic materials in the soil may effect the safety and quality of food produced on those soils. Depending on how they are managed, soils can be important sources or sinks for carbon dioxide and other gases that contribute to the greenhouse effect (greenhouse gases). Soils store, degrade, or immobilize nitrates, phosphorus, pesticides, and other substances that can become pollutants in air or water.

## Defining Soil Quality

Soil quality is best defined in relation to the functions that soils perform in natural and agroecosystems. The quality of soil resources has historically been closely related to soil productivity (Bennett and Chapline, 1928; Hillel, 1991; Lowdermilk, 1953). Indeed, in many cases the terms *soil quality* and *soil productivity* have been nearly synonymous (Soil Science Society of America, 1984). More recently, however, there is growing recognition that the functions soils perform in natural and agroecosystems go well beyond promoting the growth of plants. The need to broaden the concept of soil quality beyond traditional concerns for soil productivity have been highlighted at a series of recent conferences and symposia.

Johnson and colleagues, in a paper presented at a Symposium on Soil Quality Standards hosted by the Soil Science Society of America in October 1990, suggested that soil quality should be defined in terms of the functions of soils in the environment and defined soil function as "the potential utility of soils in landscapes resulting from the natural combination of soil chemical, physical, and biological attributes" (Johnson et al., 1992:77). They recommended that policies to protect soil resources should protect the soil's capacity to perform several functions simultaneously including the production of food, fiber, and fuel; nutrient and carbon storage; water filtration, purification, and storage; waste storage and degradation; and the maintenance of ecosystem stability and resiliency.

Larson and Pierce defined soil quality as "the capacity of a soil to function, both within its ecosystem boundaries (e.g., soil map unit boundaries) and with the environment external to that ecosystem (particularly relative to air and water quality)" (Larson and Pierce, 1991:176). They proposed "fitness for use" as a simple operational definition of soil quality and stressed the need to explicitly address the function of soils as a medium for plant growth, as a means to partition and regulate the flow of water in the environment, and as an environmental buffer. Parr and colleagues, in a paper presented at a Workshop on Assessment and Monitoring of Soil Quality hosted by the Rodale Institute Research Center in July 1991, defined soil quality as "the capability of a soil to produce safe and nutritious crops in a sustained manner over the long term, and to enhance human and animal health, without impairing the natural resource base or harming the environment" (Parr et al., 1992:6). Parr and colleagues (1992) stressed the need to expand the notion of soil quality beyond soil productivity to include the role of the soil as an environmental filter affecting both air and water quality. They suggested that soil quality has important effects on the nutritional quality of the food produced in those soils but noted that these linkages are not well understood and that research is needed to clarify the relationship between soil quality and the nutritional quality of food.

The growing recognition of the importance of the functions of soils in the environment requires that scientists, policymakers, and producers adopt a broader definition of soil quality. Soil quality is best defined as the capacity of a soil to promote the growth of plants, protect watersheds by regulating the infiltration and partitioning of precipitation, and prevent water and air pollution by buffering potential pollutants such as agricultural chemicals, organic wastes, and industrial chemicals. The quality of a soil is determined by a combination of physical, chemical, and biological properties such as texture, water-holding capacity, porosity, organic matter content, and depth. Since these attributes differ among soils, soils differ in their quality. Some soils, because of their texture or depth, for example, are inherently more productive because they can store and make available larger amounts of water and nutrients to plants. Similarly, some soils, because of their organic matter content, are able to immobilize or degrade larger amounts of potential pollutants.

Soil quality can be improved or degraded by management. Erosion, compaction, salinization, sodification, acidification, and pollution by toxic chemicals can and do degrade soil quality. Increasing the protection the soil is afforded by crop residues and plants; adding organic matter to the soil through crop rotations, manures, or crop residues; and carefully managing fertilizers, pesticides, tillage equipment, and other elements of the farming system can improve soil quality.

Management of soil resources should be based on a broader concept of the fundamental roles that soils play in natural and agroecosystems. The implications of this broader concept of soil on policy development become clearer if one examines in more detail the ways that soils affect agricultural productivity, water quality, and the global climate.

## Importance of Soil Quality

Changes in agricultural productivity, water quality, and global climate are linked to soil quality through the chemical, physical, and biological processes that occur in soils.

### Agricultural Productivity

Damage to agricultural productivity from soil degradation has historically been the major concern about soil resources. Agricultural technologies have, in some cases, improved the quality of soils or have masked much of the yield loss that could be attributed to declining soil quality, except on soils vulnerable to rapid and irreversible degradation. Studies have predicted that losses in crop yields because of soil erosion will be less than 10 percent over the next 100 years, assuming high levels of inputs (Crosson and Stout, 1983; Hagen and Dyke, 1980; Pierce et al., 1984; Putnam et al., 1988). Those studies have begun to shift the emphasis of federal policy to the off-site damages caused by erosion.

*Conservation of soil productivity should remain an important long-term goal of national soil resource policy.*

The effect of soil degradation on agricultural productivity has been underestimated. Estimates of the agricultural productivity lost because of soil erosion have not accounted for the damages caused by gully and ephemeral erosion, sedimentation (Pierce, 1991), or reduced water availability as a result of decreased infiltration of precipitation. Those studies also assume that the optimum nutrient status is maintained by using fertilizers to replace the nitrogen, phosphorus, and potassium lost from eroding lands. Larson and colleagues (1983) estimated that the value of the nitrogen, phosphorus, and potassium lost through erosion in 1982 was $677 million, $17 million, and $381 million, respectively. The total mass of nitrogen and phosphorus estimated to be lost in eroded sediments in 1982 was equal to 95 and 39 percent, respectively, of the total nitrogen and phosphorus applied to all U.S. croplands in that same year (Larson et al. [1983] estimates of nitrogen and phosphorus applications were from Vroomen [1989]). In addition to erosion, compaction, salinization, and acidification, other deleterious forces can also cause yield losses and increase costs. More important, erosion, compaction, salinization, and acidification may interact synergistically to accelerate soil degradation. Losses in yields and increases in costs will be greater than those projected if investigators consider all degradation processes and their interactions. In addition, reductions in yield can be severe where soil degradation is serious but are obscured in estimates of U.S. average soil erosion or yield reductions. Other analysts have suggested that losses in potential productivity will occur sooner and be of a larger magnitude than absolute losses in yields from soil degradation.

Crosson (1985) pointed out that it is the cost of erosion, not the predicted yield losses, that is of interest. Similarly, the cost of compensating for reduced soil quality because of degradation by compaction, acidification, salinization, and loss of biological activity, as well as erosion, is of the most importance when assessing the effects of soil degradation on soil productivity.

Evidence of gully and ephemeral erosion is a clear indication that soil quality is threatened. Credit: U.S. Department of Agriculture.

To date, improvements in agricultural technology have kept the costs of compensation for losses in soil quality low enough or increases in yields high enough to offset the costs of soil degradation on most croplands. Given the multiple processes of soil degradation, however, and the probable underestimation of the full cost of soil degradation on the cost of production, on-site changes in soil quality may have significant effects on society's ability to sustain a productive agricultural system. The increased amounts of fertilizers, pesticides, and other inputs used to compensate for declining soil quality have themselves become a problem when they pollute surface water and groundwater.

### Water Quality

*Policies to prevent water pollution caused by agricultural production should seek to enhance and conserve soil quality as a fundamental step to improving water quality.*

Soil and water quality are inherently linked; conserving or enhancing

---

## SOIL QUALITY AFFECTS AGRICULTURAL PRODUCTIVITY

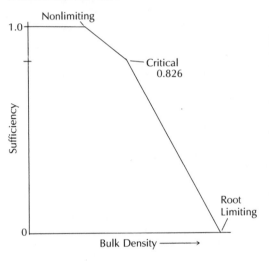

The potential of a soil to produce crops is largely determined by the environment that the soil provides for root growth. Roots need air, water, nutrients, and adequate space in which to develop. Soil attributes such as the capacity to store water, acidity, depth, and density determine how well roots develop. Changes in these soil attributes directly affect the health and productivity of the crop plant.

The figure illustrates how changes in one soil attribute, bulk density (bulk density is a measure of the compactness of a soil), affects agricultural productivity. When the bulk density of a soil increases to a critical level, it becomes more difficult for roots to penetrate the soil and root growth is impeded. As bulk density increases beyond this critical level, root growth is more and more restricted. At some point, the soil becomes so dense that roots cannot penetrate the soil and root growth is prevented. Heavy farm equipment, erosion, and the loss of soil organic matter can lead to increases in bulk density. Similarly, these processes of soil degradation can lead to reduced soil depth, reduced water-holding capacity, and increased acidity. At some critical point, these changes in soil quality affect the health and productivity of the crop plant, leading to lower yields and/or higher costs of production.

Source: *Derived from F. J. Pierce, W. E. Larson, R. H. Dowdy, and W. A. P. Graham. 1983. Productivity of soils: Assessing long-term changes due to erosion. Journal of Soil and Water Conservation 38:39–44.*

soil quality is a fundamental step toward improving water quality. Reducing losses of nutrients, pesticides, salts, or other pollutants will be impossible or difficult if soil degradation is not controlled. Indeed, use of nutrients, pesticides, and irrigation water to compensate for declining soil quality may be an important cause of water pollution.

Soil quality degradation causes both direct and indirect degradation of water quality. Soil degradation from erosion leads directly to water quality degradation through the delivery of sediments and attached agricultural chemicals to surface waters. Clark and colleagues (1985) estimated that the cost of sedimentation from eroding croplands on recreation, water storage facilities, navigation, flooding, water conveyance facilities, and water treatment facilities, among other damages, was $2.2 billion annually (in 1980 dollars based on 1977 erosion rates).

The indirect effects of soil quality degradation may be as important as the direct damages from sediment delivery, but they are often overlooked. Soil erosion and compaction degrade the capacities of watersheds to capture and store precipitation, altering stream flow regimes by exaggerating seasonal patterns of flow; increasing the frequency, severity, and unpredictability of high-level flows; and extending the duration of low-flow periods. The increased energy of runoff water causes stream channels to erode, adding to sediment loads and degrading aquatic habitat for fish and other wildlife.

Erosion, compaction, acidification, and loss of biological activity reduce the nutrient and water storage capacities of soils, increase the mobilities of agricultural chemicals, slow the rate of waste or chemical degradation, and reduce the efficiencies of root systems. All of these factors can increase the likelihood of loss of nutrients, pesticides, and salts from farming systems to both surface water and groundwater (Figure 2-1).

Improvements in soil quality alone, however, will not be sufficient to address all water quality problems unless other elements of the farming system are addressed. Improving soil quality, for example, will not reduce nitrate damages to surface water and groundwater if producers apply excessive amounts of nitrogen to the cropping system. If nitrogen applications are too high, changes in soil quality may change the proportion of nitrates delivered to surface water rather than to groundwater, but total nitrate losses may remain excessive. In this example, improved soil quality must be linked to improved nitrogen management.

Even if soil quality is very high, producers who mismanage inputs may still have unacceptable losses of nutrients, pesticides, and other

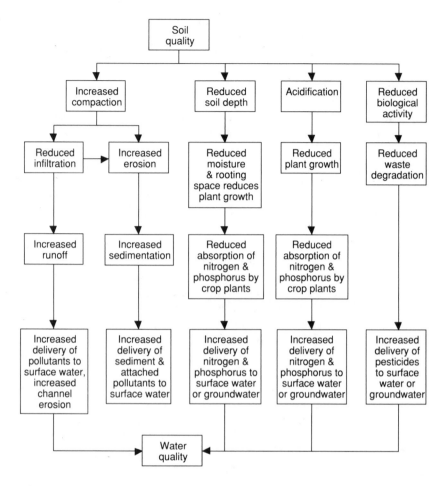

FIGURE 2-1   Changes in soil quality affect water quality.

pollutants from their farming systems. High soil quality is not a substitute for careful management of all components of the farming system.

### Global Climate

*The effect of soil management on global climate change should receive more attention in environmental policy.*

Soils can serve as a source or sink of carbon, depending on how they are managed. Lal and Pierce (1991), for example, estimated that if 1 percent of the organic carbon stored in predominantly tropical soils is mineralized per year, 128 billion metric tons (141 billion tons) of carbon will be released into the atmosphere. This figure compares with an estimated 0.325 billion metric tons (0.358 billion tons) of carbon emitted each year from fossil fuels and 1.659 billion metric tons (1.828 billion tons) emitted from deforestation (Brown et al., [1990] as cited by Lal and Pierce [1991]). Little is known, however, about the contribution of soil-related processes to greenhouse gas emissions under different soil and crop management systems. What is known suggests that soil may play an important role in regulating greenhouse gas concentrations.

### Soil Policy Goals

*The long-term goal for soil management should be the conservation and enhancement of soil quality.*

To date, the national debate over the appropriate goals and objectives of soil management has been driven by estimates of the effect of erosion on soil productivity. Conserving soil productivity alone, however, is not a sufficient objective for national soil resource policies and programs. The cost of soil degradation is greater than simply the effect on agricultural productivity. The direct and indirect effects of soil degradation on water quality and global climate change, may, in many circumstances, be more important than the effect of soil degradation on agricultural productivity.

#### *Erosion Control Alone Is Not Sufficient*

*Soil management policies should explicitly address compaction, salinization, acidification, loss of biological activity, and soil pollution as well as erosion.*

In the same way that soil productivity has framed the debate about soil management policies, efforts to control erosion have dominated programs and policies for protecting soil resources. Erosion control is an important means of conserving and enhancing soil quality, but it is not the only means. Other processes of soil degradation may, in some circumstances and regions, be more important threats to soil quality than erosion.

Not all forms of soil degradation are equally damaging. Erosion, salinization, and compaction by, for example, wheeled traffic, are most worrisome because their effects are often not easily reversible. Acidification can have important effects, but in most cases it is reversible through proper management. Biological degradation is difficult to define, but it is closely related to organic matter content. The soil's biological activity has important effects on all other soil quality attributes and on the capacities of soil to function as an environmental buffer and water regulator. More important, processes of degradation interact to accelerate soil degradation. Soil compaction, for example, reduces the soil's water-holding capacity, in turn increasing surface runoff and thereby accelerating erosion, which reduces the soil's biological activity by stripping away organically enriched topsoil.

In the United States, slightly more than 20 million ha (49 million acres) (13 percent) of cropland were estimated to be eroding at greater than the soil loss tolerance level in 1987 because of sheet and rill erosion (U.S. Department of Agriculture, Soil Conservation Service, 1989a), and wind-caused erosion greater than the soil loss tolerance level was estimated to be occurring on about 15 million ha (37 million acres) (9 percent) of U.S. cropland in 1982 (U.S. Department of Agriculture, Soil Conservation Service, 1989b). Salinization or sodification affected nearly 20 million ha (49 million acres) (9 percent) of cropland and pastureland; about 5.6 million ha (13.8 million acres), or 21 percent of irrigated cropland, was slightly saline or sodic in 1982 (U.S. Department of Agriculture, Soil Conservation Service, 1989b). Little information is available for other forms of soil degradation. No national estimates of the extent or severity of soil compaction or acidification exist, and researchers have made little attempt to estimate the loss of biological activity.

### Soil Degradation as an Environmental Problem

*Protecting soil quality, like protecting air and water quality, should be a fundamental goal of national environmental policy.*

Currently available data underestimate the severity and extent of soil degradation and overlook many of the costs that soil degradation impose on the environment. The Science Advisory Board of the U.S. Environmental Protection Agency (EPA) (1990) recently recommended that the ecological risk imposed by human activities, including soil degradation, should receive much greater attention by the agency and the nation because they pose relatively high-risk problems to the natural ecology and human welfare.

A new effort is needed to reevaluate the relative importance soil degradation should receive in national environmental policy. This effort should have three components. First, new criteria are needed to quantify soil quality. Second, national soil and water resource assessments need to be redirected to provide the information needed to determine the extent and seriousness of soil degradation. Third, soil management at the farm level needs to receive greater attention as a fundamental component of efforts to improve soil and water quality.

## Measurement of Soil Quality

If the conservation and enhancement of soil quality are to be the primary objectives of soil resource policies, methods for measuring changes in soil quality and predicting the effects of farming systems on soil quality are needed. Key indicators of soil quality need to be identified and used as the basis for monitoring and predicting changes in soil quality. A great deal is known about the general relationship of specific soil attributes to soil quality, and several authors have recently recommended various soil attributes as indicators of soil quality (Alexander and McLaughlin, 1992; Arshad and Coen, 1992; Granatstein and Bezdick, 1992; Griffith et al., 1992; Hornsby and Brown, 1992; Larson and Pierce, 1991; Larson and Pierce, in press; Olson, 1992; Pierce and Larson, 1993; Reagnold et al., 1993; Stork and Eggleton, 1992; Visser and Parkinson, 1992; Young, 1991). More work needs to be done, however, to develop more quantitative methods of estimating change in soil quality. Over time, changes in these soil quality indicators will provide the information needed to assess the effects of current farming systems and land use on soil quality, to develop new farming systems that improve soil quality, and to guide the development of national policies to protect soil and water quality.

*The Secretary of the U.S. Department of Agriculture (USDA) and the Administrator of the EPA should initiate a coordinated research program to develop a minimum data set of soil quality indicators, standardized methods for their measurement, and standardized methods to quantify changes in soil quality.*

The development of methods to quantify changes in soil quality will require measurable indicators that are relatively easy to sample and not subject to extreme variation in time or space. Models that can integrate measurements of multiple soil attributes into quantitative estimates of change in soil quality with reasonable confidence given the spatial and temporal variability of soils will also be needed. This task will require integrating research from many scientific disciplines and scientists from universities, industries and government agencies. This research effort should include

• identification of the soil attributes that can serve as indicators of change in all three soil functions (promotion of plant growth, regulation and partitioning of infiltration and runoff, and environmental buffering) and development of simplified models that relate changes in the selected attributes to changes in soil quality;

• standard field and laboratory methodologies to measure changes in indicators of soil quality;

- a coordinated monitoring program that can quantify changes in the indicators of soil quality; and
- a coordinated research program designed to support, test, and confirm the models used to predict the impact of management practices on soil quality.

Such a coordinated research effort would be comparable to efforts that have been undertaken to improve erosion simulation models for use in resource assessments, conservation planning, and program implementation. A comparable effort will be needed to develop the data and models required to estimate changes in soil quality.

### National-Level Assessments of Soil Quality

*The National Resources Inventory should include quantifiable measures of changes in selected soil quality indicators and should be broadened to produce estimates of compaction, salinization, sodification, acidification, and biological degradation in addition to erosion.*

The 1977, 1982, and 1987 National Resources inventories (U.S. Department of Agriculture, Soil Conservation Service, 1989a,b,c), were by far the most extensive and quantitative inventories of soil resources in the United States. These inventories and assessments, however, are limited by their focus on quantifying rates of erosion and related processes of soil degradation rather than a focus on assembling and assessing the information needed to monitor changes in soil attributes that can be related to changes in soil quality.

Measures of changes in soil quality indicators should be included as part of national-level resource inventories such as the National Resources Inventory. A system that enables more direct quantification of actual changes in soil attributes will allow policymakers to direct policies and programs more specifically toward monitoring actual damages to soil quality. Quantifiable estimates of soil degradation processes in addition to erosion are also needed to direct national soil management efforts comprehensively and to set priorities for soil management and conservation programs.

### Assess Currently Available Data

*The Resource Conservation Act appraisal process should assemble all currently available information to assess the current state of and trends in soil quality.*

Currently available data on rates of erosion, salinization, sodification, acidification, and compaction should be assembled and interpreted to make preliminary estimates of the full extent and severity of soil degradation. Much information on the quality of U.S. soils could be assembled from state and private soil testing laboratories; local, state, and regional soil conservation programs; and other sources to supplement the data already collected as part of the National Resources Inventory.

Currently available models should be used to predict the effects of erosion, compaction, and other forces of degradation on those soil attributes related to soil quality. Information on crop yields and cropping patterns, for example, could be used to predict trends in soil organic matter content, an important and integrating indicator of soil quality (see Table 5-1 in Chapter 5). Other models could be used to predict the effects of current tillage, harvest, and other machinery on compaction to make preliminary estimates of the extent and severity of compaction. Data in SOILS-5 (a data base maintained by USDA that contains information on the attributes of different soils) on surface soil horizons could be used with existing models to estimate the locations and geographic extent of soils particularly vulnerable to different forms of soil degradation.

Such an appraisal would identify the utility of current data and models for soil quality assessment and would clearly identify gaps in the data and understanding needed to complete comprehensive assessments of the quality of U.S. soil resources.

## Soil Management at the Farm Level

Public policies for soil management at the farm level have, to date, focused primarily on erosion control. The major thrust of programs such as Conservation Compliance, Sodbuster, and the Conservation Reserve Program has been to reduce erosion rates to the soil loss tolerance level or to adopt farming practices that result in a specified reduction in erosion rates. The measures used to evaluate management of soils at the farm level need to be refined to reflect a broader concern for the protection of soil quality.

### Soil Quality Thresholds

*Soil quality indicators and models should be used to set threshold levels of soil quality that can be used as quantitative guides to soil management.*

Once in place, these threshold values should be used as the basis for conservation planning and programs such as Conservation Compliance.

Shown is a farmer using the stubble mulch tillage system. The stubble that remains after harvest reduces erosion and enriches the soil when it is incorporated into the soil as mulch during tillage. Credit: U.S. Department of Agriculture.

Trend in soil attributes toward or away from threshold values will indicate whether current management is improving, degrading, or maintaining soil quality. In the long term, changes in soil quality should replace the soil loss tolerance level as the standard used to determine acceptable rates of erosion. In the short term, however, the soil loss tolerance level is the best standard available. Quantitative standards to quantify the effect of erosion, compaction, salinization, biological degradation, and other processes of soil degradation on the minimum data set of soil quality indicators are needed to enable comprehensive and cost-effective management of soil resources.

*A set of soil quality indicators should be added to Soil Conservation Service field office technical guides, and the effects of recommended practices on soil quality should become an integral part of the development of integrated farming system plans.*

The Soil Conservation Service should add to their field office technical guides a simple checklist of soil attributes that can serve as soil quality indicators. This checklist should be developed and used while a more rigorous effort goes forward to develop soil quality indicators suitable for national-level monitoring and quantification of changes in soil quality. The checklist of indicators could serve as an important integrating concept while a more holistic approach to resource management systems is implemented. The effect of alternative management systems on soil quality indicators could be depicted in simple matrices, as could the significance of these effects on soil and water quality.

Measures of soil attributes in the checklist of soil quality indicators should be added to routine soil test reports, and the significance of the measured levels of indicators should become part of the routine interpretations issued with soil test reports. For example, estimates of organic matter, leaching potential, crop yield potential, erosion potential, and other interpretive information should be added. A rapidly increasing number of counties throughout the United States have computerized soil survey information. Given the legal description of the farm, soil test reports could be enhanced by also reporting the soil mapping units and interpretive information from the soil survey (see Table 5-1 in Chapter 5).

## Soil-Specific Management

Tailoring the ways that farming systems are managed to differences in soil quality is a way to improve soil quality, water quality, and profitability simultaneously. Soils vary greatly over the landscape and often within the same field. Soils in one part of the field may be much more vulnerable to erosion or compaction. Similarly, soils within the same field or farm differ in their capacity to hold water, nutrients, and pesticides. Because the irregular soil quality distribution over the landscape does not match up well with the regular geometric pattern of crop fields, most of these differences in soil quality are ignored during farming operations (Larson and Robert, 1991). Ignoring the differences in soil quality between and within fields leads to soil degradation and water pollution by agricultural activities.

*Promise of New Technology*

New technology can link digitized soil maps with the exact position of equipment in the field to tailor applications of agricultural chemicals to differences in soil quality (see, for example, Reichenberger and Russnogle [1989] and Robert and Anderson [1986]). This new technology holds the promise of allowing producers to make on-the-go changes in rates of nutrient and pesticide applications as soil quality changes over the landscape.

A similar technology can be used to vary tillage and residue cover for controlling erosion and compaction, the planting rate, the crop variety selection, and many other facets of the crop production system. Robert and colleagues (1992) recently recounted the advances and the research needed in guidance systems, field equipment, soil and terrain mapping, environmental protection, and the economic consequences of this promising and rapidly developing technology.

*Better Use of Available Information*

The soil maps and soil information available in county Soil Conservation Service and extension offices can, if used properly, help tailor management practices to gross differences in soils between and within fields. When linked with soil test results for fields that are appropriately sampled, this information can lead to better management, even in the absence of new technology.

Computers offer great opportunities for combining all of the information available for a field or a farm. Soil surveys can be digitized and made available along with interpretive information such as crop yield potential, erosion potential, nutrient status, and leaching potential. This interpretive information—when combined with soil test data; records of actual crop yields and pest problems; tillage practices; nutrient, pesticide, and irrigation water applications; and crop rotations over a period of years—builds the information base needed to greatly refine management of the farming system. Crop-soil consultants and some crop producers have already developed and implemented these computer-based systems. Expert system software can be used to make farming system information management easier and more useful. Such software is becoming more commonly available. A great deal of information about soils is currently collected as part of soil surveys, research projects, or from routine soil tests that could be used to improve soil management. Current data collection protocols and systems of storing and processing these data may need to be updated to facilitate their use.

Tailoring farming system management practices to differences in soil quality can reduce the potential for runoff or leaching of nutrients and pesticides and can reduce the potential for erosion and other soil degradation processes. In some cases, increases in profitability are also possible when the increased costs required to obtain the information are offset by reduced spending on nutrients, pesticides, or other inputs.

Development and implementation of these technologies could lead to increases in the total efficiency and effectiveness of agricultural production. Widespread use of such technologies could also lead to the development of extremely accurate data bases that link actual production practices on croplands to their effects on productivity, soil degradation, and water pollution. These data bases would be extremely useful in developing and implementing new farming systems.

## INCREASING INPUT USE EFFICIENCIES

Nitrogen, phosphorus, and pesticides are important inputs to agricultural production systems. They are also important pollutants when they are delivered to air and water. The drainage from irrigated fields transports salts, pesticides, and nutrients to both surface water and groundwater, and the management of irrigation water has important effects on the kinds and amounts of pollutants carried in drainage water.

### Mass Balance between Inputs and Outputs

The nitrogen, phosphorus, and pesticides introduced into the environment during crop production follow various pathways that determine their eventual fates in the environment. Figures 2-2 through 2-4 show simplified pathways of the nutrients (nitrates and phosphorus), pesticides, and irrigation water, respectively, used as inputs to agricultural production systems. The nutrients introduced in fertilizers or manures or fixed by legumes become part of the nutrient cycle in the

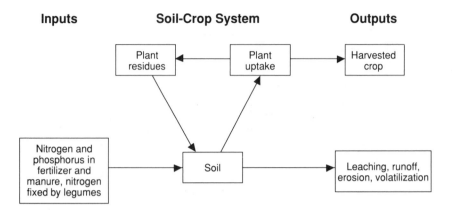

FIGURE 2-2   Nutrient cycle and pathways in agroecosystems.

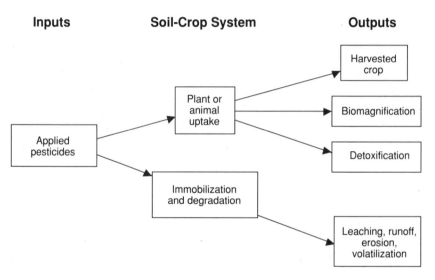

FIGURE 2-3   Pesticide pathways in agroecosystems.

soil-crop system. Nitrogen and phosphorus can leave the soil-crop system in the harvested crop, or they can be lost through erosion, runoff, or leaching (Figure 2-2).

The fate and transport processes for pesticides are more complicated. There is no natural pesticide cycle comparable to nutrient cycles. The pesticides added to the soil-crop system can be immobilized or degraded in the soil or can be taken up by plants or animals. Pesticides taken up

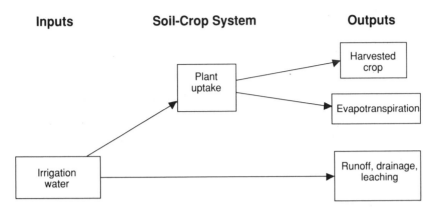

FIGURE 2-4   Irrigation pathways of water in agroecosystems.

by plants and animals can be removed with the harvested crop, passed on to other animals through the food chain, or detoxified by biological processes within the crop or pest organism. Pesticides can be lost from the soil-crop system through runoff, erosion, leaching, and volatilization (Figure 2-3).

Irrigation water flows through the soil-crop system, and irrigation drainage water carries salts, nutrients, pesticides, and trace elements. Much of the applied irrigation water that is taken up by plants is transpired back into the atmosphere. A smaller portion is incorporated into plant tissues, and the remainder leaves the soil-crop system by leaching, runoff, or subsurface drainage (Figure 2-4).

In farming systems, mass balances can be used to review the balance between inputs and outputs to assess where the opportunities lie for preventing pollution. Although the nature of the inputs and outputs vary among farming systems, regions, and fields, mass balances provide a conceptual framework that can be applied across a diversity of farming systems and geographic scales.

### Increased Input Efficiency

*Increasing the efficiency with which nutrients, pesticides, and irrigation water are used in farming systems should be a fundamental objective of policies to improve water quality.*

Despite the complexity and regional diversity of the fate and transport processes that determine how agricultural production affects soil and water quality, two general approaches can control loadings of nutrients, pesticides, salts, and trace elements. One approach is to reduce the total residual masses of nutrients, pesticides, salts, and trace elements in the soil-crop system by increasing the efficiency with which nutrients, pesticides, and irrigation water are used. The second approach is to keep these residuals in the soil-crop system by curtailing the transport processes (leaching, runoff, erosion, volatilization) that carry pollutants out of the soil-crop system or by increasing the mass of inputs immobilized or degraded in the soil-crop system.

To date, the effects of most agricultural conservation programs on water quality have been to reduce erosion and runoff, thereby reducing the transport of nutrients, pesticides, salts, and trace elements to surface water. The nutrients and pesticides retained in the upper portion of the soil profile are subject to additional management efforts that can biologically or chemically reduce their concentrations through microbial degradation or chemical reactions or through subsequent uptake by crops. Once pesticides, nutrients, and salts have moved out of the upper soil profile to deeper levels in the soil or to surface runoff, the probability that they will be delivered to surface water or groundwater increases. Keeping the applied inputs in the soil-crop system by increasing the resistance of farming system soils to erosion and runoff should continue to be an important element of programs to improve soil and water quality. (This approach is discussed in the next section.)

Reducing the total residual mass of nutrients, pesticides, salts, and trace elements in the soil-crop system by increasing the efficiency with which nutrients, pesticides, and irrigation water are used, however, is essential to preventing surface water or groundwater pollution. This approach is also promising because of the potential for financial gains for the producer by reducing the input costs per unit of production.

There are two ways to improve input efficiency. The first and most direct way is to bring the amount of nitrogen, phosphorus, pesticides, and irrigation water applied into better balance with crop needs. The second more indirect but not less promising way is to alter cropping systems through rotations, cover crops, intercrops, and other cropping patterns that both reduce the need for inputs and help to keep those inputs that are applied in the soil-crop system.

Improved management can greatly increase the efficiency with which inputs are used in crop production systems. Input management can be improved by using information and technologies that already exist. The diversity of production systems, soils, and landscapes that characterize U.S. agriculture, however, makes generalizations difficult. Part Two of this volume provides more in-depth discussions of ways to improve input management. The most important opportunities that are generally applicable to crop production systems are discussed here.

## Improving Nitrogen Management

Economically viable food and fiber production often requires additions of large amounts of nitrogen. Nitrogen cycling in the soil-crop system is complex (see, e.g., Figure 6-1 in Chapter 6). Nitrogen in the form of nitrate is very mobile in the environment, making some losses of nitrogen into the environment inevitable. Many facets of nitrogen management can be improved, however, to reduce those losses.

### Reduction of Residual Nitrogen in the Farming System

*Reducing the amount of residual nitrogen in the soil-crop system by bringing the nitrogen entering the system from all sources into closer balance with the nitrogen leaving the system in harvested crops should be the objective of nitrogen management to reduce losses of nitrogen to the environment.*

The nitrogen supplied in excess of that needed for crop requirements leaves a pool of residual nitrogen in the soil. Over time, the size of the residual nitrogen pool directly influences the magnitude of losses of available or mobile forms of nitrogen to surface water, groundwater, and the atmosphere. Nitrogen applications beyond the amount required for crop growth lead to increases in the mass of residual nitrogen that is vulnerable to loss to the environment through leaching or subsurface drainage. Nitrate losses tend to be greatest in agricultural watersheds in which nitrogen inputs from synthetic fertilizer, manure, or legumes greatly exceed the amount of nitrogen taken up by the crop (Meisinger and Randall, 1991).

The nitrogen delivered in fertilizers, manures, rainfall, and irrigation water; the nitrogen mineralized from soil organic matter and crop residues; and the nitrogen fixed by legumes all contribute to the nitrogen budget of a particular agricultural field. Nitrogen in the form of ammonium ions and nitrate are of particular concern because they are very mobile forms of nitrogen and are most likely to be lost to the environment. All forms of nitrogen, however, are subject to transformation to ammonium ions and nitrate as part of the nitrogen cycle in agroecosystems and all can contribute to residual nitrogen and nitrogen losses to the environment. The importance of any particular source depends on the type of agricultural enterprise, soil properties, geographic location, and climate. (See Chapter 6 for a discussion of the nitrogen cycle.)

### Nitrogen Mass Balances

National-, regional-, and farm-level nitrogen mass balances suggest that current nitrogen inputs from all sources usually exceed the nitrogen harvested and removed with crops. Table 2-1 presents estimates of national and regional nitrogen inputs and outputs. (See Chapter 6 for a discussion of nitrogen mass balance estimates and the Appendix for a complete discussion of how these estimates were made.) In most regions and for the United States as a whole, the nitrogen applied in synthetic fertilizers is less than that harvested in crops. The nitrogen in synthetic fertilizers is not the only nitrogen source; the nitrogen in manures (manure-N) and the nitrogen fixed by legumes (legume-N) provide nearly as much nitrogen as is applied in commercial fertilizers in these estimates (the nitrogen in crop residues is assumed to be present in inputs and outputs in equal amounts). If all sources of nitrogen are accounted for, the estimated nitrogen inputs are nearly 1.5 times as great as the nitrogen removed in harvested crops or crop residues. Very little nitrogen is applied to legumes such as alfalfa and soybeans, yet those crops account for more than 35 percent of the nitrogen harvested by all crops. If only nitrogen inputs and outputs for major commodities such as corn, wheat, and cotton are considered, only about 35 percent of total nitrogen inputs are accounted for in the harvested crop.

Peterson and Frye (1989) estimated the amount by which the nitrogen in synthetic fertilizers applied to corn fields (aggregated at the national level) replaced the amount of nitrogen removed in the grain of that year's crop. The nitrogen applied to corn in synthetic fertilizer exceeded that removed in the grain by 50 percent or more for every year since 1968. Peterson and Frye's (1989) estimates of the amount of nitrogen applied included that in synthetic fertilizers only and did not include estimates of the nitrogen provided in manures or by legume fixation.

Similarly, Peterson and Russelle (1991) estimated the amount of nitrogen applied to corn in synthetic fertilizers and that supplied by alfalfa in the Corn Belt (see Table 6-7 in Chapter 6). Their estimate of the amount of nitrogen supplied by alfalfa included estimates of the amount of nitrogen that might be applied as manure from alfalfa fed to cattle as well as nitrogen fixed by alfalfa. Depending on whether they used a low or a high estimate of how much nitrogen was supplied by alfalfa, Peterson and Russelle estimated that nitrogen applications could be reduced by between 8 and 14 percent for the region as a whole. For states such as Michigan, Minnesota, and Wisconsin, which grow more alfalfa than other states, the estimated nitrogen reductions were much

TABLE 2-1 Regional and National Estimates of Nitrogen Inputs, Outputs, and Balances on Croplands, Medium Legume-N Fixation Scenario (metric tons)

| Region | Input | | | | | Output | | | Balance |
|---|---|---|---|---|---|---|---|---|---|
| | Nitrogen Fertilizer | Recoverable Manure-N | Legume-N | Crop Residues | Total | Harvested Crop | Crop Residues | Total | |
| Northeast | 252,000 | 224,000 | 284,000 | 70,200 | 831,000 | 412,000 | 70,200 | 482,000 | 349,000 |
| Appalachia | 564,000 | 109,000 | 395,000 | 102,000 | 1,170,000 | 491,000 | 102,000 | 593,000 | 577,000 |
| Southeast | 609,000 | 78,100 | 173,000 | 43,800 | 904,000 | 236,000 | 43,800 | 280,000 | 624,000 |
| Lake States | 988,000 | 278,000 | 984,000 | 368,000 | 2,620,000 | 1,420,000 | 368,000 | 1,780,000 | 834,000 |
| Corn Belt | 2,720,000 | 240,000 | 2,750,000 | 1,220,000 | 6,940,000 | 3,860,000 | 1,220,000 | 5,080,000 | 1,850,000 |
| Delta | 468,000 | 59,400 | 552,000 | 105,000 | 1,180,000 | 444,000 | 105,000 | 548,000 | 636,000 |
| Northern Plains | 1,510,000 | 256,000 | 1,020,000 | 602,000 | 3,390,000 | 1,930,000 | 602,000 | 2,530,000 | 859,000 |
| Southern Plains | 920,000 | 183,000 | 82,000 | 129,000 | 1,310,000 | 503,000 | 129,000 | 632,000 | 682,000 |
| Mountain | 554,000 | 160,000 | 451,000 | 158,000 | 1,320,000 | 786,000 | 158,000 | 944,000 | 379,000 |
| Pacific | 798,000 | 146,000 | 177,000 | 88,700 | 1,210,000 | 498,000 | 88,800 | 587,000 | 623,000 |
| United States | 9,390,000 | 1,730,000 | 6,870,000 | 2,890,000 | 20,900,000 | 10,600,000 | 2,890,000 | 13,500,000 | 7,420,000 |

NOTE: See the Appendix for a complete discussion of the methods used to estimate the nitrogen inputs, outputs, and balances given here.

TABLE 2-2 Nitrogen Budgets for Four Farms (A, B, C, and D) in Southeastern Minnesota

| | kg of Nitrogen/ha Used | | | |
|---|---|---|---|---|
| Nitrogen Budget Item | A | B | C | D |
| Sources | | | | |
| Commercial fertilizer | 174 | 146 | 162 | 155 |
| Soybean credits | 29 | NA | NA | NA |
| Alfalfa credits | 11 | 30 | 19 | NA |
| Manure | 37 | 146 | 32 | NA |
| Total sources | 251 | 321 | 214 | 155 |
| Nitrogen needed to meet yield goals | 184 | 172 | 184 | 169 |
| Excess nitrogen per hectare | 67 | 149 | 30 | -15 |

NOTE: NA, not applicable.

SOURCE: Adapted from T. D. Legg, J. J. Fletcher, and K. W. Easter. 1989. Nitrogen budgets and economic efficiency: A case study of southeastern Minnesota. J. Prod. Agriculture 2:110-116.

larger—20 to 36 percent for Michigan, 13 to 23 percent for Minnesota, and 37 to 66 percent for Wisconsin.

Such national- or regional-level mass balances are crude generalizations of the real situation in particular crop fields. The actual balance between the nitrogen applied and that required for crop growth varies from region to region, from farm to farm, and even from field to field. Nitrogen must be applied in excess of the amount actually harvested in grain and residues because the efficiency of nitrogen uptake by the crop is less than 100 percent and because precise crop needs vary with time and weather. The magnitude of the unaccounted for nitrogen in estimated mass balances, however, indicates the underlying reason for the loss of nitrogen from crop production and illustrates the potential for improvements in nitrogen management.

Farm-level nitrogen mass balances in many instances reinforce the picture that emerges from national- and regional-level estimates of nitrogen mass balances. Legg and colleagues (1989), for example, estimated nitrogen budgets for four farms in southeastern Minnesota (Table 2-2). Generally, the amount of nitrogen applied in synthetic fertilizer was about the same as the amount of nitrogen removed in the crop. The nitrogen from all sources, however, was far in excess of the nitrogen removed in the crop on those farms where multiple sources of nitrogen were available, suggesting that the use of supplemental applications of nitrogen in synthetic fertilizer could have been reduced. Similar budgets at the farm level have been reported by Lanyon and

Beegle (1989), Duffy and Thompson (1991), Schepers and colleagues (1986), Bouldin and colleagues (1984), and Pratt (1984).

### Refining Fertilizer Recommendations

*The most important immediate opportunity for improving nitrogen management is to refine recommendations for application of synthetic fertilizers containing nitrogen.*

Although nitrogen is supplied to cropping systems from many sources, including legumes and manures, most adjustments to the total nitrogen applied to cropping systems come by refining the quantity, location, and time of year that producers apply synthetic fertilizers containing nitrogen. Applications of synthetic fertilizers containing nitrogen are much easier to manage because the amount of nitrogen applied is known with accuracy. More important, when livestock or legumes are an important part of the farm enterprise, nitrogen additions from these sources are a fixed part of the nitrogen budget for the enterprise, and adjustments in the total amount of nitrogen applied will likely be made by adjusting the amounts of synthetic fertilizers containing nitrogen that producers apply. Use of legumes and applications of manure may be needed to improve soil quality in addition to their value as sources of nitrogen. Improving the management of synthetic fertilizers containing nitrogen presents the greatest opportunity for improved nitrogen use efficiency.

Recommendations for application of synthetic fertilizers containing nitrogen can be improved by setting realistic yield goals and accounting for all sources of nitrogen when making fertilizer recommendations.

### Realistic Yield Goals

As a crop's yield increases, the crop's need for nitrogen increases, at least initially. The dilemma for producers is that nitrogen must be applied before the crop yield is known. Nitrogen recommendations, therefore, must be based on some expectation of crop yield. For many crops, nitrogen requirements and recommendations are based on yield goals, that is, the yield expected by the producer under optimum growing conditions. The importance of setting realistic yield goals as the basis for making both economically and environmentally sound recommendations has often been highlighted (see, for example, Bock and Hergert [1991]; Peterson and Frye [1989]; University of Wisconsin-Extension and Wisconsin Department of Agriculture, Trade and Consumer Protection [1989]; U.S. Congress, Office of Technology Assessment [1990]). An unrealistically high yield goal will result in nitrogen applications in excess of what is needed for the yield that is actually achieved and will contribute to the mass of residual nitrogen in the soil-crop system.

*Yield goals should be established on the basis of the historical yields achieved on a field-by-field basis.*

The actual yield in any year depends on the weather and the inherent soil quality. The importance of weather in determining yields means that yields will vary from year to year, even under the best management conditions. If the producers apply nitrogen after planting, there is more opportunity to adjust applications to weather during the growing season. If most or all of the nitrogen is applied before planting, the yield goal set by the producer has important effects on the potential for water pollution. In such cases, the best way to establish yield goals is to obtain an average yield for the field during the previous 5 years.

If a producer uses the yield from a bumper crop as the goal, the result will be overapplication of nitrogen during most years. This practice increases production costs and the amount of residual nitrogen. In addition, many soils, though not those with very low levels of organic matter, typically supply the additional nitrogen needed during a bumper crop year because optimal weather and soil conditions that lead to a bumper crop also increase the amount of nitrogen mineralized from the organic matter in the soil (Schepers and Mosier, 1991).

No national data are available to estimate how realistically producers now set their yield goals. A number of local studies (Schepers and Mosier, 1991; Schepers et al., 1986) suggest that the yield goals set by producers are not often achieved. The data reported by Schepers and colleagues (1986) showed that producers in Hall County, Nebraska, generally set their yield goals at 2,688 kg/ha (40 bu/acre) more than the yield that they actually achieved. These unrealistically high yield goals meant that nitrogen was applied at rates averaging more than 45 kg/ha (40 lbs/acre) over the rate recommended by the University of Nebraska soil testing laboratory.

*Accounting for All Sources of Nitrogen*

The amount of nitrogen that needs to be applied to cropland depends on the amount already available in the soil from all sources. Producers must account for the nitrogen available from manure applications, legumes, soil organic matter, and other sources before recommendations for supplemental nitrogen applications can be made. The importance of carefully accounting for all nitrogen sources has repeatedly been stressed as a way to improve nitrogen management (see, for example, Bock and Hergert [1991]; Peterson and Frye [1989]; Schepers and Mosier [1991]; University of Wisconsin-Extension and Wisconsin Department of Agriculture, Trade and Consumer Protection [1989]; U.S. Congress, Office of Technology Assessment [1990]).

The nitrogen balances estimated in Table 2-1 reinforce the importance of accounting for all sources of nitrogen when making decisions about rates of nitrogen fertilization. The nitrogen that is recoverable in manure and legumes supplies roughly the same amount of nitrogen applied to crop fields in synthetic fertilizers containing nitrogen. In some regions where producers grow large amounts of soybeans or alfalfa, the nitrogen credits from legumes alone may exceed the amount supplied by synthetic fertilizers.

Although no comprehensive data are available on how well producers credit the nitrogen available from manures, legumes, and other sources when making fertilizer application decisions, those data that are available suggest real opportunities for improvement. Peterson and Russelle (1991), for example, estimated that fertilizer applications to corn in the Corn Belt could be reduced by between 237,000 and 435,000 metric tons (261,000 and 479,000 tons) by properly accounting for the nitrogen supplied by alfalfa (see Table 6-7 in Chapter 6). If the nitrogen supplied from soybeans had also been included in their analyses, the possible nitrogen application reductions would be greater. Two statewide surveys of producers in Iowa found that although 80 and 76 percent of producers took some credits for the nitrogen value of soybean and 92 percent of producers took some credits for the nitrogen value of alfalfa, the credits taken were inadequate and only 50 percent of the producers took credit for the nitrogen in manures (Duffy and Thompson, 1991; Kross et al., 1990). Nitrogen balances for individual farms also indicate the importance of accounting for the nitrogen in manures and fixed by legumes (Bouldin et al., 1984; Lanyon and Beegle, 1989; Legg et al., 1989; Padgitt, 1989).

El-Hout and Blackmer (1990) evaluated corn fields that followed alfalfa in rotation in northeast Iowa. Fertilizer application rates ranged from 6 to 227 kg/ha of nitrogen (5 to 203 lb/acre) and 59 percent of the fields sampled also received manure applications. Of the fields sampled, 86 percent had greater soil-nitrate concentrations than needed for optimal yields, 56 percent had at least twice the optimal amount, and 21 percent had at least three times the amount of soil-nitrate needed. Similarly, farm assessments completed as part of a nutrient and pesticide management program in Wisconsin showed that more than half of the farms were applying 50 percent more nitrogen than recommended for optimal crop production (Nowak and Shepard, 1991).

*Recommendations for applications of fertilizers containing nitrogen should be made by taking a full accounting of and by considering all internal sources of nitrogen, including the nitrogen fixed by legumes and in manures.*

The single most important way to improve nitrogen management is to reduce supplemental applications of nitrogen to account for nitrogen supplied by legumes and manures. Most states publish estimates for the nitrogen replacement value of alfalfa, soybeans, and other legumes. Widespread use of even these estimates when making fertilization application decisions would result in immediate improvements in nitrogen management.

Uncertainty in estimating the nitrogen contents of manures is an important constraint to an accurate accounting of the nitrogen contributed by manures (Schepers and Mosier, 1991). (See the section on manures later in this chapter and in Chapter 11.) Published estimates and ranges of the nitrogen contents of manures are available, however, and could be used as rules of thumb for improving manure management. If results of statewide surveys in Iowa are representative, more than one-half of the producers would benefit from taking even a conservative manure credit, since they now take no credits (Duffy and Thompson, 1991; Hallberg et al., 1991; Padgitt, 1989).

### Synchronizing Fertilizer Applications with Crop Needs

Supplying the nitrogen needed for crop growth during the period when it is most needed can be an important way to improve nitrogen management. Nitrogen is needed most during the period when the crop is actively growing. The nitrogen applied before that time is vulnerable to loss through leaching or lateral subsurface flow because of the mobility of nitrates in the soil system. Larger nitrogen applications are required if the nitrogen is applied in the fall or early spring before planting to make up for the nitrogen lost or that becomes unavailable in the soil during the period between application and crop growth. Use of a nitrification inhibitor that slows down the rate at which nitrogen is converted to mobile nitrates produces an intermediate effect but may not reduce losses to the environment over the long term.

*Fertilizers containing nitrogen should, whenever possible, be applied during and/or after planting.*

The opportunity to increase the efficiency with which nitrogen is used by synchronizing applications with periods of crop growth has often

TABLE 2-3    Crops Receiving Fertilizer Nitrogen Before, During, and After Seeding

| Crop | Area Planted ($10^3$ ha) | Percent of Planted Area Fertilized | Average Application Rate (kg/ha) | Percent of Fertilized Area Receiving Nitrogen | | | |
|---|---|---|---|---|---|---|---|
| | | | | Before Seeding | | At Seeding | After Seeding |
| | | | | Fall | Spring | | |
| Corn | 23,800 | 97 | 148 | 28 | 57 | 43 | 26 |
| Cotton | 3,940 | 80 | 96 | 35 | 44 | 8 | 51 |
| Rice | 729 | 97 | 128 | 10 | 16 | 4 | 92 |
| Soybeans | 19,500 | 17 | 27 | 24 | 47 | 23 | 8 |
| Winter Wheat | 16,300 | 84 | 69 | 73 | NA | 22 | 42 |
| Spring Wheat | 6,400 | 69 | 59 | 38 | 34 | 63 | 2 |

SOURCE: H. H. Taylor. 1991. Fertilizer application timing. Pp. 30-38 in Agricultural Resources: Inputs Situation and Outlook. Report No. AR-24. Washington, D.C.: U.S. Department of Agriculture, Economic Research Service, Resources and Technology Division.

been highlighted (Ferguson et al., 1991; Jokela and Randall, 1989; Peterson and Frye, 1989; Randall, 1984; Russelle and Hargrove, 1989; University of Wisconsin-Extension and Wisconsin Department of Agriculture, Trade and Consumer Protection, 1989; U.S. Congress, Office of Technology Assessment, 1990). Similarly, increased losses of nitrogen from nitrogen applications in the fall or early spring have also been noted. Applications of manures without incorporation in fall, winter, and spring can be a particularly important source of surface water pollution (University of Wisconsin-Extension and Wisconsin Department of Agriculture, Trade and Consumer Protection, 1989). Data on the timing of nitrogen applications, however, suggest that preplant applications are the rule, not the exception, for most major commodity crops (Table 2-3). Only 26 percent of corn, 51 percent of cotton, and 8 percent of soybeans, received fertilizer containing nitrogen following seeding.

Some improvement in nitrogen management could be achieved by increasing the percentage of crops treated with nitrogen postplanting rather than preplanting except for small applications at planting that may be needed as starter fertilizer. Corn production alone consumes more than 40 percent of the nitrogen applied to commodity crops (Vroomen, 1989). Opportunities to improve nitrogen management by adjusting the timing of nitrogen applications appear to be particularly great for corn production, for which fall and spring applications are both common and application rates are high (Table 2-3). The advantages of changing application times may be specific to the climate and site conditions (Killorn and Zourarakis, 1992). When coupled with efforts to account for all sources of nitrogen and to set realistic yield goals, significant financial and environmental benefits may be achieved.

### New and Improved Tools

Current nitrogen management data suggest that substantial improvements in nitrogen management could be achieved through the more widespread use of current technology for setting yield goals, making and implementing fertilizer recommendations, and increasing postplanting nitrogen applications. Weather and crop yield variabilities, however, create uncertainties about a crop's nitrogen requirements and the amount of residual nitrogen available from the soil that makes refining nitrogen management difficult.

*The development, testing, and implementation of improved methods of estimating crop nitrogen requirements following planting should be the highest research priority for improving nitrogen management.*

Typical, currently used methods of testing soils are not suitable for supplying the information needed to reduce the uncertainty in estimating a crop's nitrogen needs. The initial results from testing new methods of estimating the nitrogen content of soils or crop tissues appear promising. Practical and accurate testing methods that would allow nitrogen fertilizer recommendations to be made following planting is the single most important technical innovation needed to improve nitrogen management. The inadequacy of current methods for reducing this uncertainty is a serious impediment to improving nitrogen management.

Various plant and tissue tests have proved to be valuable tools for more efficient nitrogen management in vegetable and citrus crops, but such methods must be refined and implemented for the major row crops such as corn, to which most of the nitrogen used in the United States is applied. Many methods are being tested across the Corn Belt (Binford et al., 1992; Blackmer et al., 1989; Cerrato and Blackmer, 1991; Fox et al., 1989; Magdoff, 1991a; Motavalli et al., 1992; Piekielek and Fox, 1992; Tennesse Valley Authority, National Fertilizer Development Center, 1989). The presidedress soil nitrate test developed in Iowa, for example, measures the amount of nitrogen available in the upper 0.3 to 0.6 m (1 to 2 feet) of the soil profile and has been used to refine recommendations for supplemental fertilizer applications. In a project in which fertilizer dealers used the test to refine fertilizer recommendations, nitrogen applications were reduced an average of 42 percent while maintaining crop yields (Blackmer and Morris, 1992; Hallberg et al., 1991).

Development of monitoring and modeling systems to help estimate the nitrogen available to the crop from the soil and from carryover of nitrogen applied the previous year are also needed. Models that integrate climatic, soil, and crop conditions to predict the nitrogen available from the previous year could help producers refine their annual plans for nitrogen application.

*Research to refine crop yield response models for use in estimating optimal nitrogen application rates should be undertaken.*

Fertilizer recommendations are made on the basis of models of crop responses to various rates of nitrogen application. These models are normally developed by each state for the crops grown in that state and for the soils and climatic regions of that state. The accuracies of these models in large part determine the level of refinement possible in fertilizer recommendations.

Various studies have noted that different models that use the same data from crop response field tests predict very different optimal rates of nitrogen application. Cerrato and Blackmer (1990) evaluated the five most widely used response models. All of the models predicted very similar maximum obtainable yields, but the optimal nitrogen rates differed by 250 kg/ha (223 lbs/acre). In addition, the crop response models in use have often been developed on the basis of studies of plots with only two to four different rates of nitrogen application. Blackmer (1986) found that much greater refinements in the estimation of crop requirements were obtained from crop response models developed from a greater variety of application rates.

Refining current crop response models to allow greater precision in estimates of optimal nitrogen application rates is an important way to improve nitrogen management.

## Improving Phosphorus Management

Phosphorus, like nitrogen, is both an important plant nutrient and a serious pollutant when delivered to surface water. Most forms of phosphorus compounds are bound more tightly to most soils than nitrogen compounds, creating important differences in the approaches taken to control phosphorus losses from farming systems.

### Phosphorus Cycle

Like nitrogen and other plant nutrients, the phosphorus added to the soil-crop system goes through a series of transformations as it cycles through plants, animals, microbes, soil organic matter, and the soil mineral fraction. Unlike nitrogen, however, phosphorus is tightly bound in most soils and only a small fraction of the total phosphorus found in the soil is available to crop plants.

Most of the phosphorus in soil is found as a complex mixture of mineral and organic materials. Organic phosphorus compounds in plant residues, manures, and other organic materials are broken down through the action of soil microbes. Some of the organic phosphorus can be released into the soil solution as phosphate ions that are immediately available to plants. Much of the organic phosphorus is taken up by the microbes themselves. As microbes die, the phosphorus held in their cells is released into the soil. A considerable amount of organic phosphorus is held in the humic materials that make up soil organic matter. A portion of this organic phosphorus is released each year as these humic materials decay. The phosphate ions released from the decomposition of organic phosphorus compounds or added directly in fertilizers containing inorganic phosphorus readily react with soil minerals and are immobilized in forms that are unavailable for plant growth. (Figure 7-1 in Chapter 7 provides an illustration of the phosphorus cycle in the soil-crop system.)

### Transport Processes

Phosphorus can be lost from the soil-crop system in soluble form through leaching, subsurface flow, and surface runoff. Particulate phosphorus is lost when soil erodes. Phosphorus loss by leaching to groundwater, in most regions of the United States, is not a problem (Gilliam et al., 1985). The majority of phosphorus lost from agricultural lands is with surface flow, both in solution (soluble phosphorus) and bound to eroded sediment particles (particulate phosphorus). Most of the total phosphorus loss from cropped land is in the sediment-bound form (Gilliam et al., 1985; Sharpley and Menzel, 1987; Viets, 1975). Soluble phosphorus is more readily available to stimulate eutrophication, but particulate phosphorus can be a long-term source of phosphorus once it is delivered to surface water (Gilliam et al., 1985; Sharpley and Menzel, 1987).

### Phosphorus Mass Balance

Phosphorus is added to agricultural lands in crop residues and manures, in synthetic fertilizers, and from phosphorus-bearing minerals in the soil. Part of the phosphorus entering the soil-crop system is removed with the harvested crop; the balance is immobilized in the soil, incorporated into soil organic matter, or lost in surface or shallow subsurface flow, primarily to surface water.

Table 2-4 provides estimates of regional and national phosphorus inputs and outputs in 1987. (See Chapter 7 for a discussion of phosphorus mass balance estimates and the Appendix for a complete discussion of how these mass balance estimates were made.) The difference between phosphorus inputs and outputs in crops and crop residues is reported as the phosphorus mass balance.

The phosphorus in synthetic fertilizers is the single most important source of phosphorus added to croplands in the United States (Table 2-4). Approximately 3.6 million metric tons (4.0 million tons) of phosphorus was added to croplands in 1987. The amount of synthetic fertilizer applied represents 79 percent of phosphorus inputs. The amount of recoverable phosphorus voided in manures is small compared with that supplied in synthetic fertilizers at the national level. Locally, the proportion of phosphorus supplied by manures can be large. The recoverable phosphorus in manure (manure-P), for example, supplies 65 percent of total phosphorus inputs in Vermont (see Table 7-3 in Chapter 7).

Approximately 1.3 million metric tons (1.4 million tons) of phosphorus—or 29 percent of total phosphorus inputs—was harvested along with crops in the United States in 1987 (Table 2-4). Another 272,000 metric tons (300,000 tons) of phosphorus—or 6 percent of total phosphorus inputs—was contained in crop residues. About 1.6 million metric tons (1.8 million tons)—or 36 percent of total phosphorus inputs—can be accounted for in harvested crops and crop residues, leaving an unaccounted for balance of 2.9 million metric tons (3.1 million tons)—or 63 percent of total phosphorus inputs.

The fraction of total phosphorus inputs lost in eroded soil and in surface runoff can be substantial, but it is difficult to estimate. Larson and colleagues (1983) estimated that 1.74 million metric tons (1.92 million tons) of phosphorus—or about 50 percent of the estimated total phosphorus balance in Table 2-4—was lost in eroded sediments in 1982. Additional phosphorus can be lost in solution.

The majority of the unaccounted for phosphorus balance on croplands is immobilized in the soil's mineral or organic fractions. The actual magnitude of the unaccounted for balance of phosphorus added to farming systems varies from region to region, soil to soil, and farm to farm. The national mass balance in Table 2-4, however, suggests that the potential for buildup of phosphorus levels in cropland soils over time is large. The buildup of phosphorus in soil increases the

TABLE 2-4 Regional and National Estimates of Phosphorus Inputs, Outputs, and Balances on Croplands, 1987 (metric tons)

| Region | Input | | | | Output | | | Balance |
|---|---|---|---|---|---|---|---|---|
| | Fertilizer-P | Recoverable Manure-P | Crop Residues | Total | Harvested Crop | Crop Residues | Total | |
| Northeast | 155,000 | 76,700 | 6,660 | 239,000 | 48,300 | 6,660 | 55,000 | 184,000 |
| Appalachia | 334,000 | 42,300 | 9,130 | 385,000 | 60,500 | 9,130 | 69,600 | 315,000 |
| Southeast | 264,000 | 32,400 | 3,960 | 300,000 | 27,800 | 3,960 | 31,800 | 269,000 |
| Lake States | 442,000 | 98,500 | 34,100 | 575,000 | 168,000 | 34,100 | 202,000 | 373,000 |
| Corn Belt | 1,130,000 | 115,000 | 112,000 | 1,350,000 | 481,000 | 112,000 | 593,000 | 761,000 |
| Delta | 119,000 | 22,200 | 10,600 | 151,000 | 54,600 | 10,600 | 65,200 | 86,200 |
| Northern Plains | 446,000 | 100,000 | 58,800 | 605,000 | 261,000 | 58,800 | 319,000 | 286,000 |
| Southern Plains | 274,000 | 59,900 | 14,300 | 348,000 | 72,400 | 14,300 | 86,700 | 262,000 |
| Mountain | 186,000 | 53,300 | 14,500 | 254,000 | 93,500 | 14,500 | 108,000 | 146,000 |
| Pacific | 224,000 | 54,300 | 8,450 | 287,000 | 57,700 | 8,450 | 66,200 | 221,000 |
| United States | 3,570,000 | 655,000 | 272,000 | 4,500,000 | 1,320,000 | 272,000 | 1,600,000 | 2,900,000 |

NOTE: See the Appendix for a complete discussion of the methods used to estimate the phosphorus inputs, outputs, and balances given here.

amount of phosphorus lost in runoff water and sediments from croplands.

### Control Phosphorus Buildup in Soil

Because phosphorus is tightly bound to the soil, most efforts to reduce the amount of phosphorus lost from farming systems have focused on reducing erosion. Reducing erosion alone, although essential, will not be sufficient to control phosphorus losses if the phosphorus levels in soil buildup to high levels. Excessive levels of phosphorus in the soil increases the amount of soluble phosphorus lost in surface runoff and the concentration of phosphorus in sediments, thus counteracting some of the reductions in phosphorus pollution gained by controlling erosion.

*Efficient management of phosphorus inputs to prevent the buildup of excess phosphorus levels in soil while providing adequate phosphorus for crop growth should be a fundamental part of programs to reduce phosphorus loadings to surface water.*

The level of phosphorus in surface soil is a critical factor that determines the phosphorus loads in runoff water and the relative proportions of phosphorus lost in solution and attached to soil particles. Increased residual phosphorus levels in the soil lead to increased phosphorus loadings to surface water, both in solution and attached to soil particles. Policies and programs to reduce phosphorus losses from farming systems should pay much more attention to improving the management of phosphorus inputs to reduce the buildup of phosphorus in soil. Erosion control should remain an important objective for reducing the amount of phosphorus lost from farming systems, but it should be coupled with efforts to reduce the buildup of phosphorus in soil.

Because of a history of phosphorus applications in excess of that harvested or naturally high phosphorus levels in soil, or both, phosphorus levels have increased in many U.S. soils (Thomas, 1989) and many now have high phosphorus levels. The results of tests for the levels of phosphorus in the soil are reported as being very low, low, medium, high, or very high. These results are based on the probability that crops grown on that soil will respond to an application of phosphorus fertilizer rather than on the absolute amounts of extractable phosphorus that were detected in the soil. A crop grown on a soil testing very low for phosphorus, for example, has a high probability (90 to 100 percent) of responding to supplemental applications of phosphorus fertilizer. Conversely, a crop grown on a soil testing very high for phosphorus has a low probability (0 to 10 percent) of responding to supplemental appli-

TABLE 2-5 Percentage of Soil Tests Reporting High to Very High Levels of Soil Phosphorus

| State | Percent | State | Percent |
|---|---|---|---|
| Alabama | 35 | Nebraska | 31 |
| Arizona | 51 | Nevada | 48 |
| Arkansas | 14 | New Hampshire | — |
| California | 41 | New Jersey | — |
| Colorado | 43 | New Mexico | — |
| Connecticut | 51 | New York | 38 |
| Delaware | 65 | North Carolina | 67 |
| Florida | 45 | North Dakota | 30 |
| Georgia | 38 | Ohio | 68 |
| Idaho | 60 | Oklahoma | 48 |
| Illinois | 63 | Oregon | 49 |
| Indiana | 78 | Pennsylvania | 44 |
| Iowa | 56 | Rhode Island | — |
| Kansas | 39 | South Carolina | 40 |
| Kentucky | 42 | South Dakota | 56 |
| Louisiana | 37 | Tennessee | 49 |
| Maine | 51 | Texas | 37 |
| Maryland | 74 | Utah | 60 |
| Massachusetts | — | Vermont | 25 |
| Michigan | 73 | Virginia | 58 |
| Minnesota | 76 | Washington | 54 |
| Mississippi | 34 | West Virginia | — |
| Missouri | 35 | Wisconsin | 66 |
| Montana | 41 | Wyoming | 38 |

NOTE: Dashes indicate no data were reported.

SOURCE: Adapted from Potash and Phosphate Institute. 1990. Soil test summaries: Phosphorus, potassium, and pH. Better Crops with Plant Food 74(2):16-18.

cations of phosphorus fertilizer. Table 2-5 reports the percentage of soil tests in each state that reported soils testing high to very high for phosphorus.

The phosphorus level in many U.S. soils is high enough that applications of additional phosphorus would not increase crop yields (McCollum, 1991; Novais and Kamprath, 1978; Yerokun and Christenson, 1990). Mallarino and colleagues (1991) cited several studies reporting that increases in soybean or corn yields are small or nonexistent when phosphorus levels in soil are in the medium category (Grove et al., 1987; Hanway et al., 1962; Million et al., 1989; Obreza and Rhoads, 1988; Olson et al., 1962; Rehm, 1986; Rehm et al., 1981). Phosphorus additions to soils that test high for phosphorus typically do not increase corn or soybean yields in the Corn Belt (Bharati et al., 1986; Hanway et al., 1962; Olson et al., 1962; Rehm, 1986). This suggests that applications of additional phosphorus to 56, 63, 78, 68, and 35 percent of the soils tested in Iowa, Illinois, Indiana, Ohio, and Missouri, respectively, would not be expected to increase yields.

Similar situations exist in the southeastern United States. Kamprath (1967, 1989) and McCollum (1991) have shown that corn and soybeans grown on Piedmont and Coastal Plain soils testing high in available phosphorus do not respond to phosphorus fertilizer additions. On the basis of the soil test data presented in Table 2-5, no response to phosphorus would be expected on approximately half of the soils that were tested in this region. In North Carolina, the amount of phosphorus recommended for use on soybeans grown in soils that tested medium for phosphorus is higher than the amount of phosphorus removed in the grain (Kamprath, 1989). Thus, current recommendations will lead to phosphorus levels in soil higher than those needed for corn or soybean production. The magnitude of the potential reduction in application of phosphorus, however, depends on the soil, climate, and crop planted.

### Thresholds for Phosphorus Levels in Soil

*Threshold levels of phosphorus in soil—beyond which no crop response from added phosphorus except for small starter applications would be expected— should be established.*

Most states have soil testing procedures and facilities that could be used to establish threshold levels of phosphorus in soil beyond which no crop response would be expected. Application of phosphorus to soils that contain phosphorus in excess of threshold levels should be discouraged or disallowed in extreme cases in which phosphorus loadings are causing severe damage. Once established, such threshold levels should be routinely reported as part of soil test results and fertilizer recommendations made by public and private organizations.

Reducing or suspending phosphorus applications to soils already testing high or very high for phosphorus is an important way to improve both the economic and environmental performance of farming systems. Mallarino and colleagues (1991), for example, studied the effect on yields of phosphorus additions to a soil testing high for phosphorus. They reported occasional positive yield responses to fertilization, but these positive responses were not, in most cases, sufficient to pay for the cost of the added phosphorus. In the 11 years of the study, phosphorus applications to soils testing high for phosphorus provided appreciable positive economic returns in only 1 year for corn. Added phosphorus provided no economic benefits for soybeans. The addition of phosphorus resulted in negative returns in most years for both corn and

TABLE 2-6   Proportion of Cropland Soils Tested for Nutrient Levels, Major Field Crops, 1989

| Crop | Area Planted (ha) | Percent Soil Tested | | | |
| | | Phosphate or Potash | | Nitrogen | |
| | | 1987 | 1988–1989 | 1987 | 1988–1989 |
| --- | --- | --- | --- | --- | --- |
| Corn | 23,431 | 26 | 33 | 13 | 20 |
| Cotton | 3,417 | 20 | 29 | 16 | 23 |
| Winter wheat | 14,047 | 16 | 16 | 10 | 10 |
| Spring wheat | 6,710 | 28 | 32 | 26 | 30 |
| Durum wheat | 1,214 | 14 | 26 | 14 | 26 |
| Northern soybeans | 15,277 | 25 | 27 | NR | NR |
| Southern soybeans | 20,692 | 23 | 26 | NR | NR |
| Rice | 844 | 15 | 20 | 9 | 14 |

NOTE: NR, no data reported.
SOURCE: M. Spiker, S. Dabrekow, and H. Taylor. 1990. Soil Tests and 1989 Fertilizer Application Rates. Pp. 46-49 in Agricultural Resources: Inputs Situation and Outlook. Report No. AR-17. Washington, D.C.: U.S. Department of Agriculture, Economic Research Service, Resources and Technology Division.

soybeans, with losses in some years being greater than $49/ha ($20/acre) for corn.

Several studies have investigated the buildup of soil phosphorus under continuous phosphorus fertilization (McCallister et al., 1987; Schwab and Kulyingyong, 1989); another study documented the loss of soil under a continuous cropping system in which only residual phosphorus was available for crop uptake (Novais and Kamprath, 1978). Both the buildup and the decline of soil phosphorus phases have been studied as well (Cope, 1981; McCollum, 1991; Meek et al., 1982), but relatively few (e.g., Cope, 1981; McCollum, 1991) have been conducted over long time spans (several decades). These few studies may provide some of the best information that can be used to aid in the prediction of residual phosphorus effects and actual phosphorus fertilization needs. These long-term studies suggest that soils with high phosphorus levels can be cropped for a decade or more without the amount of phosphorus in soil reaching a level at which fertilizer additions would result in a crop yield increase.

Few comprehensive data are available on how often and how many producers currently use soil tests when deciding how much phosphorus to apply. Data assembled by the Economic Research Service of USDA (Table 2-6) suggest that immediate improvements in phosphorus management and pollution prevention could be realized simply by expanding the use of currently available soil tests by producers coupled with setting thresholds for phosphorus levels in soil.

## Improving Manure Management

Manure supplies nitrogen, phosphorus, and other nutrients for crop growth; adds organic matter and improves soil structure and tilth; and increases the soil's ability to hold water and nutrients and to resist compaction and crusting. Disposal of manure as a waste often leads to both surface water and groundwater degradation. Improved manure management can effectively capture the benefits of manure as an input to crop production and can reduce the environmental problems associated with manure disposal.

### Nutrient Value of Manures

About 124 million metric tons (137 million tons) of manure was produced by cattle, sheep, swine, and poultry in the United States in 1987 (see the appendix for a discussion of how these estimates were made). These manures contained more than 5.0 million metric tons (5.5 million tons) of nitrogen and nearly 1.4 million metric tons (1.5 million tons) of phosphorus. At 1987 prices for nitrogen in anhydrous ammonia and phosphorus in superphosphate (Vroomen, 1989), the values of these nutrients in manures were about $1.2 billion and $450 million, respectively. Only part of the total nitrogen and phosphorus voided in manures is economically recoverable for use on croplands. Some of the manure is voided on pastures and rangelands where recovery is not feasible. Nitrogen is lost from manures by volatilization, and both nitrogen and phosphorus can be lost from barnyards and feedlots in runoff water. Tables 2-1 and 2-4 estimate that nearly 1.8 million metric tons (2 million tons) of nitrogen (34 percent of the total nitrogen voided) and about 726,000 metric tons (800,000 tons) of phosphorus (49 percent of the total phosphorus voided) were available from manures in 1987. The importance of manures as a source of nitrogen and phosphorus in crop production systems varies from region to region (Tables 2-1 and 2-4). Nationally, nitrogen from manures supplies about 8 percent of the total nitrogen applied to croplands in synthetic fertilizers, legumes, crop residues, and manures but ranges from 3 to 26 percent among farm production regions. In the Northeast region, the amount of nitrogen applied to croplands in manures is nearly equal to the amount in synthetic fertilizers. Phosphorus in manures supplies about 15 percent of total phosphorus inputs nationally, ranging from 8 percent in the Corn Belt to 32 percent in the Northeast.

### Manure Is an Important Source of Water Pollution

*Improvements in manure management should be a high priority in programs to improve water quality.*

Manures pose particularly difficult environmental problems. Manures reaching surface water disrupt aquatic ecosystems by depleting dissolved oxygen. Human health can be endangered by contamination of drinking water supplies with fecal bacteria and viruses carried in manures. Losses of nitrogen and phosphorus from feedlots and barnyards can be great, and they can be an important source of water quality problems (Bouldin et al., 1984; Brown et al., 1989; Daniel et al., 1982; Pinkowski et al., 1985; University of Wisconsin-Extension and Wisconsin Department of Agriculture, Trade and Consumer Protection, 1989; U.S. Congress, Office of Technology Assessment, 1990). Losses of nitrogen and phosphorus in runoff from croplands on which manures have been applied to the soil surface can also be great (Brown et al., 1989; Moore et al., 1978; U.S. Congress, Office of Technology Assessment, 1990). In places where animal agriculture is important and manures are not well managed, manures can be a particularly important source of nitrogen and phosphorus pollution.

### Obstacles to Improving Manure Management

Several important obstacles will make improvements in manure management difficult and costly. These include the concentration of livestock, which leads to a shortage of available cropland to which manure can be applied, the buildup of nitrogen and phosphorus in soils after repeated manure applications, and high capital costs.

#### Livestock Concentration

The concentration of livestock production in areas with insufficient cropland for effective utilization of the nitrogen and phosphorus in manure is perhaps the single greatest challenge to improving manure management. The concentration of livestock production in large confinement feeding operations or the development of regional concentrations of dairy, poultry, or other animal agricultural systems has created situations in which more manure is being produced than can be used on available cropland (see Figure 11-2 in Chapter 11). In Lancaster County, Pennsylvania, for example, the number of beef cattle increased 55 percent, dairy cattle increased 61 percent, hogs increased 677 percent, poultry layers and pullets increased 193 percent, and broilers increased

540 percent between 1960 and 1986 (Lanyon and Beegle, 1989). The resulting oversupply of manure leads directly and indirectly to increased pollution of surface water and groundwater with nitrogen, phosphorus, bacteria, and other organic pollutants.

Reducing supplemental nitrogen and phosphorus inputs to account for the nitrogen and phosphorus in manures will lead to improvements even in areas where animal agriculture is concentrated. In some watersheds, however, reducing nitrogen and phosphorus loadings to surface water or groundwater will be difficult unless livestock concentrations are reduced or unless the means of processing, transporting, or marketing manures or products derived from manures are developed.

*Nitrogen and Phosphorus Buildup after Repeated Applications*

The problem of manure oversupply is exacerbated by the buildup of phosphorus and nitrogen in soil after repeated manure applications. When manure is applied to the same field year after year, each succeeding year requires less manure to maintain the same amount of nitrogen available to the crop. For example, when manure containing only 1 percent nitrogen on a dry weight basis is added to cropland, it requires about 20 metric tons (22 tons) to supply 112 kg of available nitrogen per ha (100 lbs/acre) the first year but only about 5.1 metric tons (5.6 tons) after 15 years of repeated applications (see Table 11-3 in Chapter 11).

This problem is even more acute for phosphorus. The ratio of nitrogen to phosphorus in manure applied to the land is often between 2 to 1 and 3 to 1. Therefore, when manure is applied to supply adequate nitrogen for most cropping conditions, excess amounts of phosphorus are added, leading to phosphorus buildup in the soil. Application of manures at rates that prevent the buildup of phosphorus in soil, however, dramatically increases the amount of surplus manure that needs to be used. In regions where phosphorus pollution of surface water and groundwater is not a problem, the benefits derived from using manures to enhance overall soil quality and as a primary source of nitrogen for plant growth may outweigh any potential negative effects associated with increased phosphorus levels in the soils.

*High Capital Costs*

Effective use of the nutrients in manures requires equipment for collection and application and facilities for storage. Manures must be collected and stored until they can be applied to croplands. Storing manures where they are exposed to runoff and leaching leads to large losses of nitrogen and phosphorus. Equipment is needed to inject or incorporate manures into the soil to reduce runoff losses. The rate at which manures are applied to croplands is often imprecise and better application equipment is needed. Schepers and Fox (1989), for example, reported that actual manure application rates ranged from 29 to 101 metric tons/ha (13 to 45 tons/acre) in numerous manure calibration demonstrations in Lancaster County, Pennsylvania, even though most producers thought they were applying 45 metric tons/ha (20 tons/acre). Capital, equipment, and labor costs can be an important constraint to improving the efficiency of manure management.

### Special Emphasis on Manure Management

Manure management can be improved by improving manure collection and storage facilities, using better application equipment, and testing manures for nutrient content. In many cases, good manure management with accompanying reductions in outlays for synthetic fertilizers can lead to improved profits (Bouldin et al., 1984; Hallberg et al., 1991; Lanyon and Beegle, 1989). In regions or watersheds where manures supply a significant proportion of nitrogen and phosphorus to crop production, improved manure management to protect water quality should be emphasized.

There are few comprehensive national data that can be used to judge how well producers currently manage manures. Those studies that are available for particular farms and regions suggest that there is a substantial opportunity to improve manure management by taking appropriate credits for the manures that have been applied and by improving applications and storage practices (Bouldin et al., 1984; Duffy and Thompson, 1991; Hallberg et al. 1991; Lanyon and Beegle, 1989; Padgitt, 1989).

It may be necessary to provide subsidies or impose penalties. The capital cost of improving manure collection, storage, and application equipment may be large enough to constrain adoption without providing subsidies or imposing penalties. When the amount of manure produced is greater than the amount that can be efficiently applied to the available cropland, even the best manure management may still lead to large losses of nitrogen and phosphorus, increasing the potential for water pollution. There are no easy solutions to the obstacles to improving manure management created by large concentrations of livestock (see Chapter 11 for a more complete discussion of this problem). Restricting the number of animals or the amount of manure that a producer can produce may have a substantial effect on profitability and viability of the enterprise, and the capital cost of manure management facilities can be high.

There are technologies such as composting, anaerobic digestion, and gasification that hold some promise of producing products such as soil amendments, fertilizers, feedstuffs, and fuels from manures that can be transported out of regions with concentrations of livestock. None of these alternatives are problem free, however. The cost of transportation and application of fertilizers and soil amendments derived from manures may make these products unattractive to other agricultural producers. Special markets such as homeowners, landscapers, or greenhouses willing to pay higher prices for soil amendments or fertilizer may not be large enough to absorb the supply of such products. Use of manure as fuel for power generation on a scale large enough to be profitable may require transport of manures over distances that increase costs well beyond alternative feedstocks. Preliminary evaluations of such alternatives in Lancaster County, Pennsylvania, for example, indicated that the alternatives would be expensive to implement on a scale sufficient to solve excess nutrient problems in the county (Young et al., 1985).

Refining the composition of feeds to minimize the nitrogen and phosphorus in manures may hold promise. Van Horn (1991), for example, using data from Morse (1989) and National Research Council (1989c), concluded that the amounts of both phosphorus and nitrogen voided in manure can be controlled by composition of the feed. They recommended that management of the diet of dairy cattle, with this result in mind, should become an important component of nutrient management on dairy farms.

Solutions to the problem of manure management in regions with large concentrations of livestock will require efforts on multiple fronts. Livestock producers will have to be encouraged or required to reduce supplemental applications of nitrogen and phosphorus and improve the collection, storage, and application of manures. Regulation of feedlots and confined animal feeding facilities are needed to ensure that adequate manure handling, storage, and disposal systems are in place. Research to explore the feasibility of manure processing and to enable refined management of animal diets to reduce nitrogen and phosphorus in manure is also needed.

## Improving Pesticide Management

Natural biological processes play an important role in controlling damages caused by pests and pathogens. In highly managed ecosystems such as modern agricultural systems, these natural processes are often disrupted, increasing the risk of damage to crops and animals caused by insects, weeds, pathogens, or nematodes. The advent of synthetic chemical pesticides after World War II revolutionized pest control in agriculture. Pesticides have now become an important and nearly universally used input in agricultural production systems.

### Constraints to Making General Recommendations

Pesticides create particularly difficult problems for policymakers because general recommendations are difficult to make. Nearly 50,000 pesticide products are now registered for use with EPA, although the number of pesticides used extensively is much smaller. The ways that these pesticides behave in the environment depend on the interactions among their chemical properties, the soil at the site, the cropping system in which they are used, and the way in which they are applied. Improved pesticide management is therefore a chemical- and site-specific process. This specificity makes broad generalizations on the most promising ways to improve pesticide management more difficult to establish. The list of specific best-management practices to improve the ways in which pesticides are used in agricultural production systems is extensive (University of Wisconsin-Extension and Wisconsin Department of Agriculture, Trade and Consumer Protection, 1989; U.S. Congress, Office of Technology Assessment, 1990). Some general approaches, however, can form the basis of a national policy to improve pesticide management.

### Reducing the Total Mass of Pesticides Used

*Source control to reduce the total mass of pesticides applied to cropping systems should be the fundamental approach to reducing pesticide losses from farming systems.*

Unlike nitrogen and phosphorus, there is no inherent, natural pesticide cycle comparable to the nutrient cycles in agroecosystems. (Figure 8-1 in Chapter 8 describes the fates of pesticides applied to farming systems.) Pesticides applied to cropping systems are volatilized and lost to the atmosphere, lost to surface water bodies in solution or attached to sediments, leached to groundwater, exported with harvested crops, immobilized in the soil, or degraded in soil or by plants. Pesticide properties, soil properties, site conditions, and management practices interact to determine the fate of a pesticide. These interactions are complex and often site and chemical specific. The environmental effect of pesticides is also chemical specific. A small mass of a highly toxic or active pesticide may be more damaging than a larger mass of a less active or toxic pesticide. In some cases there is uncertainty about the fate, transport, and effect of pesticide degradation products as well.

In general, however, pesticides that are not either degraded or immobilized are eventually lost to the air or water. Site-specific processes determine how that mass of pesticides that is not degraded or immobilized is partitioned among surface water, groundwater, and the atmosphere. The first step toward reducing the amount of pesticides eventually delivered to surface water, groundwater, or the atmosphere is to reduce the total mass of pesticides introduced into the environment in farming systems.

*Pesticide Mass Balance*

Because of the complexity of the processes that determine the fate and transport of pesticides used in agricultural production systems, construction of pesticide mass balances is difficult. Despite the vast knowledge base available on pesticide reactivity and transport, a complete mass balance on the fate of any pesticide applied to a field does not exist. Some studies (see below) have measured the pathways followed by pesticides applied to crop fields, and those studies provide some perspective on ways to improve pesticide efficiency.

Losses through volatilization and spray drift during pesticide application can be substantial. Spray drift accounts for 3 to 5 percent of loss of insecticides applied under low-speed-wind conditions, but under normal conditions spray drift loss is typically 40 to 60 percent for many insecticides. Loss by volatilization from spray application ranges from 3 to 25 percent for most insecticides, but it may be as great as 20 to 90 percent for the insecticide methylparathion, for example. The delivery loss to soil and peripheral nontarget foliage may be as high as 60 to 80 percent for most sprays.

The percent of pesticide losses from soil-incorporated application are much lower. Volatilization from soil-incorporated application normally ranges from 2 to 13 percent, but it can be much higher for particularly volatile pesticides. Seasonal surface runoff of pesticides is less than 1 to 5 percent (Wauchope, 1978). Losses of pesticides through leaching are more difficult to estimate. Using a simulation model, Tanji (1991a) found that only very small fractions of the dibromochloropropane (DBCP) applied to the soil surface made its way through the soil to groundwater; concentrations of 1,500 mg/L (1,500 ppm) at the soil surface translated to concentrations of 0.009 mg/L (0.009 ppm) in groundwater. Data on

DBCP concentrations from 240 wells confirmed the simulation model results. For DBCP, however, even these minuscule leaching losses were a cause for concern, since the maximum contaminant level of DBCP considered acceptable in drinking water is 0.00002 mg/L (0.00002 ppm).

It is difficult to obtain mass balances for pesticides and, therefore, to predict their fate and transport in site-specific locations. Even very small losses of pesticides to particular parts of the environment can be a cause for concern. A concerted effort at source control through increased efficiency in the use of pesticides offers the best assurance that losses of pesticides from agricultural production systems will be reduced.

### Improved Pesticide Use Efficiency

*Aggressive efforts to adopt currently available technologies, systems, and practices to reduce the total mass of pesticides used should be pursued.*

Immediate gains in reducing the total mass of pesticides used in agricultural production systems can be achieved by using currently available improved pest management practices. The opportunities to increase pesticide use efficiency can be grouped into (1) use of integrated pest management practices, (2) improvements in pesticide formulations, (3) improvements in application practices and, (4) matching pesticide characteristics to site-specific conditions. If producers integrate currently available technologies and practices into their farming systems, many will be able to reduce the amounts of pesticides they use and sustain the profitabilities of their operations. The magnitude of financially feasible reductions will vary from region to region, crop to crop, and farming system to farming system.

#### Integrated Pest Management

Integrated pest management (IPM) is an ecologically based pest control strategy that integrates all available pest control tactics— including crop rotations, tillage practices, water management, residue management, biological controls, and pesticides—to achieve an optimal level of pest control. The concept of a treatment threshold is central to IPM systems. Pest control is designed to keep pest populations below a given threshold level at which damage is expected to cause losses in yields, profits, or some other measure of damage (Zalom et al., 1992).

IPM has a proven track record of reducing the need for pesticide applications while maintaining adequate levels of pest control. The application of IPM to cotton production beginning in 1971, for example, has led to dramatic declines in insecticide use. In 1971, an estimated 6.5

TABLE 2-7    Use of Integrated Pest Management for 12 Major Crops
in the United States, 1986

| Crop | Hectares Planted (10³) | Hectares under IPM (10³) | Percent Total Hectares under IPM |
|---|---|---|---|
| Alfalfa | 10,833 | 515 | 5 |
| Apples[a] | 187 | 121 | 65 |
| Citrus[b] | 428 | 284 | 70 |
| Corn | 31,053 | 6,075 | 20 |
| Cotton | 4,068 | 1,963 | 48 |
| Peanuts | 637 | 279 | 44 |
| Potatoes | 492 | 79 | 16 |
| Rice | 972 | 379 | 39 |
| Sorghum | 6,205 | 1,606 | 26 |
| Soybeans | 24,899 | 3,603 | 14 |
| Tomatoes | 153 | 126[b] | 83 |
| Wheat | 29,173 | 4,328 | 15 |

NOTE: Integrated pest management (IPM) is defined broadly to include all lands where basic scouting and economic threshold techniques reportedly are used.

[a]Includes the area under IPM by USDA's Cooperative Extension Service, grower organizations, producer industries, or consultants.

[b]Data are based in part on conversations with IPM entomologists in major growing regions for citrus and tomatoes.

SOURCE: Adapted from National Research Council. 1989. Alternative Agriculture. Washington, D.C.: National Academy Press.

kg of insecticide per ha (5.8 lbs/acre) was applied to cotton; in 1982, insecticide applications were 1.7 kg/ha (1.5 lbs/acre). In 1976, cotton crops received 49 percent of the total mass of insecticides applied to major field crops. By 1982, this was reduced to 24 percent. Some of the reduction was due to the adoption of new pesticides that were effective at lower rates, but use of IPM was also responsible (Zalom et al., 1992).

Zalom and colleagues (1992) also cited other examples of the success of IPM in increasing the efficiency of pesticide use. Use of insecticides on peanut crops declined from 4.4 to 0.9 kg/ha (3.9 to 0.8 lbs/acre) as producers adopted IPM practices. Use of IPM techniques in California almond production reduced crop damage, increased total production, and reduced total pesticide use by 31 percent.

*IPM programs should be accelerated.*

For some crops, particularly high-value crops, the use of IPM has become common (Table 2-7). The use of IPM for major field crops such as corn, soybeans, and wheat on which the largest masses of pesticides, particularly herbicides, are used is much less common. Yet, application of IPM techniques may have great potential for increasing pesticide use efficiency in the production of major field crops. Use of crop rotations can effectively reduce the need for soil insecticides to control corn rootworms in the Corn Belt. Banding of herbicides—a practice by which an herbicide is applied only to the crop row rather than broadcast over the entire field—can reduce herbicide applications by nearly half, for example. The development and adoption of IPM techniques suitable for use in the production of major field crops is needed.

*Research to develop IPM practices for weeds should be accelerated.*

IPM has had its major successes in the control of insects. Weed control is also important, particularly in the production of major field crops; and producers use more herbicides than any other class of pesticide in their agricultural production systems. Crop rotations and herbicide banding currently have the most promise for weed management in the corn-soybean rotation system (Edwards and Ford, 1992). Improved diagnostic tools to determine when weed control is needed could also increase pesticide use efficiency (Edwards and Ford, 1992). Much more research is needed to develop, test, and implement IPM strategies that are widely adaptable to weed management, particularly in major field crops.

### Design Better Pesticides

The chemical and physical properties of pesticides have important effects on their ultimate fates when they are applied to farming systems. It is possible to design new pesticides that pose lower risks because they are, for example, less toxic, less likely to be lost to surface water or groundwater, or more effective at lower application rates. Mechanisms to encourage and facilitate the registration of more environmentally benign pesticides could help increase the options available to producers.

### Improve Pesticide Application Practices

Simple improvements in pesticide application practices—such as following the directions on the pesticide label, carefully measuring the pesticides added to spray mixtures, and calibrating and maintaining spray equipment—can reduce the amounts of applied pesticide that are lost as well as increase application efficiencies. Technologies that apply pesticides only to the target site or pest (for example, banding) can also reduce pesticide losses. Improved application technologies such as controlled droplet applicators, drift-shielded applicators, ultra-low-volume equipment, electrostatic sprayers, and computer-controlled equipment or formulations that thicken the spray can also reduce pesticide losses. Aerial application methods generally result in higher drift losses than ground application and should be done when wind speeds are low and temperatures are cooler but not when rain is likely to occur.

### Match the Pesticide to Site Conditions

The ability to predict the behavior and transport of pesticides under field conditions appears to be weak, in part because of the spatial and temporal variabilities of pesticides in field soils. On the basis of chemical-specific properties and vulnerable site conditions, however, it should be possible to assess whether or not a given pesticide will contaminate surface water or groundwater.

*Existing knowledge should be used more fully to match pesticide selections to site conditions.*

Wauchope and colleagues (1992) have recently developed a hierarchy of pesticide properties; it lists pesticides according to their surface loss and leaching potentials. These pesticide properties can be matched with soil ratings information available in soil surveys so that producers can select those pesticides with the lowest potential for loss to surface water or groundwater. The relative toxicities of pesticides can be used as collateral criteria to refine pesticide selections (Hornsby, 1992). These data should be widely used by producers, crop-soil consultants, pesticide dealers, extension agents, and others who make pesticide recommendations.

*Increased resources should be devoted to the development of sampling, monitoring, analysis, and modeling protocols for pesticides in the environment.*

Sampling, monitoring, and improved modeling of the efficacies and fates of pesticides in the environment require substantial additional resources, facilities, and time. In the long term, these investments in efficient pesticide use will provide the models and data needed to refine the management of pesticides with greater precision. In the short term, however, current understanding should be used to reduce the total mass of pesticides used, reduce runoff and erosion from cropping systems, improve the efficacy of pesticide applications, and match pesticide selection to site conditions. These efforts should go forward at the same time that understanding of pesticide behavior in the environment is improved.

*Alternative Pest Control Technologies*

*Research required to develop alternative pest control strategies and to develop farming systems based on alternative pest control practices should be accelerated.* New methods of pest control that rely on biological or natural processes may provide alternatives to current pesticides, with less potential for water pollution. A large number of new biologically based methods of pest control appear to hold promise. Such methods range from crop rotations, cover crops, intercrops, development of resistant plant varieties, the use of pheromones and other plant or animal compounds to disrupt reproduction of pests, the use of highly specific toxins produced by bacteria or other organisms, and the introduction or management of living biological control agents as diverse as insects, nematodes, fungi, viruses, and bacteria (Charudattan, 1991; DeBach and Rosen, 1991; Luna and House, 1990; McManus, 1989; University of California, Study Group on Biological Approaches to Pest Management, 1992; U.S. Congress, Office of Technology Assessment, 1992; Watson, 1991). The development of such biologically based systems of pest and disease control, if developed, implemented, and adopted, could solve many of the environmental problems currently associated with pesticide use while assuring effective pest and disease control. Long-term gains in reducing the total mass of pesticides used in farming systems can be achieved only by continued efforts in research and development of alternative pest control management and farming systems.

## Improving Irrigation Management

Irrigation can cause soil salinization and waterlogging of soils, particularly in arid environments. These problems have plagued irrigated agriculture for centuries (Tanji, 1990). Today, irrigated agriculture faces the same problems. The irrigated agricultural system in California's San Joaquin Valley, for example, is facing an economic and ecological crisis. About 38 percent of the irrigated cropland is waterlogged, and 59 percent is affected by the accumulation of salts (San Joaquin Valley Drainage Program, 1990) (see Chapter 10 for a more complete discussion).

Irrigation also inevitably requires the disposal of drainage water that carries salts, trace elements, pesticides, and nutrients. In the San Joaquin Valley, disposal of irrigation water has become a critical problem and the recent discoveries of selenium and other toxic trace elements in irrigation drainage water has increased the difficulty in managing irrigation-induced soil and water quality problems (National Research Council, 1989b).

*Disposal of Drainage Water*

Irrigation requires natural or constructed systems that dispose of drainage water. When producers apply irrigation water to crops, much of the water is taken up by crops and is evaporated. As the water evaporates, the salts and other minerals left behind buildup in the soil. Unless it is flushed out of the soil in irrigation drainage water, the resulting buildup in soil leads to yield losses and eventual soil destruction. Irrigation, then, inevitably leads to the need for disposal of drainage waters that contain salts, nitrates, pesticides, trace elements, and other pollutants.

*Reduction of the Volume of Drainage Water*

*Reducing the volume of irrigation drainage water by increasing the efficiency with which irrigation water is used should be the major objective of programs to reduce salt and trace-element loadings to surface water and groundwater.*

A combination of carefully controlled, efficient irrigation with an appropriate match between the crop grown and water quality will minimize the amount of drainage water requiring disposal and, thereby, reduce the potential for water pollution and soil degradation (National Research Council, 1989b). Efforts to reduce the damage to soil and water quality caused by irrigated agriculture should have as their primary objective a reduction of the volume of drainage water needing disposal. Reducing the total mass of applied irrigation water through increased efficiency of water use is the most promising means of achieving this objective.

Currently available technology, if used, could result in immediate improvements in the efficiency with which irrigation water is used. Improved irrigation scheduling can greatly increase the efficiency of water use by ensuring that irrigation water is applied only when and in the amounts needed for crop growth. Reusing drainage water or tailwater, blending or using alternate sources of irrigation water to match the crops' salt tolerance, and changes in the type and sequence of crops grown can also improve the management of irrigation water.

## New Cropping Systems

The preceding sections of this chapter focused on improved management of inputs—nitrogen, phosphorus, manures, pesticides, and irrigation—as ways to improve soil and water quality. In many cases, improving input management promises to improve both the financial and the environmental performances of cropping systems. The goal of improving input management is to bring input applications closer to optimal rates for crop growth, thus minimizing the total mass of residual nitrogen, phosphorus, pesticides, salts, and trace elements lost from farming systems. It is impossible to be certain, however, that improved input management alone will be enough to meet water quality standards in all regions. Both technical and economic constraints have an impact on the degree to which input management, particularly for nitrogen, phosphorus, and pesticides, can be refined.

### Technical Constraints to Input Management

The studies available are not sufficient for making comprehensive predictions of how seriously technical constraints will reduce the effectiveness of refined input management. These studies do suggest, particularly for mobile pollutants like nitrates, that in some regions changes in land use or cropping systems will be needed.

Hall (1992), for example, monitored changes in groundwater nitrate concentrations beneath heavily fertilized and manured fields in Lancaster County, Pennsylvania, following the implementation of nitrogen management practices to reduce nitrogen inputs. Fertilizer and manure inputs were decreased between 39 and 67 percent (222 to 423 kg/ha, or 198 to 378 lbs/acre), and nitrate concentrations in the groundwater decreased by 12 to 50 percent. By the end of the study, however, all wells still exceeded federal drinking water standards for nitrate. The decreases in groundwater nitrate concentrations were much greater where the initial nitrate concentration was highest, suggesting that while reductions in nitrate will be significant in cases of dramatic overfertilization, achieving reductions in more conventional situations will be more difficult.

Clausen and Meals (1989) reported that during 7 years of monitoring water quality in a dairying region of Vermont, the levels of dissolved oxygen, phosphorus, turbidity, and fecal coliform bacteria in runoff and stream water frequently exceeded water quality standards, despite implementation of best-management practices in the watershed. Addiscott and Powlson (1989) and Addiscott and Darby (1991) have argued that reducing nitrogen inputs will not result in large reductions in groundwater nitrate concentrations as long as there are extended periods when soil nitrate levels are high without the presence of actively growing plants; a common occurrence in modern cropping systems. Cartwright and colleagues (1991), for example, reported that in Baden-Württemburg, Germany, 3 years of reductions in nitrogen inputs brought no improvement in groundwater nitrate levels, while in North Rhine/Westphalia, a more aggressive program that included purchase of sensitive areas and mixing 15,000 metric tons (16,500 tons) of clay into sandy soils brought rapid improvements in groundwater quality. Studies of subsurface drainage water quality from continuous corn production in the fertile soils of Iowa suggest that only corn with zero nitrogen applied would consistently keep nitrate concentrations in drainage water below the 10 mg/L level set for drinking water. Use of no added nitrogen significantly reduced yields and is clearly not sustainable for the producer (Baker and Melvin, 1992).

These studies confirm that impressive improvements in input management, particularly for nitrogen, are possible and, in many cases, will significantly improve water quality. These studies also indicate that even dramatic improvements in input management may not result in meeting water quality standards in certain regions or quickly enough to meet legislated deadlines. There may be significant lag times between improvements in input management and changes in water quality. Phosphorus and pesticides in stream sediments, nitrates in surficial aquifers, and phosphorus or salts that have built up in soils, for example, will continue to contribute to water pollution for a period of time after input management is improved. The length of the lag time is difficult to predict. In cases where technical constraints to improving input management are large or lag times unacceptably long, new cropping systems or changes in land use will be required.

### Economic Constraints to Input Management

Producers try to apply inputs at economically optimum rates if they want to maximize their profits. Economically optimum rates of application are closely related to the rates that are optimal for crop growth, but they are not necessarily the same. Economically optimum rates can be greater than optimum rates for crop growth because of uncertainties about outcomes and the prices of inputs and the crop.

This problem is best illustrated by an example developed by Bock and Hergert (1991) for nitrogen management (Figure 2-5). Producers are often thought to apply nitrogen at rates greater than those required for optimal crop growth as insurance against making a wrong decision that leads to lower yields. Figure 2-5 shows average losses caused by the underapplication of nitrogen and the gains from the overapplication of nitrogen as insurance. Bock and Hergert concluded that economic

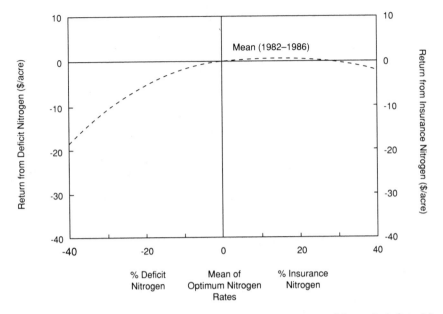

FIGURE 2-5  Economic return from insurance nitrogen (N) and deficit N applications. Source: B. R. Bock and G. W. Hergert. 1991. Fertilizer nitrogen management. Pp. 140-164 in Managing Nitrogen for Groundwater Quality and Farm Profitability, R. F. Follet, D. R. Keeney, and R. M. Cruse, eds. Madison, Wis.: Soil Science Society of America. Reprinted with permission from © American Society for Agronomy, Crop Science Society of America, and Soil Science Society of America.

incentives to nitrogen application at the optimum rate are not great, particularly when the yield response to nitrogen is highly variable and nitrogen/crop price ratios are low.

This example illustrates the general case—that the economically optimal rate of nutrients, pesticides, or irrigation can be larger than the rate that is technically optimal for crop growth. In addition, the risk of making a mistake increases as input use approaches the technical optimum application rate. The economically optimal rate, therefore, may exceed the environmentally optimal rate under current management. Technologies that afford greater precision in managing inputs can help solve this problem, but input management alone may not be sufficient to prevent water pollution. Managing cropping systems through rotations, cover crops, and multiple crops may be needed to augment efforts to improve input management.

## Managing Cropping Systems

*The use of cover crops should receive much greater attention as an integral part of soil and water quality programs.*

Much recent research has shown that the use of cover crops planted to cover the soil following harvest, or in some cases while the crop is growing, shows promise for reducing nitrogen and phosphorus losses from cropping systems, reducing the need for pesticides, and reducing soil erosion and runoff.

Cover crops have demonstrated the ability to reduce erosion, surface runoff, and leaching of nitrates to groundwater. Sharpley and Smith (1991) reported that the addition of a cover crop to farming systems that produce corn, wheat, cotton, and soybeans consistently reduced runoff, soil erosion, and the amounts of nitrogen and phosphorus transported in erosion and runoff. The use of cover crops dramatically reduced erosion and runoff when they were used in corn, cotton, and soybean cropping systems in Georgia, Iowa, Kentucky, Louisiana, Mississippi, Missouri, New York, Oklahoma, South Carolina, Tennessee, Texas, and Wisconsin (Langdale et al., 1991). Meisinger and colleagues (1991) showed that the use of nonleguminous cover crops to capture and hold residual nitrogen reduced the amount of nitrogen leached from farming systems by between 31 and 77 percent.

Currently, cover crops are widely used only in the southeastern United States (Power and Biederbeck, 1991). Langdale and colleagues (1991) reported that cover cropping systems are more well developed in the Southeast than in other parts of the United States. The drawbacks and concerns associated with cover crop use include depletion of the water in soil by cover crops, the slow release of nutrients contained in cover crops biomass, added costs of production, and difficulties in establishing and then killing cover crops, especially in northern areas of the United States (Frye et al., 1988; Lal et al., 1991; Wagger and Mengel, 1988).

*Research to develop cover cropping systems that can be used in colder and drier regions should be accelerated.*

The use of cover crops in colder and drier regions of the United States is limited by the lack of available cultivars and the lack of techniques to establish and manage cover crops where soil moisture is limiting (Power and Biederbeck, 1991). The potential benefits of cover crop use in these regions, if technical obstacles can be overcome, are great.

*Research to develop innovative cropping systems to meet long-term soil and water quality goals is needed.*

Cover crops are likely to become more important components of environmentally sound farming systems. More dramatic changes in the structure and function of farming systems, however, will be required to achieve soil and water quality goals in many areas because of the problems discussed above. In the long term, cropping systems very different from those that are most commonly used now may be needed if farming is to continue in areas with severe soil and water quality problems.

Kirschenmann (1991) suggested that there are currently two basic conceptions of what these new farming systems may entail. Some observers anticipate the development of multiple cropping systems in which companion plants are used to produce nitrogen through biological fixation and to provide weed control by acting as a living mulch; insect and disease control will rely on sophisticated webs of biological control agents. Other observers anticipate improved information-gathering devices in the crop field that will interface with farm machinery capable of varying the application of inputs to achieve higher efficiency. These two approaches are not mutually exclusive and both are important for guiding efforts to improve the environmental perfor-

---

## POTENTIAL BENEFITS OF COVER CROPS

Cover crops are legumes, grasses, cereals, or other crops that are added to crop rotations to protect the soil, reduce pest infestations, and improve water quality. Unlike other crops, cover crops are not normally harvested but are killed or plowed under when the cash crop is planted. There are drawbacks to cover crops in some regions and situations including soil moisture depletion, competition with the cash crop, and increased cost of production. When these problems can be solved, cover crops can reduce erosion and runoff, protect soil quality, suppress pests, and prevent water pollution.

**Reduced erosion and runoff.** Cover crops provide a protective vegetative layer that shields the soil from the impacts of raindrops, reduces the velocity of runoff, and increases the portion of runoff that is absorbed by the soil. Together these protective effects of cover crops can dramatically reduce erosion and runoff from farming systems. Langdale and colleagues (1991), for example, reported erosion reductions of up to 91 percent after adding a cover crop to soybean fields.

**Improved soil quality.** In addition to protecting soil from erosion, cover crops also improve soil structure, enhance soil fertility, and sustain or increase soil organic matter and soil biological activity. Cover crops may be particularly effective in restoring the quality of degraded soils. Langdale and colleagues (1992b) reported that the combination of a cover crop and conservation tillage significantly improved the structure, organic matter content, and infiltration rate of severely eroded soils in Georgia. After 5 years, crop yields from the soils that had been severely eroded were the same as those from only slightly eroded soils.

mance of agriculture. Farming systems that integrate both approaches will likely be more profitable and environmentally sound.

Guiding the research to develop new farming systems requires a long-term perspective and a vigorous imagination. The cropping systems and management practices currently used have little resemblance to the systems that were common 50 to 75 years ago. It is reasonable to expect that future systems will be equally different from current systems. Long-term, imaginative direction to current research programs is important and should result in better farming systems in the future.

## INCREASING RESISTANCE TO EROSION AND RUNOFF

*Reducing erosion and runoff should be fundamental to efforts to improve soil and water quality.*

Erosion is the single greatest threat to soil quality. Some of the most direct and serious water pollution problems result from the delivery of sediments to surface water; and the cost of dredging several million

---

**Pest suppression.** Use of cover crops may help control weeds through nutrient competition, allelopathy (suppression of growth of one plant species by another by the release of toxic substances), and physical effects. Although some living mulches also compete with row crops, compatible cover crops can provide an alternative to herbicide use without significantly decreasing productivity. A 3-year study in New Jersey showed that corn planted into growing subterranean clover (*Trifolium subterraneum*), a winter legume, produced the same or better yields than corn grown with conventional herbicides and no mulch, regardless of the type of tillage used (Enache and Ilnicki, 1990). Cover crops can also increase the diversity of insects, including species that prey on crop pests.

**Prevent water pollution.** Cover crops prevent water pollution by a combination of effects that have already been discussed. Reducing erosion and runoff also reduces the amount of sediment and agricultural chemicals that reach surface water. Improving soil quality improves the effectiveness of a the soil as a filter to capture and degrade potential pollutants, and the need to use less pesticides reduces the chance that pesticides will pollute surface water or groundwater. In addition to these effects, cover crops may capture, recycle, or immobilize residual nitrogen, phosphorus, and pesticides from crop production. The potential for cover crops to trap these potential pollutants is promising but not yet well understood.

---

SOURCE: *Adapted from R. Lal, E. Regnier, D. J. Eckert, W. M. Edwards, and R. Hammond. 1991. Expectations of cover crops for sustainable agriculture. Pp. 1–10 in Cover Crop for Clean Water, W. L. Hargrove, ed. Ankeny, Iowa: Soil and Water Conservation Society.*

cubic meters of sediments from U.S. rivers, harbors, and reservoirs is quite high. Erosion and runoff also deliver nutrients, pesticides, and salts to surface waters. Increased runoff volume and energy from croplands disrupt water flow regimes, increasing discharge peaks and stream channel erosion. Degradation of watersheds and disruption of stream channels has caused erosion runoff of stream banks and streambeds to become an important contributor to sediment loads in streams and rivers.

Controlling erosion and runoff has been the emphasis of traditional conservation programs and there is a vast body of information available on farming practices that effectively control erosion. These farming practices are well understood and a great deal of effort has been and should be expended by federal, state, and nonprofit organizations to increase the use of these practices by producers. There are, however, opportunities to make national erosion control efforts more effective by understanding the time lag between erosion control and sediment reduction, by addressing stream channel degradation, and by recognizing the importance of episodic events that suddenly and dramatically increase erosion and runoff.

## Time Lag of Sediment Load Reductions

Current sediment loads are more an indication of past rather than current erosion rates because there are long time lags between the time when soil is eroded from a field and when the sediment is finally delivered to a stream or river. Reducing erosion will not, therefore, often result in immediate reductions in sediment loads. The time lag between reduced erosion and reduced sediment loads depends on the length of time that sediments spend in storage in the watershed before they are delivered to larger streams and water bodies.

Most sediment spends most of its life in storage. In the Piedmont region, between southern Virginia and eastern Alabama, for example, an estimated 25 km³ (6 miles³) of soil was eroded from the uplands in the last 200 years (Meade, 1982; Trimble, 1975). Estimates indicate, however, that 90 percent of that soil is still stored on the hill slopes and valley floors of the region. Similarly, Trimble (1983) estimated that only 7 percent of the human-induced eroded soil has actually left the immediate watershed where the erosion occurred in Wisconsin.

In some areas, stream sediment loads are increasing, despite changes that have resulted in reduced levels of erosion on croplands in the watershed (Meade, 1982). The activities of a stream can be viewed simply as a struggle for the stream to balance its sediment load with its sediment-transport capacity. Decreased soil erosion does not immediately translate into less suspended sediment in a stream. As the sediment load in runoff water from croplands decreases because of erosion control, the capacity of the cleaner water to pick up sediment in streambeds or stream banks increases. This process continues until the stream channel and the runoff water develop a new equilibrium between the sediment delivered in runoff water and the sediment stored in the stream channel.

Many streams and rivers are still adjusting to the erosion and watershed disruption that occurred in the past. Large amounts of sediments are still in storage in streambeds, stream banks, and floodplains. Sediment loads will continue to be large in these watersheds, even if current erosion rates are reduced.

## Protecting Stream Channels

High volumes and energy of runoff water disrupt stream channels, increasing the erosion of sediments from streambeds and stream banks. In many watersheds, erosion of sediments from the streambed and stream bank contributes a large share of the sediment load. The volume and energy of runoff water is also closely related to the amount of pesticides, phosphorus, nitrates, and other pollutants delivered to surface water.

*Programs to protect water quality should seek to reduce the total volume and energy of runoff from croplands in addition to reducing total erosion.*

Erosion and runoff are closely related, but they are not the same process. Erosion reductions without comparable reductions in runoff energy and volume can cause trade-offs between the delivery of pollutants to surface water, attached to sediments, or dissolved in runoff water. Conservation tillage systems, for example, reduce total erosion, but reductions in total runoff are not as great. The decreased erosion reduces the amount of sediment-attached phosphorus that is lost, but it may increase the amount of phosphorus lost in soluble form in the runoff water (Alberts and Spomer, 1985; Angle et al., 1984; Barisas et al., 1978; Langdale et al., 1985; McDowell and McGregor, 1984; Romkens and Nelson, 1974; Romkens et al., 1973).

Erosion reduction should be linked to efforts to improve agricultural watersheds by protecting riparian zones, stream channels, and wetlands and by using other measures to manage the volume and energy of surface runoff reaching surface water bodies. Reducing erosion from croplands is a fundamental first step toward improving soil and water quality. However, managing the volume and energy of runoff water is just as important. Efforts to reduce erosion and runoff from croplands through the use of residue management systems should be coupled with efforts to protect agricultural watersheds by protecting or restoring riparian vegetation, wetlands, and grassed waterways and by using other measures that reduce the damage caused by excessive volumes and energy of runoff water. Such field and landscape buffer zones hold great promise for improving water quality in agricultural watersheds. (See the section on field and landscape buffer zones later in this chapter and in Chapter 12.)

### Resistance to Episodic Damage

Agricultural ecosystems are vulnerable to erosion and runoff during storm events because plant diversity is intentionally low in such

---

## CHANNEL INSTABILITY

Many streams experience serious instability as a result of land use changes and the adverse impacts of river management on drainage and flood control. The primary response in the fluvial system is degradation, which leads to damage to in-stream and riparian ecosystems, damage to infrastructure (bridges, dams, and roads), and the generation of heavy sediment loads, which cause aggradation problems downstream. Lowering of the water level in a stream or riverbed also increases the chance that the stream bank or riverbank may collapse. Thorne (1991) reported that when banks become unstable, the thrust of the channel instability switches from degradation to rapid stream widening. Widening involves destruction of valuable valley bottomlands, damage to infrastructure, and the prolongation of the heavy sediment supply to the system downstream. Rapid stream widening associated with bank instability results from bed scour and lateral toe (bank-bed contact) erosion in the degrading channel. The precise timing of failure and the mode of bank collapse are controlled by bank geometry, bank stratigraphy, bank material properties, as well as the hydrology above the failure areas, which in turn affect the flow hydraulics.

Thorne (1991) investigated these instability problems in the loess hills of the Yazoo Basin in Mississippi. He showed that a grade control structure could be used successfully to halt bed degradation and induce aggradation as a mechanism to produce width stabilization. The bank stabilization provided by a grade control structure was determined to be a cost-effective solution to chronic problems of retreating banks and widening channels in this highly erodible area.

---

SOURCE: *Thorne, C. R. 1991. Analysis of channel instability due to catchment land-use change. Pp. 111–122 in Sediment and Stream Water Quality in a Changing Environment-Trends and Explanations. IAHS Publication No. 203.*

ecosystems and the period during which plants are actively growing and the soil is covered is often short. Much of the soil and water quality degradation can occur during episodic climatic events, such as heavy rainfalls, droughts, or windstorms. Farming system resistance to episodic damages can be increased by lengthening the period during which growing plants or residues are present and by increasing the amount of soil cover provided by plants and residues.

### Conservation Tillage and Residue Management

*Efforts to increase the use of conservation tillage and other forms of residue management should continue to be an important component of programs to protect soil and water quality.*

Immediate gains in soil and water quality can be attained if producers adopt currently available conservation tillage and residue management systems. If producers incorporate such conservation systems into current farming systems, resistance to episodic events will increase and runoff energy and soil erosion will be reduced. A great diversity of tillage and residue management systems are available to producers (see Table 9-1 in Chapter 9), although the use of these systems may be less attractive in some situations because of unfavorable physical or economic factors. Use of these systems results in dramatic decreases in erosion and runoff from farming systems (see Table 9-2 in Chapter 9) and from agricultural watersheds (see Table 9-3 in Chapter 9).

The use of conservation tillage practices has increased over time, and considerable time and effort have been devoted to increasing the use of these systems by producers. The percentage of croplands on which various forms of conservation tillage are used varies by crop and region (see Table 9-4 in Chapter 9). In 1985, the proportion of cropland on which producers practiced some form of conservation tillage varied from about 12 to 48 percent, depending on the farm production region. Table 2-8 indicates that the use of conservation tillage also varies by crop.

The most important way to increase the soil and water quality benefits that could be realized through wider use of conservation tillage and residue management systems is to increase adoption of those systems on those lands that are most vulnerable to soil quality degradation or that contribute the most to water quality degradation. The data in Table 2-8 suggest that large percentages of highly erodible land were not farmed by conservation tillage techniques in 1990. Full enforcement of the conservation compliance provisions of the 1985

TABLE 2-8 Highly Erodible, Not Highly Erodible, and Nondesignated Lands on which Conservation Tillage Systems Are Used for Various Crops, 1990

| | Highly Erodible | | Not Highly Erodible | | Nondesignated | |
|---|---|---|---|---|---|---|
| Crop | Total Area (1000 ha) | Percent Conservation Tilled | Total Area (1000 ha) | Percent Conservation Tilled | Total Area (1000 ha) | Percent Conservation Tilled |
| Corn | 5,140 | 31 | 17,510 | 26 | 1,160 | 21 |
| Cotton | 850 | ID | 2,810 | 2 | 284 | ID |
| Winter wheat | 4,702 | 23 | 10,370 | 17 | 1,190 | 18 |
| Spring wheat | 968 | 38 | 4,860 | 23 | 567 | 25 |
| Durum wheat | 38 | NR | 1,015 | 35 | 203 | 38 |
| Northern soybeans | 2,908 | 34 | 11,120 | 25 | 717 | 20 |
| Southern soybeans | 470 | 46 | 3,710 | 17 | 620 | 9 |
| Rice | 16 | NR | 644 | 4 | 3,170 | 3 |

NOTE: Conservation tillage is any tillage system resulting in 30 percent or greater surface residue cover after planting. ID, insufficient data; NR, no data reported. Totals of percentages may not add to 100 because of rounding.

SOURCE: Adapted from U.S. Department of Agriculture, Economic Research Service. 1991. Agricultural Resources: Inputs Situation and Outlook. Report No. AR-21. Washington, D.C.: U.S. Department of Agriculture.

Food Security Act is expected to dramatically increase the amount of highly erodible land on which conservation tillage or some other form of residue management is used. A concentrated effort to increase the use of conservation tillage in watersheds where water quality degradation is greatest would lead to even greater benefits.

### Develop New Cropping Systems

Most natural ecosystems resist erosion through biotic control over the abiotic environment (Bormann and Likens, 1979). The plant canopy intercepts incoming precipitation, which greatly reduces the energy and erosive potential of rainfall. The litter layer, the forest floor, and organic soil horizons further reduce the erosive potential of rainfall, allowing gradual infiltration of rainfall water into the surface soil horizons. The presence of a litter layer also helps to maintain soil structure and prevent soil crusting. Because of these multiple layers of soil protection, most natural ecosystems have very low rates of soil erosion and are resistant to most storm events.

*Research and development of economically viable cropping systems that incorporate cover crops, multiple crops, and other innovations should be accelerated.*

As long as there are extended periods when the soil is inadequately covered, farming systems will be vulnerable to erosion and runoff events. To achieve higher resistance to the erosive power of large and small precipitation events, researchers and producers could develop farming systems that mimic the multilayer soil protection that occurs in many natural ecosystems. Increasing the number of layers of vegetative cover over the soil will further increase the resistance of farming systems to severe episodic damages. Some cropping systems already achieve this multilayer soil protection. Relay or double-cropping systems in which producers use no-tillage techniques, such as the wheat-soybean system used in the southeastern United States, have nearly continuous vegetative cover and a well-developed litter layer of crop residues. Rates of erosion in these systems are very low, even under the highly erosive conditions that exist in the southeastern United States (Langdale et al., 1992a). Further development of cropping systems that are more resistant to erosion and runoff should be a high priority. In the long term, incorporation of cover crops, multiple crops, and other changes in cropping systems hold great promise for increasing resistance to erosion and runoff.

## Probability Analysis

*Conservation systems should be designed to increase soil cover during periods when the probability of episodic damages is highest.*

Immediate gains in preventing soil degradation and water pollution can be achieved by incorporating the probability of episodic events into the design of farming systems. Current computer simulation capacities and available climatic data could be used to analyze the probability of episodic events that would lead to damaging erosion or runoff events.

Average annual soil loss is normally used for conservation planning programs by USDA. For example, cropping systems designed to meet the soil loss tolerance concept based on universal soil loss equation evaluations may not be optimum. Hjelmfelt and colleagues (1988) measured the distribution of erosion events over a 37-year period in Missouri and found that soil loss was greater than average during only 9 of the 37 years (Figure 2-6). On an individual-event basis, 4 percent of the events accounted for 50 percent of the total soil loss. Similar data from Iowa indicate that sediment yields were greater than the average for 4 years of an 18-year record. Three percent of the individual storm

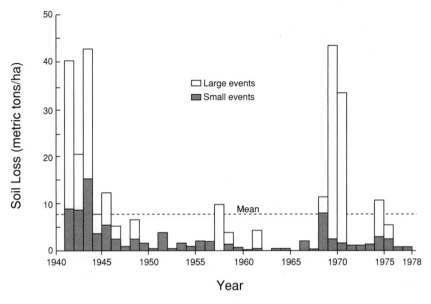

FIGURE 2-6   Distribution of erosion events over 38 years on a field in Missouri. Source: A. T. Hjelmfelt, Jr., and L. A. Kramer. 1988. Unit hydrograph variability for a small agricultural watershed. Pp. 357–366 in Modeling Agricultural, Forest, and Rangeland Hydrology. Proceedings of the 1988 International Symposium, December 12–13, 1988, Chicago, Illinois. St. Joseph, Mich.: American Society of Agricultural Engineers.

events accounted for more than 50 percent of the total erosion. Thus, in most years, conservation systems designed for average annual soil loss would be overdesigned, yet during years with severe storms, the damage might be catastrophic. These data indicate that conservation planning to increase the resistance of farming systems to erosion and runoff cannot be done only on the basis of an average year or an average event but should also be done on the basis of extreme events.

Erosion control techniques should be based on practices that result in a certain probability of controlling erosion from a storm event of a specified duration and intensity. Another possibility would be to apply conservation treatments that control soil erosion during specific types of storm events during specific periods of the year. Such approaches might assist in more precise placement of conservation practices on landscapes.

*Current sedimentation estimation technology should be improved to (1) adequately address the transport of chemicals adsorbed to soil particles, (2) produce probabilistic models that can be used to control erosion risks in a manner similar to that of flood control systems, and (3) develop wind erosion models that can predict the effects of wind erosion on water quality.*

The technology currently used to estimate and predict erosion and sedimentation suffers from weaknesses that constrain researchers' and producers' abilities to target erosion control efforts. Currently available erosion and sedimentation data are also often incompatible with modern computer simulation technology.

Fundamental understanding of the physical processes of water and wind erosion, however, is increasing as a result of recent research. USDA's Agricultural Research Service is currently developing process-based erosion prediction technology called the Water Erosion Prediction Project and the Wind Erosion Prediction System; this technology will offer agencies such as the Soil Conservation Service an opportunity to evaluate new approaches for conservation planning.

## CREATING FIELD AND LANDSCAPE BUFFER ZONES

The preceding sections argued for the need to change farming systems in ways that conserve and enhance soil quality, increase input use efficiency, and increase resistance to erosion and runoff. Farming systems, however, exist within a landscape described by the patterns of soils and slopes; the patterns of streams, lakes, and wetlands; and the adjacent ecosystems such as forests and wetlands. The ultimate effects of farming systems on soil and water quality are affected by the interaction between cropland and livestock production systems and the landscape in which production takes place.

The preceding sections have described various opportunities to improve the management of farming systems at the farm-level. These improvements will reduce the losses of sediments, nutrients, pesticides, or salts from each individual farm. There are limits, however, to the soil and water quality protection that can be achieved by working only at the farm-level. In some watersheds, even small losses from individual farms may, in the aggregate, result in water pollution or soil degradation; the cost of reducing losses from those individual farms further, however, may be great. In addition, large storms or other episodic events, as discussed in the preceding section, may result in soil degradation and water pollution even from well-managed farming systems. Managing the landscape by creating or restoring buffer zones is a promising way to increase the effectiveness and lower the cost of programs to protect soil and water quality.

A 2-acre pothole (area of low wetland) in a soybean field is being studied to determine water movement, chemical transport, and the fate of agricultural chemicals in saturated soils. Credit: Agricultural Research Service, USDA.

### Creating Managed Buffer Zones

Buffer zones can be divided into two different types: (1) field-scale buffer or filter strips, usually containing managed grasses, and (2) landscape- or watershed-scale riparian or wetland buffer zones. Buffer zones can range from a few to many meters in width. The kind of vegetation and location of the buffer zone, in addition to its size, have important effects on its pollutant-trapping capacity. Both field and landscape zones can be useful in protecting soil and water quality, but uncertainties exist in how best to design and manage buffer zones.

*The creation, protection, and management of field and landscape buffer zones should be an important objective of programs to protect soil and water quality.*

Field-by-field efforts to conserve soil quality, improve input use efficiency, and increase resistance to erosion and runoff will not be adequate where overland and subsurface movements of nutrients, pesticides, salts, and sediment are pervasive. The use of buffer zones—ranging from natural riparian corridor vegetation (vegetation along waterways) to simple, strategically placed grass strips, to sophisticated artificial wetlands—to intercept or immobilize pollutants before they reach surface water or groundwater holds promise for increasing the effectiveness of efforts to protect soil and water quality.

The purpose of creating field and landscape buffer zones is to create landscape sinks that trap or immobilize sediments, nutrients, pesticides, and other pollutants before they reach surface water or groundwater. The importance of field and landscape buffer zones in reducing the delivery of pollutants has received increasing attention (Dillaha et al., 1988, 1989b; Ehrenfield, 1987; Hayes and Hairson, 1983; Jacobs and Gilliam, 1985; Karr and Schlosser, 1978; Kovacic et al., 1991; Phillips, 1989). Grass waterways and vegetative strips have been used for erosion and runoff control in croplands for some time, but the use of these areas as sinks for nutrients and pesticides is more recent.

*Research to develop design and management standards for field and landscape buffer zones should be accelerated.*

The use of field-scale buffer or filter strips can be useful for reducing runoff, and for trapping the sediments, nutrients, and pesticides that move in surface runoff from specific fields. Unless the filter strips are relatively large, however, they can become clogged with sediments and their trapping efficiencies can decrease. Filter strips can fail during large storm events because of lateral water flow along the field-filter interface, leading to strip channelization at a low point along the interface. Moreover, although the sediment-trapping abilities of well-maintained filter strips are well established, nutrient and pesticide uptake by filter strips is poorly understood.

Forested riparian buffer zones have been demonstrated to be effective at trapping the sediments and nutrients that move in both surface and subsurface flows from crop fields (Groffman et al., 1992; Jacobs and Gilliam, 1985; Karr and Schlosser, 1978; Lowrance et al., 1984a; Peterjohn and Correll, 1984; Simmons et al., 1992). The remaining uncertainty about riparian buffer zones relates to their long-term effectiveness: will these areas be able to absorb sediments and nutrients indefinitely, or will they become saturated over time? The fates of trapped sediments, nutrients, and pesticides following fire, logging, or flooding are also unclear.

### Protection of Existing Natural Vegetation

*Federal, state, and local government programs to protect existing riparian vegetation, whether bordering major streams or small tributaries, lakes, or wetlands, should be promoted.*

Existing riparian vegetation, particularly the vegetation bordering smaller streams and tributaries, is an important resource that should be protected to serve as sinks for sediments, nutrients, and pesticides, to protect the stream bank from erosion, and to reduce excessive runoff into stream channels. Loss of these areas will increase existing water quality problems or create new ones. The cost of replacing the water quality benefits of existing vegetation with efforts on the farm level to increase soil quality, input use efficiency, and resistance to erosion may be high.

### Balance Needed

*The creation of field or landscape buffer zones should augment efforts to improve farming systems. They should not be substitutes for such efforts.*

The creation of field or landscape buffer zones cannot be seen as an alternative to efforts to improve farming systems. The capacities of field and landscape buffer zones to trap and immobilize sediments, nutrients, and pesticides are limited by the size of the buffers, the plants growing in the buffers, and the manner in which the buffers are managed. Both field and landscape buffer zones can be overwhelmed by large flows of runoff water, sediment, nutrients, and pesticides. Efforts to improve farming systems and to create field or landscape buffer zones are complementary. Emphasis on one effort to the exclusion of the other will achieve much less improvement in soil and water quality than is possible by striking a balance between the two efforts.

# 3

---

# A Systems Approach to Soil and Water Quality Management

The preceding chapter defined four broad opportunities that should be pursued by national policies to prevent soil degradation and water pollution. These opportunities are to (1) conserve and enhance soil quality as the first step toward environmental improvement; (2) increase nutrient, pesticide, and irrigation use efficiencies in farming systems; (3) increase the resistance of farming systems to erosion and runoff; and (4) make greater use of field and landscape buffer zones. Realizing those opportunities depends on the ability and willingness of producers to change their management and production practices. Producers, however, do not make isolated changes in these practices. A change in one production or management practice affects other components of the farming system that producers manage. Programs and policies that pursue these four opportunities, therefore, should also incorporate a systems perspective.

## LINKAGES AMONG OBJECTIVES

Inherent links exist among soil quality conservation, improvements in input use efficiency, increases in resistance to erosion and runoff, and the wider use of buffer zones. These links become apparent only if investigators take a systems-level approach to analyzing agricultural production systems. The focus of such an analysis is the farming system, which comprises the pattern and sequence of crops in space and time, the management decisions regarding the inputs and production practices that are used, the management skills, education, and objectives of the producer, the quality of the soil and water, and the nature of the landscape and ecosystem within which agricultural production occurs. An integrated systems approach is necessary for the development of policies and programs to accelerate the adoption of farming systems that are viable for producers, that conserve soil quality, and that do not degrade water quality (Jackson and Piper, 1989).

## LINKAGES AMONG PROGRAMS

A broad range of programs at the local, state, and federal levels seek to solve the environmental problems associated with agricultural pro-

---

## FARMING SYSTEM PLANNING

Development of an integrated farming system plan begins with an inventory of farm resources. This inventory is meant to provide the data to answer some of the following questions:

- Are there opportunities to improve pest, nutrient, or soil management through crop rotation?
- Are there livestock enterprises on the farm or nearby farms from which animal manures might be collected and used as nutrient inputs?
- What amount of pest control inputs have been used in the past?
- How are irrigation applications scheduled?
- Are land ownership or lease arrangements an obstacle to changes in farm management?
- Does the producer participate in U.S. farm programs?
- Does the equipment inventory allow or hinder the capability to improve tillage practices and residue management?
- How soon is tillage, application, or other capital equipment scheduled for replacement?
- How aware are producers of problems in their operations?
- What are producers' perceptions of the risks involved in changing their current farming systems?

Once a general picture of the farm enterprise emerges, more detailed information on production practices needs to be assembled. Often, records of input use, soil tests, crop yields, and other data are not available and must be constructed as completely as possible from memory to answer the following questions:

- What is the crop rotation history on a field-by-field basis?
- Are credits for the nitrogen fixed by legumes taken when making fertilizer applications?
- Has manure routinely been applied only to a small, particular area?
- Have particular pest problems been associated with a particular field, a particular area within a field, or a particular crop or cropping sequence?
- What do soil analyses indicate about the relative soil quality and soil fertility between fields?

duction. Ribaudo and Woo (1991) reported that various erosion control measures have been adopted by 17 states, nutrient control measures by 17 states, pesticide control measures by 16 states, land use control measures by 3 states, and input taxes by 4 states. A similar diversity of programs exists at the federal level (see Tables 1-1 and 1-2 in Chapter 1).

All of these local, state, and federal programs have specific objectives that address soil erosion; nutrient, pesticide, or irrigation water management; and protection of wetlands or other environmentally sensitive lands. The objectives of one program can conflict with, complement, or reinforce the objectives of other programs at the farm level, where the programs are ultimately implemented. Such inconsistency between the

---

- Have fertilizers and nutrient management been used on a field-by-field basis or uniformly across the farm?
- What have been the crop yields from individual fields?

At this stage improvements can begin. These might range from conservation plan improvements to input adjustments based on the results of soil tests. Fields or parts of fields where manure has been applied or legumes grown can be targeted for detailed soil sampling so that the producer can appropriately adjust fertilizer inputs.

Further refinements can be made by assessing the soil resources within each field since the soils within a field can vary dramatically. Adjustments to, for example, tillage practices and the inputs used can increase both the economic and environmental performance of the field. Fertilizer applications, for example, should be different on the top of a hill than on the side of a hill. Particular weed problems are often associated with microclimatic conditions related to the different soils located in different parts of the landscape. The yield potential can be much different on different soils in the same field; adjusting the inputs to the parts of the field with different yield potentials can increase input use efficiency.

The progression from whole-farm analysis to field-by-field and intrafield improvements is a process that takes place in steps. The producer can stop the process at any stage at which the increased cost of refined management is too high or information is not yet available to move to the next step. Movement from step to step requires better information management and improvements in the skills of the producer. Typically, implementation of such improved management requires development of a multiyear plan, which involves improved on-farm data collection, management alterations, and improved record keeping. Full implementation may be delayed until capital investment in new equipment or facilities is feasible, because of a multiyear crop rotation, or until the producer's experience with the new farming system removes doubts about its efficacy and allows the producer to overcome a perceived risk of economic loss resulting from implementation of the new farming system. In some instances, the plan is best implemented on a portion of the farm, side by side with the producer's normal management system, to increase confidence that the recommended changes will in fact work.

objectives of different programs is most likely to occur when programs promote narrow technical solutions for individual problems. Multiple programs that promote different technical solutions to different problems at the farm level increase the chance for incompatibility. The linkages between program objectives become clearer if a systems approach is used to integrate activities at the local, state, and federal levels.

## ADVANTAGES OF FARMING SYSTEMS APPROACH

Use of the farming system rather than individual best-management practices as the foundation for efforts to improve soil and water quality pays off in five ways:

1. addresses resource and enterprise variability;
2. provides a basis for targeting programs and financial support where improved soil and water quality is most needed;
3. provides a basis for coordinating local, state, and federal programs;
4. increases the chances of exploiting opportunities to simultaneously improve financial and environmental performance; and,
5. increases the flexibility to adapt programs and policies to changing resource or market conditions.

### Variability

Directing national policy toward solutions that improve soil and water quality has been made more difficult because of the geographic variability in the resources and enterprises that characterize agricultural production systems in the United States. This difficulty is exacerbated by the need to integrate the activities of local, state, and federal programs. A systems approach can be based on management principles that are applicable to the variable conditions of different farming systems and different regions. National-level programs can be based on farming system plans that can be developed by using uniform criteria. Such uniform criteria can provide a more rigorous basis for determining whether producers or programs are meeting their objectives.

### Targeting

Farming systems can be analyzed at regional scales to set national priorities. Analyzing nutrient inputs and outputs at regional scales, for example, is an effective way to target those regions where improvements in nutrient management are most likely. Figure 3-1 provides a regional breakdown of the balance between nutrient inputs and outputs

FIGURE 3-1 Proportion of national nitrogen and phosphorus inputs and balances contributed by each farm production region. Nitrogen and phosphorus balances are the differences between total nitrogen and phosphorus inputs and the nitrogen and phosphorus removed with the harvested crop or in crop residues. See the Appendix for a full discussion of the methods used to estimate nitrogen and phosphorus inputs, outputs, and mass balances.

by farm production region. Directing efforts to improve nutrient management to those regions with the greatest balance of inputs over outputs is a first step toward targeting. This kind of analysis can be done at the farm, watershed, regional, or state level to further refine targeting efforts. Similar analyses could be conducted for irrigation water, pesticides, and other inputs.

Incorporating a farming system perspective into targeting can also help identify those farming systems within those geographically defined priority areas that should be the focus of attention. Programs could be directed at farming systems that, because of their management or location, cause a disproportionate share of soil and water quality problems. The focus of targeting, then, would shift from defining geographic regions to identifying the opportunities to change farming systems within priority areas.

## Integration

Farming systems analysis provides a way to integrate the objectives of environmental programs at the local, state, and federal levels. A farming systems approach, for example, helps to make clear the relationship between programs to reduce erosion and programs to improve nutrient management. The impacts of individual programs on farming systems could be determined prior to implementation, and redundant or conflicting elements could be identified early in the policy design and implementation process.

## Win-Win Opportunities

Systematic analysis of input use, cropping systems, and tillage practices increases the likelihood that opportunities to simultaneously improve financial and environmental performance will be identified. Accounting for on-farm resources, such as nutrients from legumes or manures, can lead to improvements in nutrient management that reduce costs as well as improve soil and water quality. Similarly, a more integrated approach to analysis of weed problems can identify weedy spots in fields that need special treatment, while pest control expenditures for other parts of the field can be reduced.

Win-win opportunities also exist for program managers. A farming system approach will result in recommendations that are more appropriate to specific farms, eliminate inconsistent and conflicting recommendations, and direct the attention of program managers to those clients most in need of technical assistance. Such an approach promises to increase the effectiveness of programs and the efficiency with which recommendations can be implemented by producers.

### Adaptability

The same systems approach that is used at the enterprise level can be extended to the multiple-farm, landscape, watershed, or regional scales to direct targeting and program evaluation. Table 3-1, for example, presents the types of information and analyses that can be used at various scales to guide soil and water quality programs.

### FARMING SYSTEM AS UNIT OF ANALYSIS AND MANAGEMENT

*The farming system should be the unit of analysis and management used to direct local, state, and federal programs to protect soil and water quality.*

Environmental programs to protect soil and water quality should be evaluated on the basis of the effects of the recommended management and production practices on the total farming system. Changes in the management of farming systems rather than the adoption of individual best-management practices should be the goals of environmental programs. The linkages among soil quality, input use, erosion and runoff, and buffer zones can be managed only at the farming system level. Similarly, the linkages among different local, state, and federal programs are best understood by analyzing how these programs affect farming systems. Failure to recognize and manage these inherent linkages increases the likelihood that trade-offs between protecting soil versus water quality, protecting surface water quality versus groundwater quality, or reducing the loadings of one pollutant versus another will impede progress toward overall improvements in soil and water quality.

### Integrated Farming System Plans

Integrated farming system plans are the best mechanism available now for implementing a farming systems approach at the farm level. The current array of soil and water quality programs provides an opportunity to incorporate an integrated farming systems approach into U.S. Department of Agriculture (USDA) and U.S. Environmental Protection Agency (EPA) soil and water quality improvements efforts. The multiplicity of practices, objectives, and plans associated with these initiatives is a good example of the need for integrated farming system plans to coordinate the activities of different programs and agencies at the farm level.

TABLE 3-1  Application of Farming System Approach at Different Geographic Scales

| Geographic Scale | Farming System | Physical Resource Data | Socioeconomic Data | Inventories and Analyses |
|---|---|---|---|---|
| Farm Production Regions | Dominant farming systems in multistate regions characterized by similar soils, hydrology, and commodities produced | Soils, climate, and water availability | Income, production costs, and structure of commodity markets and agricultural sector | National Resources Inventory (SCS), production practices, costs of production, yields (ERS, NASS, Bureau of the Census), water quality monitoring (USGS, EPA, and state monitoring systems) |
| State or Regional Subdivisions[a] | Farming systems characterized by similar pattern of crop or livestock production, e.g., cash grain, cash grain-beef-hogs, or irrigated potatoes | Distribution of highly erodible land, indicators of soil quality, and surface water and groundwater quality monitoring | Fluctuations in commodity and input prices, capital requirements, off-farm employment opportunities, land tenure | Regional nutrient, pesticide, and irrigation mass balances; geographic information systems; hydrologic models; variability in production practices among producers; and segmentation of producers into target markets for program delivery |
| Farm | Pattern of input use, crop rotations, and livestock production in both space and time | Soil mapping, indicators of soil quality, drainage, and existing conservation plans | Producer attitudes, skills, goals; available equipment, capital and labor; land ownership; and participation in federal agricultural programs | History of soil test results, crop and livestock production, pest problems, irrigation scheduling, input use, and tillage practices |
| Field | Within-field patterns of pest problems, moisture, and fertility and culture practices | Soil testing keyed to soil mapping units; within-field history of nutrient, pesticide, and irrigation use; and identification of localized pest or fertility problems | Field or production unit efficiency, e.g., yield goals versus historical yields achieved, costs per kilogram of commodity produced, and year-to-year variability in costs and returns | Use of soil tests, soil maps, enterprise records of yields, pest problems, and input use to adjust production practices to variations in soils, drainage, or landscape position within the field |

NOTE: SCS, Soil Conservation Service, USDA; ERS, Economic Research Service, USDA; NASS, National Agricultural Statistics Service, USDA; USGS, U.S. Geological Survey, U.S. Department of the Interior; EPA, U.S. Environmental Protection Agency.

[a]State and regional subdivisions could include, for example, watersheds, major land resource areas, crop reporting districts, or other regions.

*The development and implementation of approved integrated farming system plans should be the basis for delivery of educational and technical assistance, should be the condition under which producers become eligible for cost-sharing dollars, and should be the basis for determining whether producers are complying with soil and water quality programs.*

Current programs, whether voluntary or nonvoluntary, are all based on a conservation planning approach. Plans are required for Conservation Compliance, Water Quality Incentive Program, the Integrated Farm Management Program option, and different elements of USDA's water quality initiatives. In addition, contracts that specify the practices that the producer should follow are required for the Conservation Reserve Program and for Agricultural Conservation Program cost-sharing agreements. Similar conditions of use and management are established in easements under the Wetland Reserve Program. Other plans will be required to comply with provisions of the 1990 Coastal Zone Management Act Reauthorization Amendments (PL 101-508).

It is possible that a single producer could be required to implement

• a conservation compliance plan stipulating erosion control measures for those fields that are highly erodible;

• a cost-sharing agreement with the Agricultural Stabilization and Conservation Service of USDA stipulating the management practices required to maintain a specific structure, such as a terrace or grassed waterway, for which the producer receives cost-sharing dollars;

• a water quality plan tied to receipt of incentive payments under the Water Quality Incentives Program; and, increasingly,

• a nutrient management plan to meet the requirements of state water quality regulations.

The effectiveness of these programs and plans will be increased if they are based on a single integrated farming system plan that balances multiple objectives and ensures that single-objective best-management practices designed to reduce erosion, improve nutrient and pest management, or improve the management of irrigation water, for example, are not working at cross purposes.

The objectives of conservation efforts are multiple; the traditional concern for reducing soil erosion has been combined with the need to reduce loadings of nutrients, pesticides, salts, and sediments to surface water and groundwater. Encouraging or requiring the adoption of single-objective best-management practices is no longer a sufficient basis for soil and water quality programs at the farm level.

Integrated farming system plans that address (1) conservation and enhancement of soil quality, (2) increased input use efficiencies, (3) increased resistance of soil to erosion and runoff, and (4) field and landscape buffer zones are needed if the multiple objectives of improving soil and water quality are to be met and trade-offs are to be minimized.

The Soil Conservation Service of USDA is beginning to use integrated farming system plans through its proposed Resource Management System. The Soil Conservation Service proposed that resource management systems address multiple objectives and the best-management practices that can be integrated into a farming system plan to improve soil and water quality.

*The first step toward implementing a farming systems approach to improving soil and water quality should be to replace current single-objective plans required to receive financial assistance through the Agricultural Conservation Program, Water Quality Incentives Program, and other programs with integrated farming system plans.*

Receipt of cost-sharing dollars should be conditional on the development of an integrated farming system plan that clearly specifies how the

---

## WIN-WIN OPPORTUNITIES: A SYSTEMS APPROACH ON A PENNSYLVANIA DAIRY FARM

Lanyon and Beegle (1989) studied a 56-ha (138-acre) dairy farm in central Pennsylvania as a model for whole-farm planning to improve nutrient management. The farm is a good example of how a farming system approach can improve soil, water quality, and profitability.

Lanyon and Beegle calculated nutrient balances using the producer's records of crop yields; the amounts of fertilizer and manure applied; sales of crops, milk, and livestock; and the amount of livestock feed purchased. Nutrient budgets for individual fields revealed that substantial reductions in the amount of nitrogen, phosphorus, and potassium were possible if inputs from manure were properly credited. The data for one corn field, for example, revealed that manure provided 277 percent as much phosphorus and 463 percent as much potassium as was removed by the corn crop. Application of purchased sources of these nutrients, except for starter fertilizers, could be suspended. The amounts of phosphorus and potassium applied to the alfalfa field were less than those removed by the crop, but soil tests revealed very high levels of phosphorus and potassium in the soil and supplementary applications of phosphorus and potassium to alfalfa were not needed. The use of on-farm supplies of nitrogen, phosphorus, and potassium from manure and legumes reduces production costs and the potential for losses of nutrients to surface water or groundwater.

The results for the nitrogen, phosphorus, and potassium balances in the livestock unit suggest that improvements in manure collection and manure storage facilities could substantially increase the efficiency with cost-sharing practice or structure supports implementation of the farming system plan. Implementation of the farming system plan, in addition to maintenance of the cost-sharing practice or structure, should be required as a condition of the cost-sharing agreement. The planning and implementation requirements for the Water Quality Incentives Program already approach this recommendation.

*In the long term, the implementation of an integrated farming system plan should be required for producers in regions where soil and water quality problems are severe regardless of their participation in federal farm programs.*

About 55 million ha (135 million acres) of U.S. cropland (about 32 percent of all U.S. cropland) will be subject to Conservation Compliance erosion control plans if producers want to receive federal farm program benefits. Full implementation of Conservation Compliance plans on these lands should help to improve soil quality and increase the soil's resistance to erosion and runoff. Soil and water quality benefits from implementation of these compliance plans could be much more comprehensive, however,

---

which producers can use the nutrients in manure. Purchased feed contributed substantially to the total nutrient flow in the farm. The nitrogen, phosphorus, and potassium supplied to the livestock enterprise from on-farm sources alone, however, provided 125 percent of the nitrogen, 87 percent of the phosphorus, and 186 percent of the potassium accounted for in livestock products and manures, suggesting that there may be substantial opportunities to refine the composition of livestock feed.

The systematic analysis of this dairy farm revealed that the best means of improving environmental and financial performance are to

- make better use of on-farm nutrient sources by redistributing nutrients to fields on the basis of soil test results, nutrient application history, and crop history;
- make better use of on-farm nutrient sources by improving manure collection and storage to reduce manure losses from the barnyard; and
- refine the feed composition, which would perhaps reduce the need for purchased feeds.

Implementation of a single best-management practice, increased soil testing, or construction of manure storage facilities, for example, would address only one component of what is required to improve nutrient management on the dairy farm. The effectiveness of soil testing or manure storage facilities will be greatly increased if they are part of a more comprehensive nutrient management approach. It is the nutrient management approach, not the practices adopted, that determine success. Similarly, it is the management approach, as reflected in an integrated farming system plan, that should be the basis of efforts to improve soil and water quality.

if they were based on integrated farming system plans that address input use efficiency and buffer zones in addition to soil erosion.

Even if fully implemented, compliance mechanisms will not rectify all soil and water quality problems because they cover only selected crops and producers (Ribaudo, 1986) and address erosion only from highly erodible lands. Evidence also suggests that compliance mechanisms, as they currently are designed and implemented, may not address many important water quality problems (Ribaudo and Young, 1989). Compliance plans, as they are currently required, do not address compaction, salinization, or other forms of soil degradation. Nutrient, pesticide, and irrigation management plans are also not addressed under Conservation Compliance plans, and the 68 percent of cropland not under Conservation Compliance plans may include important sources of soil degradation and water pollution.

Erosion control on highly erodible lands alone, while important, will not adequately control loadings of nutrients, pesticides, salts, and sediments to surface water and groundwater. Programs need to be targeted at problem areas and at problem farms (see section later in this chapter for a discussion of targeting). Croplands other than, or in addition to, those that are highly erodible will have to be included in these programs. Similarly, producers other than, or in addition to, those participating in federal farm programs will have to be added. An integrated approach that addresses all four components of the farming system (soil quality, input use efficiency, resistance to erosion and runoff, and buffer zones) will be needed.

### Rigorous Planning Standards

Rigorous planning standards are needed to increase the confidence that implementation of integrated farming system plans will result in real improvements in soil and water quality. Such standards are needed whether voluntary or nonvoluntary approaches to the development and implementation of integrated farming system plans are used.

*USDA and the EPA should convene an interagency task force to develop planning standards that can be used as the basis for implementation of the Resource Management System by the Soil Conservation Service of USDA and as guidance for state governments that meet the requirements of the Federal Water Pollution Control Act (PL 100-4) and the 1990 Coastal Zone Management Act Reauthorization Amendments (PL 101-508).*

Integrated farming system plans to improve soil and water quality should, at a minimum, specify how recommended farming practices (1) conserve or enhance soil quality, (2) increase input use efficiency, (3) increase resistance to erosion and runoff, and (4) incorporate field and landscape buffer zones into the farming system. A large body of information and models are already available to establish more rigorous standards for these criteria. That body of research should be used to prepare and evaluate integrated farming system plans.

### Soil Quality

Standards for soil quality have not yet been developed, although efforts are under way in the Soil Conservation Service and the U.S. Forest Service of USDA to develop such standards (see sections on soil quality in this chapter and in Chapter 5 for a full discussion). As a starting point, the following indicators of soil quality should be used as standards to evaluate the effects of implementing the farming practices that are recommended for an integrated farming system plan: nutrient availability, amount of organic carbon, amount of labile carbon, texture, water-holding capacity, structure, maximum rooting depth, salinity, acidity, and alkalinity.

### Input Use Efficiency

The criteria needed to evaluate input use efficiency vary depending on whether nutrients, pesticides, or irrigation water is being addressed (see sections on input use efficiency in this chapter and in Chapters 6 to 8 and 10 and 11 for a full discussion). Considerable experience has been gained in the last few years in preparing nutrient, pesticide, and irrigation management plans. Several states have already developed best-management practices and plans for nutrients, pesticides, and irrigation water. In addition, knowledge gained from the Management Systems Evaluation areas project (an interagency effort to evaluate the effect of agricultural management practices on water quality) can provide information from which to develop standards for input management plans. The management measures developed to implement the 1990 Coastal Zone Management Act Reauthorization Amendments are another source of information. Information available from these sources should be assembled to develop uniform criteria that can be applied to nutrient, pesticide, and irrigation management plans.

### Resistance to Erosion and Runoff

Erosion prediction models such as the universal soil loss equation and the revised universal soil loss equation can predict the effects that

Water runoff from a cotton field, having passed through the grass hedge (behind the agronimist), is recorded on a hygrographic chart. Credit: Agricultural Research Service, USDA.

residue management, cover crops, and other measures have on increasing the resistance of farming systems to erosion and runoff. These models are sufficient for comparing the relative effectiveness of recommended measures to increase resistance to erosion and runoff (see the section on resistance to erosion and runoff in this chapter and in Chapter 9).

### Buffer Zones

Four criteria—size, location, species selection, and vegetative management—are most important for evaluating the effectiveness of field-scale buffer or vegetative filter strips (see sections on buffer zones in this chapter and in Chapter 12). A substantial body of research and experience with vegetative filter strips within or bordering crop fields is accumulating. Models such as GRAPH (Lee et al., 1989) and GRASSF (Barfield et al., 1979; Hayes et al., 1979) have been developed to predict the amount of sediment and nutrient trapping in vegetative filter strips. This information can be used to establish criteria for size, location, species selection, and management of field-scale buffer strips. The U.S.

Forest Service has produced specific guidelines for riparian buffer zone planning, design, and maintenance (Welsch, 1991). These guidelines call for three zones, each under a different management system depending on distance from the stream and the intensity of use of adjacent uplands.

### Need for Performance Standards

Integrated farming system plans that specify a combination of production and information management practices are the best tools available now to guide efforts to prevent soil degradation and water pollution. In the long term, however, standards based on more quantitative estimates of soil degradation and water pollution caused by farming practices are needed. These standards are needed to ensure that soil and water quality goals are met and that unnecessary requirements are not imposed on producers.

In some cases of industrial pollution, polluters are required to meet performance standards based on water quality criteria applied to their discharges or to the water body that receives those discharges. Polluters are allowed to discharge a certain level or load of pollutants; or, in other cases, groups of polluters work together to achieve a given level of "receiving water" quality. In either case, water quality goals and the obligations of the polluters are unambiguous. Moreover, the performance standard approach that requires achieving a specified water quality standard rather than specifying the control technology that should be used allows polluters to develop innovative strategies for achieving compliance.

It is difficult to apply a performance standard approach to nonpoint source pollution in general and to agricultural nonpoint source pollution in particular (Abler and Shortle, 1991; Foran et al., 1991; Roberts and Lighthall, 1991). It is difficult to measure pollutant outputs from specific farm fields, and it is difficult to identify pollutants from specific areas in a degraded water body. The alternative to performance standards is a "design standard" approach. In a design standard approach, polluters come into compliance by implementing a set of approved practices. This is the approach taken in the 1990 Coastal Zone Management Act Reauthorization Amendments (PL 101-508), which requires the development of an enforceable program of management measures to control nonpoint source pollution in coastal areas.

The drawback of design standards, including integrated farming system plans, is that they do not guarantee the achievement of a given level of soil or water quality. Moreover, design standards can be confining, limiting the options of a producer to improve the environmental performance of a production system.

In the short term, design standards, such as the integrated farm management planning standards discussed here, are the best that can be done with current models and data. There is an urgent need, however, to develop a performance standard approach for improving the environmental performance of farming systems. Developing such an approach requires several steps to quantify how the agricultural management practices used on a field affect the water quality in the problem water body or aquifer. First, the lowest level of sediment, nutrient, pesticide, salt, and trace element output that can be achieved with current best-management practices in different parts of the United States needs to be determined. The current Management Systems Evaluation Area network of sites should be able to provide these data for many parts of the country. Second, there is a need for models capable of depicting edge-of-field and bottom-of-the-root-zone pollutant outputs on a field-by-field basis. Model development toward this end is proceeding slowly and may need to be stimulated (see below). Once these steps are taken, researchers can begin to evaluate agricultural nonpoint source pollutant outputs against water quality discharge and can begin to develop receiving water quality criteria on a field-by-field and watershed-by-watershed basis.

### Use of Models

Substantial advances have been made in mathematical modeling and computer simulation modeling of agricultural nonpoint source pollution problems that can help develop standards for integrated farming system plans. These modeling efforts range from simple conceptual mass balances to sophisticated research models.

#### Purposes, Advantages, and Limitations of Modeling

Investigators have used models to evaluate the extent of water pollution caused by management practices as well as to simulate water quality improvements under alternative management scenarios. Sensitivity analyses on model parameters and coefficients may identify which management practices would yield the greatest net change in soil or water quality. Modeling and models are not, however, a panacea for problem solving; they serve only as tools. Models typically reflect the model builder's perception of the problem and not necessarily that of the agronomic researcher or the resource manager. A model developed for site-specific conditions often cannot be used as a generic model, and vice versa. The precise and unambiguous language of mathematics and computer science is used in models, but in reality, models are only a substitute for the behavior of real-world systems.

Furthermore, some research models that are finely tuned require extensive input and model coefficients that are not normally measured and that are difficult or extremely costly to obtain. Simple mass balance models typically do not have the requisite characteristics to provide simulations over shorter time scales (for example, daily or monthly) or spatial scales (for example, hectares). Despite these problems, models may serve as valuable tools when properly applied and when proper recognition is given to the underlying assumptions and specificity.

### Models of Nonpoint Source Pollution

A wide array of models address agricultural nonpoint source pollution. Many of them have been calibrated and validated with soil columns in the laboratory and intensively monitored plots in the field. Fewer models have been validated under field conditions with or without cropping. In a few instances, a given model has been tested at several localities or several models have been tested comparatively on a monitored field site.

A concerted effort is being made to develop or modify currently available models to assess agricultural nonpoint source pollution problems in cropping systems. For instance, Hanks and Ritchie (1991) promoted the use of computer models as a partial substitute for experimental research to determine agronomic recommendations. In addition, many states have now developed geographic information systems for natural resource inventory and management.

The Committee on Ground Water Modeling Assessment (National Research Council, 1990) assessed the development and use of groundwater containment models for scientific and regulatory applications. That committee concluded that there is a range of capability in modeling fluid flow but had some concerns about the reliability of these models. Although prototype models exist for the reactivity and transport of contaminants, they have not yet been developed for use in practice (for example, regulatory applications).

A similar overall assessment of agricultural nonpoint source pollution models has not been made, other than to identify research needs. However, there is a growing body of literature about the utility of models that consider climate, soils, and crops for agricultural applications. The potential for improved and accurate agronomic simulation models is anticipated in the near future, and their role in assessing the effect of alternative farming systems on agricultural nonpoint source pollution problems is expected to become increasingly valuable.

## On-Farm Record Keeping

The first step toward implementing a farming systems approach is to develop information on a field-by-field basis. Many of the data needed to develop and implement integrated farming system plans are available only if producers keep good records of their management practices and yields.

*Record keeping should be an essential component of integrated farming system plans.*

The lack of good information about the farming operation can be a serious impediment to the development of integrated farming system plans. At a minimum, all producers should be encouraged to keep records of the inputs and tillage practices used, crop sequence, and crop yields on the field or enterprise level. Record keeping should be mandatory when integrated farming system plans are the basis on which financial assistance is received or for ensuring compliance with soil or water quality laws. Record keeping during and after the implementation of an integrated farming system plan is critical for providing the steady flow of information needed to evaluate and adjust the farming system plan. The systems established to manage the flow and analysis of information are as important or more important than the specific management practices specified in the plan. The development of record keeping systems that link agronomic and financial decisions should be a high priority. Policies that encourage or mandate the collection and use of information by the producer may, in the long term, prove to be more effective than encouraging or mandating the use of best-management practices.

Record keeping has important benefits for the producer as well as governments. Record keeping is essential for refining enterprise management to increase profits. The information needed to manage a farm operation to maximize profit, if properly organized, will complement the information needed to improve soil and water quality. Collection and organization of this information is a way to improve profitability as well as soil and water quality.

## Developing Capacity at the Local Level

Providing the technical assistance to develop and implement integrated farming system plans will tax the current capabilities of federal, state, and local government agencies. The development of the capacities of both the public and the private sectors to deliver technical assistance to farmers should be a primary goal of agricultural environmental policy.

### Public Sector

*The Soil Conservation Service, in cooperation with the Cooperative Extension Service, should undertake an accelerated training effort targeted at federal, state, and local government personnel and at producers to develop and implement integrated farming system management plans.*

The technical capacities of county Soil Conservation Service, Agricultural Stabilization and Conservation Service, county Soil and Water Conservation Districts, and other personnel to develop and implement integrated farming system plans are variable. Expertise at the local level has been developed primarily to provide technical assistance for erosion control. The capacity to develop and implement integrated farming system plans requires broader technical understanding of input management, the transport of agricultural chemicals to water bodies, and the economics of farm planning. Training programs are urgently needed to develop this expertise at the local level.

### Private Sector

*Mechanisms should be developed to augment public-sector efforts to deliver technical assistance with nonpublic-sector channels and to certify the quality of the technical assistance provided through these channels.*

Development and implementation of integrated farming system plans at the farm level can be facilitated by public-sector programs—through the Cooperative Extension Service, the Soil Conservation Service, Soil and Water Conservation Districts, and the Agricultural Stabilization and Conservation Service; but many of the services needed to assist producers are increasingly provided by the private sector. Table 3-2, for example, shows the results of a survey conducted by American Farmland Trust to determine the sources of information producers use to make tillage, fertility, or weed and insect control decisions. Five hundred farmers in Washington state, California, Minnesota, Illinois, and Georgia were surveyed. Fertilizer, herbicide, or insecticide dealers; other farmers; and family members stand out as the most important sources used by the surveyed farmers. Public-sector sources of information were relatively untapped, at least directly, by these farmers when making their decisions.

TABLE 3-2   Ranking of Information Sources by Surveyed Farmers

| | Percentage of Farmers Ranking Source as First or Second Most Important[a] | | | | | |
| --- | --- | --- | --- | --- | --- | --- |
| | Tillage | | Fertility | | Weed and Insect Control | |
| Information Source | First | Second | First | Second | First | Second |
| CES staff | 8 | 5 | 8 | 3 | 9 | 6 |
| CES publications, meetings, or field days | 5 | 4 | 4 | 5 | 3 | 5 |
| SCS or CD staff | 2 | 3 | <1 | 1 | 0 | <1 |
| SCS or CD publications, meetings, or field days | 1 | 2 | 1 | 2 | <1 | <1 |
| ASCS staff | 2 | 2 | 1 | 1 | 1 | 0 |
| ASCS publications, meetings, or field days | 1 | 1 | 1 | 1 | 0 | <1 |
| Staff and publications of other public agencies | 1 | 1 | 1 | 1 | <1 | 1 |
| Farm organization staff | 2 | 2 | 2 | 2 | 2 | 1 |
| Fertilizer dealer | 28 | 9 | 56 | 16 | 18 | 6 |
| Herbicide or insecticide dealer | 5 | 14 | 2 | 12 | 40 | 18 |
| Fertilizer or pesticide applicator | 1 | 2 | <1 | 4 | 3 | 4 |
| Other farmers | 17 | 20 | 8 | 23 | 6 | 27 |
| Family member | 10 | 13 | 8 | 10 | 6 | 9 |
| Nonprofit, educational, or environmental organization | 1 | 1 | <1 | 1 | 0 | <1 |
| Farm magazines, journals, and radio and television programs | 8 | 13 | 2 | 12 | 2 | 11 |
| Other | 5 | 1 | 5 | 3 | 4 | 2 |
| No response | 5 | 10 | 1 | 1 | NA | NA |

NOTE: CES, Cooperative Extension Service, USDA; SCS, Soil Conservation Service, USDA; ASCS, Agricultural Stabilization and Conservation Service, USDA; CD, county Soil and Water Conservation District; NA, not available.

[a]Mean of reported percentages of farmers surveyed in Whitman County, Washington; Butte County, California; Renville County, Minnesota; Livingston County, Illinois; and Dooly County, Georgia.

SOURCE: Adapted from J. D. Esseks, S. E. Kraft, and L. K. Vinis. 1990. Agriculture and the Environment: A Study of Farmer's Practices and Perceptions. Washington, D.C.: American Farmland Trust.

Although private-sector service-oriented programs have proved successful for producers of specialty crops in the Southeast and Pacific regions, there has been relatively little development of such services for use by producers of the major commodity crops such as corn, wheat, and soybeans. The private sector may require some encouragement to develop the willingness and capability to deliver these services. Quality control and quality assurance are also needed to ensure that the technical assistance delivered by the private sector is adequate. Spiker and colleagues (1990), for example, found it difficult to determine whether the use of soil tests affected rates of fertilization because of the widely varying nutrient and fertilizer recommendations and applications that follow testing. Efforts to develop certification procedures and to increase the capacities of nonpublic-sector channels should go hand in hand.

## TARGETING PROBLEM AREAS AND FARMS

The importance of targeting—that is, attempting to direct technical assistance, educational efforts, financial resources, or regulations to those regions where soil and water quality improvements are most needed, or to those farm enterprises that cause a disproportionate portion of soil and water quality problems—is difficult to overstate. Finding ways to target programs to well-defined regions and farm enterprises has become even more important as the problems that these programs address has expanded from soil erosion to water quality. The need for refined targeting has been made more urgent as federal, state, and local policymakers have struggled to stretch sometimes shrinking budgets to keep up with the increasing list of items on the environmental problem agenda.

Chapter 2 recommended an ambitious set of objectives for a national effort to improve soil and water quality, and the previous section of this chapter emphasized the need to expand efforts at the farm level by taking a systems approach to soil and water resource management. Such an expanded agenda cannot be implemented unless it is targeted at well-defined problems and farm enterprises. The inability or unwillingness to target policies, whether voluntary or nonvoluntary, only at areas where the need to improve soil and water quality is greatest or only at those farm enterprises responsible for soil and water quality damages is a major obstacle to efforts to make soil and water quality programs more effective.

Targeting of programs to regions and farm enterprises where those programs are most needed requires information about soil degradation, water pollution, and producers' production practices. Ideally, the decision to target programs at particular regions or enterprises should be based on an

- articulation of national or state goals for soil and water quality;
- identification of regions where the benefits from achieving the goals per dollar invested are greatest;
- identification of the linkages among farm practices, soil quality, and water quality; and
- identification, within a targeted region, of those enterprises that contribute to the problem as well as their barriers to changing their farming systems.

Figure 3-2 illustrates how this information could be arrayed three dimensionally to identify those regions and producers at which programs should be targeted. The highest priority for programs to change farming systems would be where soil and water quality degradation and the potential to improve producer's management are greatest.

Unfortunately, the data needed to construct the three-dimensional targeting scheme proposed in Figure 3-2 often are not available. The need to improve targeting, however, is urgent, and new approaches are needed to identify the regions and enterprises that should receive the greatest attention. The information available for guiding targeting efforts will, most likely, always be less than ideal. It is urgent that means be found to move ahead with the information that is available now. The targets identified by using this information may not be as refined as policymakers and program managers might like, but even crude targeting will help reduce the costs and increase the effectiveness of current programs to improve soil and water quality.

## Soil and Water Quality Monitoring

Most efforts to target programs where they are most needed have been based on identifying those geographic regions where soil degradation and water pollution are most severe. These efforts have produced considerable amounts of information that could be used by policymakers and program managers to target soil and water quality programs at better-defined geographic regions. This information, however, has been collected by different agencies and for different purposes and has been based on different measures of soil or water quality. This information needs to be assembled and synthesized to identify priority regions.

*The Secretary of USDA and the Administrator of EPA should undertake a coordinated interagency effort to identify regions or watersheds that should be the*

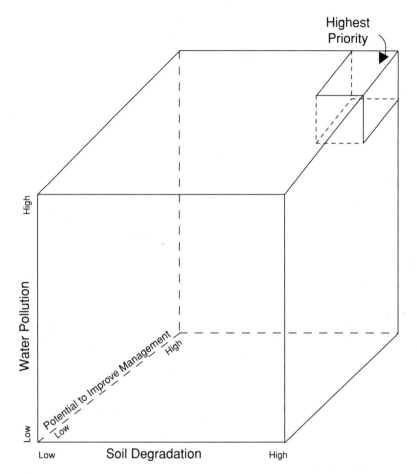

FIGURE 3-2   Conceptual diagram of three-dimensional targeting.

*highest priorities for federal, state, and local programs to improve soil and water quality.*

Federal, state, and local governments have identified priority areas for various soil and water quality problems and for various programs to improve soil and water quality. A few of these efforts are described below.

### Soil Quality

The USDA has developed criteria and data to identify highly erodible lands, that is, lands that are most vulnerable to accelerated rates of erosion if they are not properly managed. The definition of these lands has been central to the implementation of Conservation Compliance, Conservation Reserve Program, and Sodbuster. Although erosion alone does not address all of the forces affecting soil quality (see Chapters 2 and 5 for a full description), the distribution of highly erodible land should be an important criterion used to identify priority regions. The extent of salinization of soils has also been monitored by the Soil Conservation Service as well as by state governments and international organizations. These data are also available to help define priority regions. As discussed in Chapters 2 and 9, the volume and energy of runoff from agricultural lands are closely related to but are not the same as rates of erosion. The Soil Conservation Service and the Agricultural Research Service should assemble the data and models available now to identify agricultural lands that should be included as priority areas based on the expected volume and energy of runoff caused by a lack of proper management.

### Water Quality

Several efforts have been made by federal agencies and state and local governments to identify priority areas for water pollution control. The EPA is currently assembling data from assessments of nonpoint sources of water pollution that were conducted by each state under the provisions of the 1987 amendments to the Federal Water Quality Protection Act (PL 100-104) and is expected to use these data to identify priority areas for the control of nonpoint source pollution of surface waters. Similarly, the USDA identified 74 hydrologic unit areas to receive priority attention under the USDA's water quality initiative (Mussman, 1991). Congress has also identified the Chesapeake Bay watershed and the Great Lakes watershed as priority areas for water pollution control and, through the 1990 Coastal Zone Management Act (PL 100-508), it has mandated that watersheds that contribute to degradation of coastal wetlands and estuaries should be priorities for water pollution control. It is essential that the definition of priority areas under these programs be coordinated, and the data used to define these areas should be used to help define national priorities for soil and water quality improvements.

## Monitoring Production Practices

The benefits of targeting efforts on the basis of soil and water quality data that define where resource damages are most severe are widely accepted. The benefits, however, of using data on producers' production practices and enterprise characteristics have often been overlooked. Adding the third dimension—information on production practices and enterprise characteristics (Figure 3-2)—to the process of identifying targets will help in three ways: by identifying problem farms within priority areas identified by soil and water quality criteria alone, by identifying changes in the management of farming systems that should be sought within priority areas, and by identifying the barriers to adoption of improved farming systems that need to be overcome.

### Problem Farms

*The Secretary of USDA and the Administrator of EPA should initiate a multiagency effort to assemble currently available data on production practices and enterprise characteristics to identify problem farms within priority areas for soil and water quality improvements.*

Although systematic data on production practices, input use, and management systems are scarce, those studies that are available clearly suggest the benefits of targeting programs to those farm enterprises that cause the most damage to soil and water quality. Padgitt (1989) found, for example, that about 25 percent of the Iowa farmers surveyed applied fertilizer at a level of 28 kg/ha (25 lb/acre) above recommended levels. Similarly, Schepers and colleagues (in press) found that 14 percent of the land in the Central Platte Natural Resource District in Nebraska that they studied received nitrogen in excess of 100 kg/ha (89 lb/acre) of the recommended amounts. Other investigators have also found that some producers apply excessive nutrients: Hallberg and colleagues (1991) report on Iowa, and Bosch and colleagues (1992) report on two regions of Virginia.

Setia and Magleby (1988) found that targeting conservation tillage practices to the 4,452 ha (11,000 acres) that cause the most damage within the targeted watershed could reduce the cost of improving water quality from $139,000 to between $9,000 to $32,000 for each percentage point reduction in the amount of sediments. Targeting the farms that contribute the largest nutrient loadings in the watershed could reduce cost of improving water quality from $151,000 to between $11,000 and $43,000 for each percentage point reduction in nutrient loads. Similarly, Lee and colleagues (1985) found that directing improvement efforts to critical areas within a targeted watershed could reduce costs of improving water quality 5- to 10-fold.

In 1985, USDA's Agricultural Research Service and Economic Research Service concluded an analysis of the previous years' targeting efforts. Nielson (1986), drawing on that analysis, recommended that USDA concentrate its efforts on problem farms. The study indicated that few county Agricultural Stabilization and Conservation committees gave such targeting high priority.

Data from the studies mentioned above demonstrate the importance of recognizing that there are problem farms, that is, farm enterprises that because of their location and production practices and management techniques cause more soil and water problems than others. It is just as important to recognize that many farms cause no problems at all and that some are probably improving soil and water quality. Targeting programs at the set of farms that are responsible for most soil and water quality degradation will reduce the cost and increase the effectiveness of soil and water quality programs. Targeting can also prevent placing unnecessary burdens on those producers who are not causing damages and recognize those producers who are making positive contributions to improving soil and water quality.

### Monitoring Progress

Tracking changes in production practices provides a way to monitor programs in the absence of adequate soil and water quality monitoring. Nutrient mass balances, for example, can be calculated and monitored in the absence of adequate data on nutrient loadings to surface water or groundwater. Over time, improvements in nutrient management should be indicated by changes in the relative proportion of nutrient inputs and their balance with crop outputs (see Hallberg et al. [1991] for examples of programs in Iowa). This approach can be linked to existing or planned soil and water quality monitoring to further refine targeting and program evaluation. Gianessi and colleagues (1986), for example, used a large data base that described discharges to the nation's waters from approximately 32,000 point and 80,000 nonpoint sources to identify regions that would show significant improvement in phosphorus concentrations as a result of upland erosion control. As such data bases become available and better refined, they can be combined with nutrient mass balance information to identify those regions where potential water quality benefits from improved farming systems are the greatest.

Changes in the use and distribution of irrigation water, pesticides, tillage systems, or other farming system components could also function as measures of success in improving farming systems. In a survey of Iowa, Duffy and Thompson (1991) showed that 88 percent of corn and soybeans were cultivated an average of 1.3 times, indicating a significant potential to reduce herbicide use through banding rather than broadcast applications. Yet, the survey indicates that banding is used only on about 15 percent of the land planted in corn and soybeans.

### Refine Strategies to Change Producer Behavior

Analysis of data on production practices and enterprise characteristics when programs are being designed is essential for adapting national policies to local realities as the national policies are implemented (Rogers, 1983). Analysis of production practices and enterprise characteristics helps program managers understand the diversity of reasons that may account for a producer's decision not to adopt an improved farming system (Nowak, 1983, 1985; Nowak and Schnepf, 1987). When local program personnel begin to understand this diversity, they can begin to make use of program implementation tools that match the diversity of reasons for producers' nonadoption of management practices (Kelly, 1984; Lake, 1983).

An example of such an approach is provided by the Farm Practices Inventory (FPI) developed in Wisconsin. The FPI measures specific crop nutrient behaviors on a corn field identified by the respondents as being the most productive in that year. It also measures differences, if any, between the most productive and other corn fields. In addition, it contains an inventory of pesticide use and management practices, livestock inventories and manure management practices, and a series of items that measure farmstead design (for example, wells, storage, and waste disposal). The FPI also has a knowledge "test" that measures the levels of knowledge and perception of the major attributes of a series of recommended practices. Finally, it measures a limited set of farm, personal, and communication items. The FPI was designed to meet three objectives in natural resources management in Wisconsin:

- provide an accurate assessment of the agronomic behaviors of producers in Wisconsin relative to the patterns of nutrient and pesticide management;
- segment target audiences, design appropriate educational strategies, and provide direction for allocating limited fiscal resources; and
- evaluate the effectiveness of soil and water quality programs and actions.

Analysis of production practices and enterprise characteristics are needed before programs are implemented. Knowledge of prevailing agronomic practices and producers' belief systems, knowledge levels, and other characteristics helps in designing improved farming systems that are relevant to the producers affected by the problem (Grunig et al., 1988). This type of knowledge can be used to determine exactly

- what farm management and production practices contribute to soil and water quality degradation;
- what changes in management are suitable given existing knowledge levels, the nature of current farming practices, and the flexibility to invest human capital or fiscal resources in new practices; and
- what information and assistance mechanisms currently have high credibility and use among target populations.

Answers to these questions prior to program implementation will allow a level of targeting that is not now being used.

### Regional and National Data Collection

Full implementation of an integrated approach to planning and directing programs to prevent soil degradation and water pollution will only be possible if information of the appropriate density and quality is available. Providing this information will require coordinated efforts at local, state, and national levels.

*Concerted efforts at the state and local levels should be undertaken to collect new data and find ways to link data that is already collected for other purposes to provide the foundation for more integrated approaches to preventing soil degradation and water pollution.*

In many cases, existing data are not well suited to integrated approaches to program planning and direction. National data on soil and water resources such as the National Resources Inventory provide useful information for large regional scales but are not dense enough for use in the county, watershed, or smaller scale applications required to implement a systems approach at the local level. Data available at the local level, such as that found in soil surveys, are often difficult to link with other data sets that have been assembled for different purposes, such as participation in federal farm programs or cropping histories assembled by county offices of USDA's Agricultural Stabilization and Conservation Service.

The lack of systematic data on production practices is a particularly serious obstacle to targeting, monitoring, and designing soil and water quality programs. When such information is available, it is often not geographically based or linked to physical information about soil and water quality degradation. This lack of linkage between relevant natural resource data, production practices, and socioeconomic data limits the ability to realize improved targeting and program direction from an integrated approach based on farming systems (Fletcher and Phipps, 1991).

Geographic information systems (GISs) have the potential to greatly increase the usefulness of existing and provide new data to implement a systems approach to soil and water resource programs (Fletcher and Phipps, 1991). GISs are designed to collect, manage, analyze, and display data spatially; they can be used in combination with other models as a way to enhance targeting, planning, and directing programs.

For example, Prato and coworkers (1989) used a GIS to assemble and retrieve physical measures of erosion. The GIS was linked with a linear programming model to determine an economically efficient system for reducing pollution. The water quality effects of such economically efficient solutions were evaluated by using the Agriculture Nonpoint Source Model. A farm practices inventory obtained economic data that were combined with a microcomputer budget management system and an erosion planning model (Figure 3-3). Ultimately, the researchers designed a resource management system that would obtain the most income while making the desired reductions in pollutants. Reports by Tim (1992) and Hamlett and colleagues (1992) are also good examples of the potential to use GIS to target problem watersheds.

The collection of data and the development of GISs will greatly increase the ability to implement integrated approaches at the state and local levels. Similar improvements in data collection, particularly the collection of systematic data on production practices, are needed to implement a systems approach to developing and directing national policy.

*The Economic Research Service, the National Agricultural Statistics Service, and the Soil Conservation Service should assemble currently available information to provide baseline information about production practices and agronomic behaviors.*

The ability to target and direct programs is seriously constrained by the lack of comprehensive and representative data on the production practices and agronomic behaviors of agricultural producers. Few comprehensive and representative data are available on producers' nutrient, pesticide, and irrigation water management practices. Better, but still limited, information on tillage systems and erosion control practices is available. This lack of information makes it difficult to set realistic goals, identify the changes in farming practices that should be sought through environmental programs, or evaluate how effective programs have been and what remains to be done.

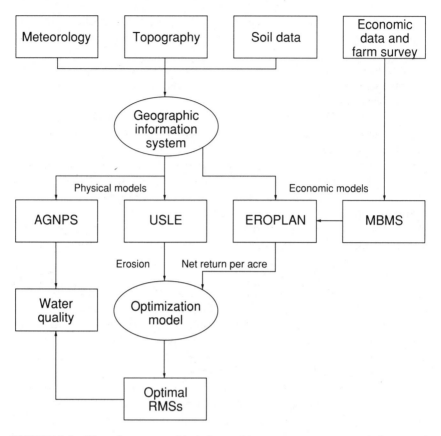

FIGURE 3-3 Use of a geographic information system to target and direct soil and water quality programs. AGNPS, agriculture nonpoint source model; USLE, universal soil loss equation; EROPLAN, erosion planning model; MBMS, microcomputer budget management system; RMSs, resource management systems. Source: T. Prato, H. S. R. Rhew, and M. Brusuen. 1989. Soil erosion and nonpoint source pollution control in an Idaho watershed. Journal of Soil and Water Conservation 44:323–328. Reprinted with permission from © Journal of Soil and Water Conservation.

The Economic Research Service, the National Agricultural Statistics Service, and the Soil Conservation Service should assemble all currently available data on production practices and agronomic behaviors. This information, if assembled in one place, would be very helpful for the direction of policy. The effort to assemble these data would also be the first step toward identifying the gaps in current data collection that need to be filled.

*The Economic Research Service, the National Agricultural Statistics Service, and the Soil Conservation Service, in coordination with the Bureau of the Census, should develop, test, and implement ongoing surveys of production practices and agronomic behaviors.*

The Economic Research Service and the National Agricultural Statistics Service are expanding current surveys of production practices. It is essential that such surveys continue over time to allow monitoring of changes in production practices. The value of production practice and agronomic behavioral data will be greatly enhanced if they can be linked to soil and water quality problems. The return on the current investment in data collection would be much greater if methods were developed to geographically link the data already collected in current and ongoing surveys. Such linkage should have as its goal improved policy formulation and implementation, particularly targeting.

The topographically integrated geographic encoding and referencing (TIGER) system, developed by the U.S. Bureau of the Census, could serve as a model for integrating agricultural census and farming system data with various land and water resource data bases. In addition, this spatial data base could contain information on factors such as the primary and secondary types of farming systems, the production activities that cause the most soil degradation or water pollution, the use of remedial production practices, and other factors that may influence the implementation of policies.

## IMPLEMENTING A SYSTEMS APPROACH

USDA, EPA, and state and local programs provide important opportunities to implement a systems approach to preventing soil degradation and water pollution. These programs, however, will have to be restructured and redirected, in some cases, to implement a systems approach. Increasing the resistance of farming systems soils to erosion and runoff has historically been the overriding objective of USDA soil and water conservation programs. More emphasis is now being placed on protection of water quality in USDA programs. Tables 1-1 and 1-2 in Chapter 1 list the soil and water quality programs administered by the USDA, and the new initiatives passed as part of the 1990 Food, Agriculture, Conservation and Trade Act (PL 101-624). New USDA programs such as the Water Quality Incentives Program, the Wetland Reserve Program, and the Environmental Easement Program signal the increasing importance of water quality in USDA programs.

The EPA's programs are also increasingly affecting agriculture (see

"Most important and least done about it" (February 6, 1936). Credit: Courtesy of the J.N. "Ding" Darling Foundation.

Table 1-1). In 1990, states began implementing the management plans they were required to prepare under section 319 of the 1987 amendments to the Federal Water Pollution Control Act (PL 100-4) (U.S. Environmental Protection Agency, 1992). The 1990 Coastal Zone Management Act Reauthorization Amendments (PL 101-508) require states to develop nonpoint source control programs within the coastal zone. The state programs are required to include enforceable policies and mechanisms to implement pollution control practices, called management measures. Seven management measures—including erosion and sediment control, wastewater and runoff control from confined animal facilities, nutrient management, pesticide management, grazing management, and irrigation water management—will have direct effects on agricultural production in the coastal zone (U.S Environmental Protection Agency, Office of Water, 1993).

State nonpoint source control programs, coastal zone programs, and the initiatives in the 1990 Food, Agricultural, Conservation, and Trade Act are important opportunities to address all four objectives proposed in this report: conserving and enhancing soil quality, improving input efficiency, increasing resistance to erosion and runoff, and making greater use of field and landscape buffer zones. Improved management of nutrients, pesticides, animal waste, or irrigation water is listed as an objective in all 16 demonstration projects and 74 hydrologic unit area projects that are part of USDA's Water Quality Initiative (U.S. Department of Agriculture, Working Group on Water Quality, 1991). These initiatives, along with the Water Quality Incentives Program represent a significant new commitment by the USDA to improve input management. Implementation of the management measures under 1990 Coastal Zone Act Reauthorization Amendments will also address the need to improve input management, and EPA's Office of Water (1982) reports that 24 percent of the management activities included in state nonpoint source pollution management plans address agricultural sources of pollution. The Wetland Reserve Program and the Environmental Easement Program created in the 1990 Food, Agriculture, Conservation and Trade Act are clear opportunities to make greater use of field and landscape buffer zones.

## Limited Funding

Funding for these initiatives, however, has been limited. Significant new commitments of general revenues to agricultural soil and water quality programs have been made since 1985. Figure 3-4 shows that expenditures by USDA and related state and local programs have

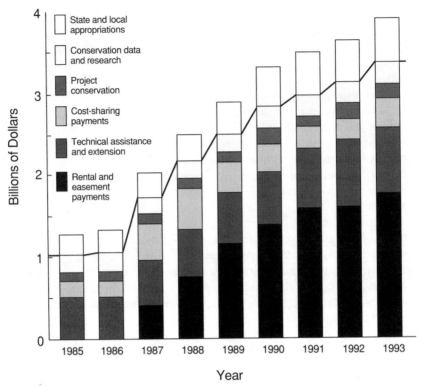

FIGURE 3-4   Conservation expenditures by the U.S. Department of Agriculture (USDA) and related state and local programs, 1983 to 1990. CRP, Conservation Reserve Program. Source: U.S. Department of Agriculture, Economic Research Service. 1990. Conservation and Water Quality. Pp. 28–41 in Agricultural Resources: Cropland, Water, and Conservation Situation and Outlook Report. Report No. AR-19. Washington, D.C.: U.S. Department of Agriculture.

increased 2.5-fold to more than $3.4 billion since 1986. Almost all of the increase in expenditures, however, was for the Conservation Reserve Program. Spending for other purposes increased much less.

In 1991, $51 million was provided by a grant through the EPA to help states implement plans to control nonpoint source pollution from all sources (U.S. Environmental Protection Agency, Office of Water, 1992). Only part of those funds were expended to control agricultural sources of pollution. The Agricultural Water Quality Protection Program was implemented under the Agricultural Conservation Program as a cost-shared practice called the Water Quality Incentives Program. The program is expected to expend $6.8 million in 1992 (U.S. Department of

Agriculture, 1992). Expenditures for water quality incentives are about 3.5 percent of total Agricultural Conservation Program expenditures in 1992 and projected to be 8 percent in 1993. Limits on the amount of cost-share dollars that can be received by an individual producer participating in the Agricultural Conservation Progam may, at times, be too low to cover a substantial share of the cost of adopting improved management practices. The Wetland Reserve Program was budgeted about $46 million expenditures in 1992 to enroll about 50,000 acres, and was budgeted for $160.9 million in fiscal year 1993 to enroll 381,000 acres (U.S. Department of Agriculture, 1992). Congress, however, failed to appropriate any funds in fiscal year 1993 for the Wetland Reserve Program. No funds have been budgeted for the Environmental Easement Program.

The historical emphasis on controlling erosion and runoff remains the focus of programs to control soil degradation and water pollution from agricultural production. In 1991, for example, 62 percent ($111.5 million) of the expenditures for cost-sharing the implementation of best-management practices in the Agricultural Conservation Program were for erosion control (U.S. Department of Agriculture, Agricultural Stabilization and Conservation Service, 1992). Seventeen percent ($30.5 million) of the cost-share expenditure was for water quality improvement, and about $15.9 million of cost-share expenditure for water quality was for one practice—agricultural waste control facilities (U.S. Department of Agriculture, Agricultural Conservation and Stabilization Service, 1992). Technical assistance provided by the Soil Conservation Service to producers is the single largest expenditure of federal funds for agricultural programs and totaled about $427 million in 1991 (U.S. Department of Agriculture, 1992). About 10 percent or $44 million of technical assistance was allocated to the implementation of the water quality initiative (U.S. Department of Agriculture, 1992). Most of the technical assistance provided by the Soil Conservation Service to producers since passage of the 1985 Food Security Act has been dedicated to helping producers determine whether their croplands are subject to Conservation Compliance, Sodbuster, or Swampbuster and to helping producers plan and implement conservation practices required under these programs (U.S. Department of Agriculture, 1992).

More money has been invested in the Conservation Reserve Program and the Wetland Reserve Program than all other conservation programs combined (U.S. Department of Agriculture, 1992). The Conservation Reserve Program was budgeted $1,642.1 million in 1992 to enroll another 1.1 million acres (U.S. Department of Agriculture, 1992).

## New Sources of Funds

*Taxes on nutrient or pesticides inputs or reallocation of commodity program expenditures should be explored as ways to increase the funding available to support and sustain soil and water quality improvement programs over the long term.*

Substantial reallocation of existing funds or a significant new source of funds will be needed if current initiatives and programs are to comprehensively address soil quality, input efficiencies, resistance to erosion and runoff, and field and landscape buffer zones. Reallocation of existing funds to priority areas will help supply the funds required to undertake the more intensive and refined efforts needed to improve farming systems. New sources of funds, however, will be needed to sustain these efforts over the long term. Relatively low taxes on nutrient and pesticide inputs or transfers of funds from commodity programs have the potential to generate large new sources of revenue.

Table 3-3 lists the 1992 expenditures for agricultural soil and water quality programs administered by USDA and EPA. Annual expenditures on technical assistance are less than 10 percent of annual expenditures for either pesticides or fertilizers on major commodity crops and 3 percent of Commodity Credit Corporation expenditures. All Agricultural Conservation Program expenditures for cost-sharing agreements with producers are about 3 percent of expenditures on either fertilizers or pesticides and about 1 percent of Commodity Credit Corporation expenditures. Expenditures for the Water Quality Incentives Program in 1992 represented only 0.1 percent of expenditures for either pesticides or fertilizers and 0.03 percent of Commodity Credit Corporation expenditures.

Special consideration should be given to revenue sources such as taxes on agricultural chemicals, fuel, heavy tractors, moldboard plows, irrigation water, and other inputs that can be related to soil and water quality degradation from agricultural production practices or to transfers from Commodity Credit Corporation programs to soil and water quality programs.

These sources should be explored as ways to generate the new funds needed to sustain soil and water quality programs. One percent ($128 million) of the annual 1990 expenditures of $12.8 billion on pesticides and fertilizers, for example, is more than 65 percent of the total 1992 expenditures on cost-sharing under the Agricultural Conservation Program, and more than 18 times the total 1992 expenditures on the Water Quality Incentive Program.

New sources of funds will be needed to implement and sustain efforts

TABLE 3-3   Expenditures for Soil and Water Quality Programs as a
Percentage of Expenditures on Pesticides, Synthetic Fertilizers, and
Commodity Programs

| | Expenditures as Percentage of Spending on: | | |
|---|---|---|---|
| Program | Pesticides[a] | Fertilizers[a] | Commodity Programs[b] |
| USDA programs[b] | | | |
| Soil Conservation Service technical assistance | 8.0 | 7.0 | 3.0 |
| Agricultural Conservation Program | | | |
| All Cost-Share Spending | 3.0 | 3.0 | 1.0 |
| Water Quality Incentive Program | 0.1 | 0.1 | 0.03 |
| Conservation Reserve Program | 30.0 | 24.0 | 10.0 |
| Wetland Reserve Program | 0.8 | 0.6 | 0.3 |
| EPA programs[c] | | | |
| Nonpoint Program Grants | 0.9 | 0.7 | 0.3 |

[a]The 1990 farm production expenditures for pesticides and for fertilizers and lime (pesticide, $5,727 million; fertilizers and lime, $7,137 million) were taken from U.S. Department of Agriculture, Economic Research Service and National Agricultural Statistics Service. 1992. Statistical indicators—farm income. Pp. 58–61 in Agricultural Outlook. Rockville, Md: U.S. Department of Agriculture.
[b]Estimated fiscal year 1992 expenditures in U.S. Department of Agriculture. 1992. 1993 Budget Summary. Washington, D.C.: U.S. Department of Agriculture. (Soil Conservation Service Technical Assistance, $478.0 million; Agricultural Conservation Program Cost-Share, $194.4 million; Water Quality Incentives Program, $6.8 million; Environmental Conservation Acreage Reserve Program, $1,786 million; Conservation Reserve Program, $1,740 million; Wetland Reserve Program, $46.4 million; commodity programs, Commodity Credit Corporation, $18,300 million).
[c]Fiscal year 1991 expenditures ($51 million) in U.S. Environmental Protection Agency. 1992. Managing Nonpoint Source Pollution: Final Report to Congress on Section 319 of the Clean Water Act (1989). Washington, D.C.: U.S. Environmental Protection Agency.

to protect soil and water quality. Soil and water quality programs may achieve greater continuity if they are funded through a mixture of both general revenues and new revenues generated by taxes on agricultural inputs.

# 4

# Policies to Protect Soil and Water Quality

Chapter 2 described four major opportunities to prevent soil degradation and water pollution caused by farming practices and outlined the technologies and scientific knowledge available to take advantage of those opportunities. These four opportunities are

1. to conserve and enhance soil quality as the first step to environmental improvement;
2. to increase the efficiency with which nutrients, pesticides, and irrigation water are used in agricultural production;
3. to increase the resistance of farming systems to erosion and runoff; and
4. to make greater use of field and landscape buffer zones.

These opportunities should be the goals that policies to protect soil and water resources seek to achieve. Chapter 3 recommended that soil and water quality programs target resources at problem areas and problem farms and use a farming system, rather than a best-management practice approach, to take advantage of the technical opportunities to prevent soil degradation and water pollution. Chapter 3 also outlined the improved tools and information that producers and program managers will need to implement a farming system approach to managing soil and water resources. Federal, state, and local programs could be made much more effective if these steps were taken.

There is considerable scientific and technical information on how to prevent soil degradation and water pollution, as the chapters in Part Two of this report demonstrate. Although gaps remain to be filled in technology and information, the more important obstacle to improving soil and water quality is the lack of incentives for producers to use the knowledge and technology that already exists.

Ultimately, it is the millions of management decisions producers make each year that determine the effect of farming systems on soil and water quality. The purpose of national policy should be to create the proper incentives that induce producers to change the way they manage their farming systems. There is, however, much less known about the factors that influence producers' choices of cropping, livestock, and enterprise management practices than there is about the technologies and management methods that will protect soil and water quality. Empirical information on the costs of changing farming systems is often lacking or is anecdotal.

Many factors influence the decisions that producers make—including market prices for inputs and products, the cost of new technologies, the labor and capital available to the producer, agricultural policy, environmental regulations, and the goals of the individual producer or enterprise (see Chapter 1, Figure 1-1). The agricultural sector is not made up of a homogeneous collection of uniform farms managed by producers with similar skills, resources, and goals. Instead, farming enterprises differ widely in the commodities they produce, the quality of their soils, and their topography. Ownership patterns and the labor or financial resources the producer can tap vary just as widely. Also, producers are a diverse set of people who have a variety of goals: profit maximization, minimization of management time, maintenance of a certain life-style, protection of personal independence, desire to obtain a certain social status, and observation of a particular environmental or religious ethic. This variability means there are many different reasons why producers choose to adopt or reject new farming systems (Table 4-1)—no single policy or program will influence all the producers whose behavior those policies seek to change.

The inadequacy of empirical data and predictive models of producer behavior and the diversity of enterprises that make up the agricultural sector make it difficult to pinpoint the precise effect of alternative policies on the behavior of producers. General understanding of the factors that influence producers' decisions, however, can guide the development of national policies to change the way producers manage their farming systems.

## ENVIRONMENTAL AND AGRICULTURAL POLICY

Environmental objectives have historically been closely linked with the larger goals of agricultural policy to support and stabilize the prices

TABLE 4-1   Constraints to Adopting New Technologies and Program
Responses to Nonadoption

| Constraint | Program Response |
|---|---|
| *Inability* | |
| Basic information needed for a sound economic and agronomic analysis is lacking or scarce. | Generate and distribute information to those who need it. |
| Time, expense, or difficulty of obtaining site-specific information is excessive. | Reduce the costs of obtaining the necessary information by increasing accessibility to the information. |
| Complexity of a technology is inversely related to the rate and degree of adoption. | Redesign or simplify the technology. |
| Investment costs, costs of operation, or profit loss are too high. | Subsidize the adoption, or design a less expensive system. |
| Labor requirements are excessive. | Redesign the practice to reduce labor requirements or subsidize the hiring of adequate labor. |
| *Unwillingness* | |
| Inconsistencies or conflicts exist in the recommendations of public sources (e.g., land-grant universities, USDA agencies), private sources (e.g., agribusinesses, financial institutions), or other sources (e.g., producer-to-producer referral networks, family members). | Work to develop a consistent set of recommendations. When legitimate differences between alternative recommendations exist, offer producers explanations of these differences. |
| Available information is not applicable or relevant to the producer's farm firm. | Generate and distribute relevant information on a local basis. |
| New technologies are not compatible with existing production systems or policies. | Develop flexible management methods and production practices capable of being altered to meet unique farm conditions. |
| Producer does not understand basic agronomic or economic aspects of a new technology, or agents who promote a new technology do not understand the basic needs of a potential adopter. | Determine the actual, not assumed, assistance needs and knowledge levels of potential adopters relative to those factors critical to adoption. Then, design education and assistance programs on the basis of producers' needs, not agency or business expertise. |
| Current planning horizon—relative to the time associated with recouping initial investments, learning costs, or depreciation of the present equipment line—is too short. | Redesign the system or subsidize a short-term unprofitable decision. |

*(continued)*

TABLE 4-1 (*Continued*)

| Constraint | Program Response |
|---|---|
| Support from local equipment or agrichemical dealers, other producers, or USDA is lacking and information and assistance networks capable of answering producers questions are inadequate. | Build the capacity of local assistance networks to meet local demands. |
| Managerial skills are inadequate. | Develop skill-building opportunities for producers. |
| A decision cannot be made without the approval of a partner, the source of financial credit, the landlord, or some other third party. | Determine who can make the adoption decision and focus efforts on those persons or organizations. Also, recognize that an adoption decision is often a family decision, and therefore, persuasion or assistance efforts need to address relevant family members. |
| Technology is inappropriate for the physical setting. | Specify the physical applicability of the technology or design the technology to be more adaptable to different physical settings. |
| Complexity of a practice, importance of the timeliness of operations, and interdependence of inputs increase the perceived or real uncertainty and risk. | Risk can be addressed in two basic ways: either increase information so that probabilistic outcomes can be calculated or subsidize the producer so that he or she can take a risk. |

SOURCE: Adapted from P. Nowak. 1992. Why farmers adopt production technology. Journal of Soil and Water Conservation 47:14–16.

of agricultural commodities and, indirectly, support and stabilize the income of producers. Effective policies and programs to achieve long-term protection of soil and water quality, however, cannot simply be adjuncts of income and commodity policies. The problem areas and problem farms that should be the focus of soil and water quality policies may be very different than the areas and farms of producers requiring income support; and the commodities produced in problem areas or problem farms may not be the same commodities that are the target of programs to support and stabilize prices.

The objectives of commodity and environmental policy are different, and the mechanisms used to achieve those objectives also will be different. It is essential that these policies not create conflicting incentives, and reform of agricultural commodity policy to reduce incentives that lead to soil degradation and water pollution is important. In the long term, however, policies to protect soil and water quality cannot

"Some folks don't know how to appreciate good news" (September 16, 1927).
Credit: Courtesy of the J.N. "Ding" Darling Foundation.

depend on incentives tied to price and income support programs. The importance of creating soil and water quality policies that are independent of commodity and income policies is best understood by briefly reviewing the historical linkage of soil conservation and income support in agricultural policy.

## A Brief History

Soil erosion problems in the United States were recognized by a few people early in the nation's history. Generally, however, new lands made available by westward expansion meant that producers and policymakers gave little attention to erosion. By the 1890s, exploitation of land and abandonment of farms when the land became "exhausted" was so commonplace that one of the first bulletins issued by the newly formed U.S. Department of Agriculture (USDA) urged producers to conserve the land they owned (Rasmussen, 1983).

The general public was alerted to soil erosion problems during the next few decades, mainly as the result of the efforts of one man, Hugh Hammond Bennett. Bennett was a soil scientist who was appointed to head the USDA's Soil Conservation Service at its creation in 1935. Bennett led a messianic campaign to convince farmers and legislators of the dangers of soil erosion. The initial federal response, in 1930, was to allocate funds to be used for soil erosion research (Kramer and Batie, 1985). Propelled by the Great Depression, stronger legislation soon followed.

When President Franklin Roosevelt took office in 1933, U.S. agricultural producers and their urban counterparts were in serious financial stress. The prices of farm products had fallen more than 50 percent since 1929. The Agricultural Adjustment Act of 1933 (PL 73-10) represented a major shift in agricultural policy toward direct government involvement in markets and farm-level decision making. The act authorized USDA to enter into voluntary agreements with producers to take land out of production for compensation (Kramer and Batie, 1985).

Simultaneously, but independently, the U.S. Congress had created the Soil Erosion Service and authorized money to be spent to combat erosion. Conceivably, the two programs of farm income support and soil conservation would have developed separately. On January 6, 1936, however, the U.S. Supreme Court ruled the Agricultural Adjustment Act unconstitutional, ruling that the production control provisions of the act were coercive. Policymakers, anxious to continue the supply adjustment program despite the ruling, found a way to use soil conservation as a vehicle for income support.

"Don't blame factories for all the unemployment" (December 22, 1939). Credit: Courtesy of the J.N. "Ding" Darling Foundation.

On February 29, 1936, Congress enacted amendments to the Soil Conservation and Domestic Allotment Act (PL 74-461), which required producers to submit conservation-oriented adjustment plans and to enroll in the Agricultural Conservation Program to participate in acreage adjustment contracts (Kramer and Batie, 1985). Producers thus were paid to set aside the acreage from "soil-depleting crops" and replace them with "soil-conserving crops" such as grasses. Since soil-depleting crops were also those crops that existed in surplus, the supply adjustment goals were accomplished by this reorientation. Soil conservation legislation was a legal vehicle for pursuit of farm relief and recovery.

From their inception, soil conservation programs were designed to support farm income and production as well as to reduce soil erosion. They remained popular because they lowered producers' operating costs, improved yields, and provided for compensation for idling lands from production.

During the 1950s and 1960s, the emphasis of soil conservation programs was on cost-sharing and technical assistance to encourage the adoption of soil-conserving practices. By the 1970s, rising commodity prices and reduced production controls encouraged producers to bring more land into production and to intensify production on their existing croplands. Some of this new cropland was highly erodible, and the push for full production led some producers to abandon conservation practices. A 1977 report by the Comptroller General to the U.S. Congress warned that soil erosion was still a serious problem, despite 40 years of soil conservation efforts (U.S. General Accounting Office, 1977).

Not until the 1985 Food Security Act (PL 99-198) was there an emergence of erosion control and water quality as independent objectives of agricultural policies. The receipt of income support was again linked to conservation practices, as in the Agricultural Adjustment Act. Income support, however, became conditional on the adoption of conservation practices on certain highly erodible lands. Soil conservation objectives took precedence over income support objectives, at least on the most highly erodible lands.

For the first time, to be eligible for farm program benefits, agricultural producers were required to implement a soil conservation plan for their highly erodible croplands. A conservation plan was required for highly erodible land converted to cropland from other uses, and Congress also established the Conservation Reserve Program (CRP) to pay producers to take highly erodible land out of production. However, many critics claim that the CRP, at least as initially implemented, was intended more to control supply and stabilize land values than to take the most highly erodible lands out of production (U.S. General Accounting Office, 1993).

## Incremental Redesigning of Agricultural Policy

Because of the history of multiple, competing objectives of agricultural policies and because of a recognition that agricultural policies are a major influence on commercial producers of commodity crops, there has been increasing attention focused on redesigning agricultural policies to remove any barriers to achieving environmental goals.

Recent research has pointed out that the current structures of price support and supply control programs erect barriers to the adoption of farming systems that improve soil and water quality. Where such barriers exist, they can seriously impede efforts to induce producers to change the way they manage their farming systems to protect soil and water quality.

### Incentives Are Perverse

*Price support, deficiency payment, and supply control policies should be reformed to remove the barriers to voluntary adoption of improved farming systems.*

The structure of U.S. farm programs induces a bias toward intensive farming practices to boost yields and to expand the base acreage of the cropland that can be enrolled in the price support programs. Deficiency payments are directly proportional to a farmer's historical yield, which is used to establish the *program yield*, and the historical cropland, which is used to establish the *base acreage* for the crop. These features create incentives for producers to increase plantings and boost yields to capture higher government payments in the future. The 1985 Food Security Act (PL 99-198) and the 1990 Food, Agriculture, Conservation and Trade Act (PL 101-624) have moderated this bias by freezing program yields at 1986 levels and by applying constraints on the expansion of base acres.

In a study of potential farm bill influences, however, Dobbs and colleagues (1992) found that current agricultural policies still pose barriers to the adoption of more sustainable farming systems for some farms, despite the modifications made in the 1990 Food, Agriculture, Conservation and Trade Act. For example, simulated reductions in target prices appeared to make farms that practice sustainable agriculture more profitable than farms that use conventional agricultural practices in northeastern and southeastern South Dakota. Dobbs and colleagues (1992) reported that giving producers who participate in agricultural programs more flexibility with respect to the choice of crop rotations did not consistently favor producers who practice sustainable agriculture.

Runge and colleagues (1990) reviewed recent farm-level studies that explored the difficulties that farmers confront when they attempt to participate in government programs and pursue environmentally sound practices simultaneously. They summarized their findings as follows:

> These case studies suggest that farmers are currently confronted by a confusing set of signals that make it difficult to remain both profitable and environmentally responsible. In one case study in southwest Minnesota, farmers describe the current government programs as putting them in a "vise grip," resulting in cropping practices that distort the allocation of fertilizer and chemical inputs, and discourage crop rotations. While such practices, if changed, would not in themselves solve all of the environmental problems affecting agriculture, they would at least not aggravate them, as current policy appears to do. A second case study, conducted in Iowa, documents a similar set of problems, showing that under current federal farm legislation, crop rotations are discouraged in favor of continuous corn, using the highest levels of nitrogen fertilizer. A third case study, in southwestern Minnesota, an area similar to many other parts of the upper Midwest with vulnerable soils and groundwater, shows both commercial fertilizer and live-stock waste must be closely accounted for if total nitrogen use is to reflect best-management practices. It also suggests that some land areas are simply more vulnerable to environmental damages than others, implying the need for more targeted environmental policies. Together, these case studies suggest that government policies and on-farm decisions are closely linked, and that better management practices will require both a different set of signals from Washington, and a renewed commitment to careful and precise farming methods that account for off-farm effects (Runge et al., 1990:v).

These studies reinforce some of the findings of a National Research Council study on alternative agriculture:

> Federal policies, including commodity programs, trade policy, research and extension programs, food grading and cosmetic standards, pesticide regulation, water quality and supply policies, and tax policy, significantly influence farmers' choices of agricultural practices. As a whole, federal policies work against environmentally benign practices and the adoption of alternative agricultural systems, particularly those involving crop rotations, certain soil conservation practices, reductions in pesticide use, and increased use of biological and cultural means of pest control (National Research Council, 1989a:6).

The lack of crop diversity either within a field or over time (in rotation) appears to be a major constraint in achieving high soil microbial activity necessary for high soil quality and attendant benefits for water quality (Harwood, 1993). The barriers created by current commodity policy may also constrain the development of innovative cropping systems to improve input use efficiency or resist erosion and runoff that were outlined in Chapter 2. Thus, the disincentives in agricultural programs for rotations and crop diversity are an important barrier for improving soil and water quality.

The incentives embedded in agricultural programs to increase production of certain crops are also a problem. The lion's share of government payments goes to producers of feed grains—especially corn, food grains (wheat and rice), and cotton—and indirectly to dairy products. Growers of most livestock products, fruit, vegetables, hay, and nearly all specialty crops are excluded from the direct influence of government programs. Although many vegetable and fruit crops that are not part of government programs receive high agrichemical applications, these crops occupy relatively small areas on a national level. In contrast, corn, cotton, soybeans, and wheat received an estimated 65 percent of total agrichemical applications (Fleming, 1987). Reichelderfer (1985) also concluded that program crops were more soil eroding on average than nonprogram crops.

### Incremental Reform

The current incremental process of policy reform, such as the increased base flexibility provided by the 1990 Food, Agriculture, Conservation and Trade Act (PL 101-624) and the freeze on established yields and the gradual decline in real target prices initiated in the 1985 Food Security Act (PL 99-198) have probably helped to encourage crop rotations and discourage excessive use of nutrients, pesticides, and irrigation water. These incremental reforms have reduced the barriers to adoption of improved farming systems erected by the selectivity and structure of U.S. farm programs and have probably achieved modest improvements in soil and water quality. At the same time, the reformed policies have remained reasonably effective in meeting price support and stabilization objectives. Gradual reform along these lines will help to remove barriers to the development and implementation of improved farming systems and permit time for farmers to adjust to the new incentives.

### Increasing Planting Flexibility

*Current price support and supply control programs should be redesigned to increase the flexibility participants have to plant different crops in order to permit greater use of crop rotations, cover crops, and other changes in cropping systems.*

Calculation of deficiency payments on the basis of program yields and base acres has created inadvertent disincentives to diversify crop mixes and make other improvements in current farming systems. To producers, these features of the price support system send signals that conflict with policies to accelerate voluntary adoption of improved farming systems. Short-term reform of price support programs should reinforce current efforts to reduce the effects of price support programs on production decisions and to increase planting flexibility.

Of all the features of traditional commodity programs that have imposed barriers on environmentally sound farming practices, the rigid base acreage structure has probably been the most influential. (Base acres are the acres of a producer's cropland that can be planted to a crop for which deficiency payments are received; base acres cannot be planted to a different crop without incurring a penalty.) Consequently, this goal can be pursued by increasing the percentage of acres that can be planted to any base crop the producer chooses, along the lines of the 1990 Food, Agriculture, Conservation and Trade Act (PL 101-624), or by giving producers complete 100 percent flexibility to plant crops on their base acres, as in the 1990 Normal Crop Acreage proposal of USDA.

Important economic factors, however, will constrain the degree to which producers diversify their farming systems even if the barriers erected by current commodity policy are relaxed. Increased flexibility will not result in wider use of crop rotations, intercropping, or multiple cropping unless markets exist for the crops added to the farming system. Specialty crops, such as canola or buckwheat that might be used in more diverse farming systems may have limited markets, thereby restricting the number of producers who can profitably incorporate those crops into their farming systems. In addition, widespread adoption of forages into farming systems requires diversification of the enterprise to include livestock. The effects of large-scale changes in the crops or forages used to feed livestock, however, could be significant. A model used in the Second RCA Appraisal, for example, predicted that given the model's flexibility to use least-cost combinations of grains and forages to meet the demand for livestock production, planted cropland would be reduced by about 26 million hectares (65 million acres) (U.S. Department of Agriculture, Soil Conservation Service, 1989a).

## Nonincremental Reform of Agricultural Policy

Dissatisfaction with the efficiency and the fiscal and environmental effects of traditional farm programs has led some policymakers and scholars to propose redesigning or eliminating the current structure

(Boschwitz, 1987; Cochran, 1986; Cochrane and Runge, 1993; Harrington and Doering, 1993; Kramer and McDowell, in press; and Tweeten, 1993). In addition, the United States has sought, since 1987, to reduce subsidies to agricultural producers that distort production and trade as part of the negotiations under the General Agreement on Tariffs and Trade.

The nature of nonincremental reform includes proposals such as the substitution of producer-paid crop insurance, the decoupling of farm income support from crop yields, producer-financed price stabilization funds, and similar proposals. The arguments for elimination of agricultural programs are captured in the following quote of Tweeten:

> Agriculture is no longer an industry of low income or low returns on resources nor would it be without commodity programs. Commodity programs transfer income from lower income/wealth taxpayers to higher income/wealth producers. Given the pressing need to promote economic efficiency and to reduce federal outlays and thereby the national debt, a strong case can be made for a transition program to end government intervention in agricultural markets.
>
> This paper reviewed the numerous justifications for continued government intervention in farm markets. Problems of instability, the environment, poverty, cash flow, loss of family farms, and competitive challenges from abroad are real. However, commodity programs, as currently structured, do not respond in a cost-effective manner to any of these problems. By simultaneously trying to address all of these problems plus a nonexistent commercial farm welfare problem, none of the problems are properly addressed. It is time to *disassemble* commodity programs. Environmental problems need to be addressed by an environment program and the poverty problem needs to be addressed by a welfare program—though not unique to agriculture (Tweeten, 1993:28-29).

### Limitations of Commodity Program Reform

Current research suggests that although price and supply programs exacerbate soil and water problems, they are not the cause of those problems. Even if these programs are eliminated, the need for programs that specifically address soil and water quality problems will remain. Incremental changes in agricultural commodity policies will most likely not result in major changes in farming practices and will likely result in only modest gains in environmental quality. The research evidence, as discussed below, also suggests that a decoupled "free market" agriculture that does not include mechanisms to address agricultural pollutants in the environment will not result in improved protection of soil and water quality and may increase soil degradation or water pollution in some regions.

### Effects of Program Elimination

Recently, attempts have been made to estimate the effects of eliminating price and supply management programs on soil and water quality. Doering (1991) suggests that there is little reason to believe that environmental quality would have been better had there been no government agricultural policies. Doering (1991) concluded that "changes in existing farm programs or even the elimination of these programs will not result in basic changes in the way farmers farm" (Doering, 1991:i). Hrubovcak and colleagues (1990) and Carlson and Shui (1991) reached essentially the same conclusion. They argued that the improvement in soil and water quality caused by the acreage reduction and supply control components of the farm programs offset the soil and water quality damage induced by the incentives to boost program yields and restrictions on planting flexibility.

In a model of the U.S. wheat sector, for example, Hertel and colleagues (1990) estimated that keeping federally established farm program yields at their 1985 levels would have reduced the use of nonland inputs (including fertilizers and pesticides) by 22 percent in 1986, a year when the target price of wheat greatly exceeded the market price. (The target price of an agricultural commodity is set by the federal government; the difference between the target price and the market price is the deficiency payment that producers receive from the federal government.) In 1982, a year when the market price of wheat was nearly the same as the target price, freezing of yields would have resulted in a 1 percent decline in the use of nonland inputs.

Faeth and colleagues (1991), however, recently modeled multilateral program decoupling and four other policy scenarios with respect to their impacts on producer incomes and off-site and on-site soil erosion costs. Multilateral program decoupling was projected to increase substantially the incomes of U.S. producers in response to significant world price increases as inefficient producers in other countries reduced their levels of production. The cost to U.S. taxpayers was also projected to plummet with multilateral program decoupling. On the other hand, off-site and on-site damages from soil erosion were projected to increase under the fencerow-to-fencerow farming practiced under the decoupling scenario. In order to make environmentally sound farming practices economically attractive to producers, it was necessary to add the social costs of erosive practices into the production costs paid by farmers (Faeth et al., 1991).

Shoemaker and colleagues (1989) estimate that today's agricultural policies increase total chemical use by only about 12 percent on program crops but that the increase is less for nonprogram crops. Such results are similar to those of Doering and Ervin (1990), who showed that even with 100 percent flexibility, nitrogen use would decline only by 4 percent and pesticide use by 2 to 3 percent. Doering (1991) concluded that even a drastic change in the nature of traditional programs was likely to produce only modest changes in cropping patterns and input use.

Some studies suggest that there are stronger linkages between eliminating price and supply management programs and improved environmental quality. For example, Tobey and Reinert (1991) reported that national general equilibrium modeling predicted a decline in the use of fertilizer as well as off-site damages if the deficiency payments made to producers were reduced. Similarly, early studies by Dixon and colleagues (1973) and Richardson (1975) showed that the farm programs of the 1960s substantially boosted agrichemical demand. Dixon and colleagues (1973) estimated that in 1965, free market agricultural policies could have satisfied food and fiber demands with one-half the pesticides and fertilizers used under the prevailing farm program structure.

### Environmental Policies for Environmental Goals

*Long-term protection of soil and water quality should be based on policies and programs that are independent of price support, supply control, or income support mechanisms; policies that target problem areas and problem farms, regardless of participation in federal commodity support programs, are needed.*

The studies cited above differ in their estimates of the effects of federal agricultural commodity programs on soil and water quality. All of the studies, however, suggest that simply eliminating these programs will not solve soil and water quality problems. Programs that have as their primary objectives soil and water quality protection are needed now, and they will be needed regardless of how price and supply management policies are reformed. Society clearly has a stake in both the production of agricultural commodities and the protection and enhancement of soil and water quality. Soil and water quality programs need to become more independent of efforts to control supply or to support commodity prices and farm income.

Agriculture now faces an environmental agenda that has expanded beyond the historic concerns over erosion control to conserve soil productivity to include concerns over the loadings of nutrients, sediments, pesticides, salts, and trace elements to both surface water and groundwater (Hamilton, 1993). Soil and water quality improvements have become important objectives of agricultural policies. Programs to improve soil and water quality can no longer be seen as adjuncts to programs that support prices or income.

## FACTORS AFFECTING PRODUCERS' DECISIONS

The overriding objective of any soil or water quality program is to induce change among producers. The design of policies, then, should be based on an understanding of the factors that affect the decisions made by producers.

Many factors affect the decisions that producers make (see Chapter 1, Figure 1-4). Most important are those factors that determine the likelihood that producers will use new technologies and information to prevent soil degradation and water quality. Producers must first be aware that new technologies and information relevant to their farming system are available. Although obvious, this first factor is easily overlooked. The availability of information, however, is one of the most important factors cited in studies of the adoption of new technologies (Esseks et al., 1990; Nowak, 1992; Nowak and Korsching, 1983; Padgitt, 1989).

Even if producers are aware of new technologies, they may fail to adopt them because they are either unable or unwilling to do so (Nowak, 1992). These reasons are not mutually exclusive. Producers may be able but unwilling, willing but unable, or both unwilling and unable. The kind of technical assistance, education, or regulation required to influence a producer who is unwilling is very different from that required to influence a producer who is unable. Recognition of this difference is crucial when designing the appropriate way to increase the use of new knowledge or technology. Table 4-1 lists the reasons why producers may be unable or unwilling to adopt new technologies or farming systems and also suggests changes in programs that might help to address those reasons. Three general observations from the lists in Table 4-1 are important.

First, programs should address obstacles that make producers unable to adopt improved farming systems. Once these obstacles are removed, it may be possible to induce an unwilling producer to adopt an improved farming system. The removal of obstacles to adoption must precede persuasion for adoption.

Second, many factors that make producers unable or unwilling to adopt new technologies or systems are beyond their control. For

A chemist and a soil scientist with the Agricultural Research Service are studying the effect of different tillage systems on the movement of pesticides into groundwater. Here they take groundwater samples from a test site next to cropland at Beltsville, Maryland. Credit: Agricultural Research Service, USDA.

example, some changes may have exceptionally negative impacts on profits. In such cases, changes in policies, prices, or technology may be needed before a producer will adopt new farming systems. In many cases it is not so much a producer failure as it is a system failure.

Third, the remedial strategies outlined in Table 4-1 are closely linked to the social and economic contexts of the farm enterprise. No single strategy to encourage or mandate the adoption of improved farming systems can be universally applied, and programs restricted to one or a few strategies will fail. Integration of a range of programs from technical assistance to nonvoluntary or regulatory programs will be needed to influence producers. Technical assistance, for example, may be sufficient to induce adoption by producers who are willing but unable to change their farming systems. The threat of regulation or penalities or the use of market-based incentives may be needed for producers who are able but unwilling.

## CONTINUUM OF POLICIES

A continuum of policies ranging from research to regulation will be required to "get the incentives right" so that producers are both willing and able to adopt new technologies and improve the management of their farming systems. Policies will need to specify exactly who will fund and direct research, provide technical assistance, provide market-based mechanisms, provide mechanisms to change land use, and use regulatory approaches. Policies must be clear about legal responsibilities of landowners and land users as regards soil and water quality. This range of policies describes a continuum from purely voluntary to purely compulsory approaches to change the behavior of producers.

In the real world, there are no absolute demarcations between policies along the voluntary-to-compulsory continuum. Research and development of new technologies supports policies on either end of the continuum, and both voluntary and regulatory programs create a demand for research and development. Similarly, providing technical assistance or facilitating a change in land use may be essential components of either a regulatory or a voluntary approach. Integration along the continuum is essential to ensure that producers receive more lasting and consistent signals to adopt and sustain farming systems that improve soil and water quality.

### Research and Development

Many recommendations for research and development of new information, technologies, and methods of managing farming systems have been made in Chapters 2 and 3 of this report. The application of science and technology to agricultural production has had a revolutionary effect on agricultural productivity. The gains that public and private investments in research and technology transfer have made possible in agricultural productivity are now cited as models for other sectors of the economy. Scientific understanding, coupled with improvements in production and information technology present agriculture a second opportunity to revolutionize production to meet the twin goals of productivity and environmental compatibility. This second revolution is already under way—producers, researchers, and educators in many localities have developed farming systems that solve local soil and water quality problems.

The importance of research and the development of new production technologies and management methods is difficult to overstate. Often overlooked, however, is the importance of research to develop a better and more empirical understanding of the factors that influence the decisions that producers make. New technologies are successful only if they meet the needs of producers. New programs and policies are effective only if they are based on an understanding of the critical factors producers consider when they decide to purchase new technologies or adjust the way they manage their farming systems. Economic and social research should be a fundamental component of both the development of new technology and new policies.

### Understanding Producers

*Policy-relevant research directed at identifying the nature and magnitude of factors influencing producers' choices of cropping and livestock practices should be a high priority for USDA and EPA research programs.*

The design of programs to protect soil and water quality is hampered by an inadequate answer to a fundamental and broad question: Why do producers make the choices they do? As discussed in Chapter 1, the agriculture sector is exceptionally diverse with diverse farms and diverse people. The major obstacle to the design of successful environmental policies will not be inadequate technical and scientific information; rather, it will be the lack of information as to the effective incentives for achieving change.

There is good general understanding of the factors that affect a producer's ability or willingness to adopt new technologies. Much more work, however, is needed to provide the empirical data needed to predict the effect of policies and programs on producers. Research is needed to measure and analyze the diversity of reasons why producers may be unable or unwilling to adopt a new system or technology. Such understanding will allow the design and implementation of programs that influence the most important factors affecting producers' decisions. Such an approach could significantly increase the rates of adoption of improved farming systems.

*Technical Innovation*

    *Research leading to the development and implementation of new technologies, cropping systems, and methods to manage farming systems that are profitable*

---

## THE NARROWS CREEK-MIDDLE BARABOO PRIORITY WATERSHED PROJECT

The Narrows Creek-Middle Baraboo watershed, located within Salk County, Wisconsin, is 453 km² (175 miles²) in size. It includes all the lands draining to the Baraboo River between Reedsberg in the northwest, Lime Ridge in the west, and West Baraboo in the east.

The Wisconsin Department of Natural Resources and the Department of Agriculture, Trade and Consumer Protection named the Narrows Creek-Middle Baraboo a Wisconsin Priority Watershed during July 1990 because of degraded water quality and the impact of sedimentation on the aquatic habitat in the Baraboo River. The watershed is almost entirely rural, with croplands, pasturelands, and woodlands dominating the land use patterns. Dairying is the major agricultural activity, making manure runoff from barnyards and fields a major concern. Finally, much of the cropping occurs on steep slopes, which facilitates the transport of nutrients and pesticides to adjacent water bodies.

Local watershed staff identified all operators in the watershed who operated at least 16 ha (40 acres) of land and who had at least 15 beef or dairy cattle. Some 53 of the 261 operators met these two criteria but refused to participate in the initial interview (79.6 percent response rate). The Farm Practices Inventory was used to collect information from the 208 operators who agreed to participate prior to implementation of the watershed project.

The results of the Farm Practices Inventory were used to establish four priorities for the entire watershed project: (1) appropriate use and interpretation of soil tests, (2) nitrogen crediting from legumes, (3) nitrogen crediting from manures, and (4) construction of manure storage structures. The inventory, however, revealed that differences in the production practices that producers used within the watershed were great. Three regions within the watershed were defined on the basis of those differences. Each region required different emphasis among the four general priorities.

### Region 1

**Problems** High levels of nitrogen (355 kg/ha [317 lb/acre] from all sources) are being applied; yet, corn yields average 7.96 metric tons/ha *and protect soil and water quality should be a high priority for USDA and EPA research programs.*

Technical innovation is already expanding the alternatives available to producers to protect soil and water quality while sustaining productive and profitable farming enterprises. Indeed, the development of new methods to manage crop residues, nutrients, pesticides, and irrigation water has helped some producers make dramatic progress in protecting soil and water quality. The potential for technical breakthroughs leading to farming systems very different from those in use today is great. New

---

(127 bushels/acre), the lowest yield of the three regions. This region has the lowest number of operators who actually credit manure nitrogen (29 percent) and legume nitrogen (47 percent) and the highest manure application rates (230 kg/ha [205 lb/acre]); and those crediting manure are under crediting available, first-year nitrogen by 39.9 percent (for example, crediting 67.3 kg/ha [60.1 lb/acre] when 112 kg/ha [100 lb/acre] are available). The sandy soils in this region make the manure storage facilities difficult to design and expensive to construct. Moreover, the overall nitrogen and phosphorus application rates on corn are not significantly different between those few livestock or dairy producers with manure storage structures and those who use a daily manure haul system. Finally, the future plans indicate that 10 percent of the dairy producers will downsize or stop farming within the next 5 years.

**Solutions** Implementation strategies for region 1 need to emphasize proper crediting and other crop nutrient management issues largely through education and use of various information transfer mechanisms. A major theme could be the $26.60 operators are spending on commercial nutrients per hectare of corn ($10.77/acre) when on-farm nutrient sources are available. Educational efforts directed toward private agrichemical dealers are also important. The majority of operators in the region rely on the dealer for soil testing, interpretation, and recommendations based on soil test results. Synthetic fertilizer recommendations need to be adjusted more realistically for on-farm nutrient sources. The use of cost-sharing dollars for manure structures is not warranted in this area.

## Region 2

**Problems** About one-third of the producers (34 percent) apply nitrogen at more than 50 kg/ha (45 lb/acre) above the recommended level. The total nitrogen applied in this region averages 237 kg/ha (212 lb/acre) compared with the recommended level of 179 kg/ha (160 lb/acre). Yet, less than one-third (32 percent) of the producers with animals credit the manures from these animals. Of those who do, they under credit the available first-year nitrogen by an average of 52.2 percent. This region also has significantly more manure storage structures than the other two regions.
*(continued)*

tests to determine the need for additions of nutrients, pesticides, or irrigation water; methods to adjust applications of inputs and tillage operations to changes in soil quality; methods that link computer-based decision systems with simple data collection methods or remotely sensed data that can be easily used by producers; and the development of imaginative cropping systems that alter the pattern or sequence of crops to protect soil and water quality are a few of the most promising developments. Many of these opportunities have been discussed in Chapters 2 and 3 and in Part Two of this report.

---

Region 2 is likely to see a small increase in livestock (12 percent) and cash crops (6 percent) over the next 5 years, whereas dairy and forage production will remain static.

**Solutions** Implementation in region 2 needs to focus on two issues. First, there needs to be proper application and crediting of manures from existing manure storage facilities. Second, most farmers do not differentiate between corn fields in relation to nitrogen and manure application rates, even though their fields differ greatly in slope and soil characteristics and drainage.

## Region 3

**Problems** Some 35 percent, or 29 of 84 ha (72 of 207 acres), of all tillable land receives manure from an average of 92 dairy cattle, 23 beef cattle, and 5 swine per farm. More producers in this region credit manures (34 percent) and legumes (67 percent), although the process is not wholly accurate. Only eight of the farms have manure storage structures. Some 12 percent of the producers in this region have wells that test for nitrate-nitrogen in excess of 10 mg/ml (10 ppm), and another 74 percent have wells that test for nitrate-nitrogen in the range of 2 to 10 mg/ml (2 to 10 ppm). Yet, the topography, soils, and location of the farm wells indicate that much of this pollution is derived from point sources (for example, farmstead design issues). Dairy, livestock, and the supporting forages will continue to grow in this region for the next 5 years. Cash crops, on the other hand, will decrease.

**Solutions** Dairy and livestock are the major focus of the farmers in region 3. Implementation strategies focusing on field nutrient management issues will have lower salience than those associated with herd management. Farmstead design, manure management, and runoff structures need to be oriented to the implications for herd management (for example, lowering of somatic cell counts or reducing conditions conducive to mastitis). The number of animals combined with the rolling topography of this region indicate that phosphorus management is a critical issue. Education on the role of manures in phosphorus management is also a critical issue. Education on the role of manures in phosphorus maintenance, possible cost-sharing of soil testing, and building of structures need to be emphasized in this region.

The development of innovative technologies is, in the long term, the most promising way to achieve lasting protection of soil and water quality while sustaining profitable production of food and fiber. A sustained program of research and development is an essential component of policies to prevent soil degradation and water pollution.

### Technical and Financial Assistance

Voluntary change has been the dominant approach used in the past to improve the farming practices used by producers. These programs have been characterized by the following:

- reliance on the development of conservation plans for individual farmers with free technical assistance provided through an extensive network including the Soil and Water Conservation Districts, Soil Conservation Service, USDA, and the Cooperative Extension Service of the USDA;
- reliance on voluntary adoption of conservation plans, with incentives provided through cost-sharing arrangements and education; and
- reliance on self-regulation through local Agricultural Stabilization and Conservation Service of the USDA and county Soil and Water Conservation District committees.

#### *New Approaches*

The voluntary approach to change through technical and financial assistance has achieved improvements in farming practices, particularly when there have been opportunities to improve environmental and financial performances simultaneously. The success of programs to encourage the adoption of conservation tillage, which reduces both soil quality damages and tillage costs, is a good example.

*There are clear opportunities to improve farming systems in ways that improve both environmental and financial performances; policies should, in the short term, first seek to take advantage of these opportunities.*

Chapter 2 outlined a diverse set of technical opportunities to implement farming systems that prevent soil degradation and water pollution. In many cases, these opportunities will have minimal or no negative effects on profitability. In some cases, the implementation of new farming practices and better farm management may increase profitability. The magnitude and nature of these opportunities will vary from region to region, crop to crop, and farm enterprise to farm enterprise. The tools and knowledge needed to implement these oppor-

A technician with the Agricultural Research Service measures the height of a grass hedge at the lower end of a cotton plot while the agronomist records the data. Hedges protect fields by holding back soil that would otherwise move off-site with water or wind. Credit: Agricultural Research Service, USDA.

tunities are, in many cases, already available. The short-term goal of agricultural policy should be to accelerate the rate at which these improved farming systems are implemented.

*The Cooperative Extension Service of the USDA should develop the information and methods needed to segment target audiences and tailor an accelerated educational program to target audiences including producers, crop-soil consultants, dealers who sell agricultural inputs, and others who affect producers' decisions.*

The crops grown, production practices, and management intensities vary widely among producers. Chapter 3 emphasized the need to recognize these differences and target efforts to improve farming systems to those problem farms that cause more soil degradation or water pollution. Similarly, the techniques used to deliver new knowledge and technology to producers should be tailored to differences in the socioeconomic characteristics of producers and to differences in the structures of farm enterprises.

An extensive body of research demonstrates how to link the dissemination of new knowledge and technology to the socioeconomic characteristics of the segmented, target audience as well as to the stage of the decision process affected by new knowledge and technology. This type of segmentation and refinement in disseminating new knowledge and technology is largely nonexistent in the implementation of current soil and water quality programs.

Procedures and methods need to be developed for program managers at the local level so that they can integrate more sophisticated marketing techniques into the dissemination of new knowledge and technology. Program managers need rational techniques to augment their familiarity with the local conditions gained through experience in a particular location and to tailor the dissemination to local social and economic characteristics.

### Potential for Change

The potential for programs based on technical and financial assistance is illustrated by efforts in Iowa to improve nitrogen management. In 1982, a consortium of state and federal organizations began implementing a coordinated set of programs in Iowa to improve soil and water quality. One of the primary objectives was to improve nitrogen management because of widespread detections of nitrates in both surface water and groundwater in the state. Although improved nitrogen management was the primary goal, an integrated approach to farming systems was the basis of the program. The program was implemented with a network of demonstration and implementation projects that attempted to accelerate the adoption of known technologies that would result in immediate improvements in nitrogen management. These projects were coupled with an aggressive marketing and educational effort designed to reach those producers who could improve nitrogen management in their operations (Hallberg et al., 1991)

The results of this program are promising. In the Big Spring Basin area that was targeted by the program, 52 percent of the 200 area producers reported in 1990 that they had reduced their applications of synthetic nitrogen fertilizer since 1981. The amount of nitrogen applied to corn was reduced 21 percent—a 0.454 million-kg (1 million-lb) reduction in nitrogen loading to the watershed and a cost savings of $200,000 per year for area producers. Statewide demonstration projects that used an improved late spring soil test for nitrogen reduced nitrogen applications by 62 percent in 1989 (23 sites) and 21 percent in 1990 (41 sites) with no differences in yields. The greater reductions in 1989 were due to drought-induced crop failures in the preceding year that left large amounts of nitrogen in the soil for use by the crops in 1989.

Statewide data show that since 1985 Iowa producers have reduced the amount of nitrogen they use, despite declines in fertilizer prices and contrary to trends for the Corn Belt as a whole. Since 1986, these reductions total more than 363 million kg (800 million lb) of nitrogen and represent a cost savings of more than $120 million. The experience in Iowa suggests that aggressive, coordinated efforts can accelerate the voluntary adoption of improved farm management techniques, at least when improved management results in financial as well as environmental benefits. It is too early to tell, however, whether these voluntary improvements in nitrogen management will be sufficient to meet water quality goals.

## Market-Based Incentives

Past experience with point source control has shown that environmental regulation can be expensive to enforce and expensive for firms to adhere to. William K. Reilly, who was the administrator of the EPA during the Bush administration, wrote

> It is becoming increasingly clear the reliance on the command and control approach to environmental regulation will not, by itself, allow EPA to achieve its mission or many long established environmental goals. A number of persistent, seemingly intractable problems remain. Whereas in the past we focused mainly on controlling pollution from large, industrial sources, we are now confronted by environmental concerns that stem from a diverse range of products and activities. . . .

> To maintain progress toward our environmental goals, we must move beyond a prescriptive approach by adding innovative policy instruments such as economic incentives. Properly employed, economic incentives can be a powerful force for environmental improvement (U.S. Environmental Protection Agency, 1991:iii).

The economic incentives considered by EPA include refundable deposits for pesticide containers, changes in water prices, and fees on the carbon contents of fossil fuels (U.S. Environmental Protection Agency, 1991). Such incentives could be extended to include tradable permits for groundwater withdrawals or taxes based on the leaching or runoff properties of agrichemicals (Jacobs and Casler, 1979; Shortle and Dunn, 1986; Tatenberg, 1985). EPA has also experimented with trading pollution permits between point and nonpoint sources, but with limited success (Carpenterier, 1993).

*Research should be directed at the design of market-based incentives to protect soil and water quality.*

Already used in the Clean Air Act (U.S. General Accounting Office, 1992) marketable permits, have also been used in agriculture to allocate water in the West (Wahl, 1989) and could conceivably be used to allocate nutrients or pesticides (Atkinson and Tietenberg, 1982; Bartfeld, 1992; Bower, 1980; Eheart et al., 1983, 1987; Krupnick, 1989; Letson, 1992; Malik et al., 1993; O'Neil, 1983a,b; Taylor, 1975). Marketable permits would require that a permit system be established in a targeted region. The permits would limit the total use of pesticides or nutrients within the region or would base permits on environmental quality standards. Producers who wish to use pesticides or nutrients would need to have a permit specifying the amount and use; the permits could be sold to others if unused. There is an increasing body of literature addressing market-based incentives for environmental goals (see, for example, Malik et al., 1993). Capalbo and Phipps (1990) note that many questions need to be answered before marketable permits for chemicals could be viewed as a feasible alternative to existing approaches; however, the flexibility of the approach, the year-to-year consistency of chemical use regardless of market agricultural policy signals, and the gravitation of permits to where they yield the highest return suggest that further analysis of the feasibility of using marketable permits is warranted.

### Facilitating Changes in Land Use

Voluntary or nonvoluntary adoption of improved farming systems, in itself, may prove to be insufficient to meet soil and water quality goals. Increased nutrient use efficiencies, for example, may not be sufficient to control nitrogen and phosphorus losses from watersheds where the concentration of livestock outstrips the available cropland on which to apply manures. Similarly, reducing soil erosion may not be sufficient to reduce sediment damage in streams unless riparian areas are protected or restored. In addition, there are lands that, because of their soils, landscape position, or hydrogeological setting, cannot be profitably farmed, even using improved farming systems, without degrading soil or water quality. Long-term changes in land uses in such cases are needed to protect soil and water quality.

### Long-Term Easement Program

*A program to purchase selective use rights from producers through long-term easements (an easement program) should be developed to provide incentives to producers to use environmentally sensitive lands sustainably so that they do not threaten soil and water quality.*

The intent of an easement program should not be to retire land from all productive uses but, rather, to prevent its use in ways that result in damage to soil or water quality. The program should serve as a way to make the transition to farming systems that are more appropriate to these sensitive lands. The specific set of rights purchased would depend on the environmental problem being addressed and the farming systems currently in use. Such easements might simply purchase the rights to grow row crops on the land covered by the easement; all other economic uses would be allowed. More restrictive easements could be used to protect riparian zones and wetlands.

The use of land set-asides has long been a component of programs to control both supply and soil and water quality damage. Figure 4-1 illustrates the history of such set-aside programs. The area of land involved in set-aside programs has varied dramatically over the past 40 years. Most recently, the Conservation Reserve Program has retired about 15 million ha (36 million acres) of highly erodible cropland. Although soil and water quality protection is great while set-aside lands are out of production, gains in soil and water quality can be lost when set-aside programs are terminated.

*The easement program should be designed to support rather than replace efforts to accelerate voluntary change or to initiate regulations.*

Accelerated voluntary and regulatory programs should be pursued on those lands that can be profitably and sustainably farmed by using available farming systems. The purchase of easements should be limited to those lands that cannot be sustainably farmed by using improved

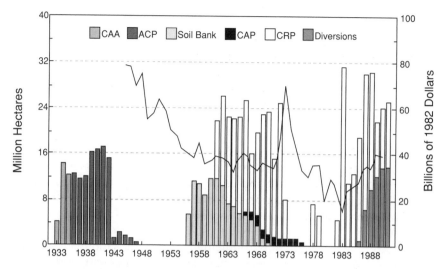

FIGURE 4-1  History of land set-aside programs in the United States as cropland area reductions by type of program (1933–1991) and net farm income (1945–1990). CAA, Conservation Adjustment Act; ACP, Agricultural Conservation Program; CAP, Cropland Adjustment Program; CRP, Conservation Reserve Program. Source: W. M. Crosswhite and C. L. Sandretto. 1991. Trends in resource protection policies in agriculture. Pp. 42–46 in Agricultural Resources: Cropland, Water and Conservation Situation and Outlook. Report No. AR-23. Washington, D.C.: U.S. Department of Agriculture, Economic Research Service, Resources and Technology Division.

farming systems or those lands that society would rather be used as habitat for fish and wildlife or recreational opportunities rather than to produce food and fiber.

Selling use rights then becomes one of several options a producer might use to meet soil and water quality goals. Long-term easements are written on a case-by-case basis, are flexible, and can be targeted to the needs of both the land and the landowner. Several alternative legal methods or instruments can be considered, including the use of easements (conveyances of property), long-term contracts (personal promises), covenants (promises connected to the land), or maintenance agreements. Landowners, however, may prefer contracts rather than options such as easements that are more permanent.

*The easement program should be designed so that state and local governments can supplement the program with efforts of their own and so that the program does not interfere with local efforts to control soil erosion or protect water resources.*

Easement programs can be integrated with other local environmental programs concerning issues such as farmland preservation, expansion of recreational opportunities, and water quality protection to maximize the environmental benefits acquired with public funds and to ensure coordination between various environmental protection efforts.

### Advantages of Easements

The most important advantage of an easement program is its focus on environmental improvement and its flexibility in being able to purchase use rights only for those lands that present environmental problems. The program could be tailored to local environmental and agricultural conditions.

For carefully selected, environmentally sensitive lands, long-term easements, or their equivalent, may offer a permanent and more cost-effective form of soil and water quality protection from a public standpoint. Appropriately designed and implemented, a system of long-term easements may be more attractive to landowners because it would allow for the recovery of partial compensation for the land while also allowing continued economic activity on the land, which would remain in private ownership.

The use of economic incentives, such as through the public acquisition of easements, is an attractive intermediate alternative to sole reliance on either voluntary or nonvoluntary approaches to protecting soil and water quality. Voluntary programs may not create a mechanism for achieving soil and water conservation on all lands, and nonvoluntary measures may create economic burdens on landowners. Easements offer the benefit of being voluntary and providing partial compensation (Hamilton, 1993). The approach may be more acceptable politically and may be more attractive to farmers and landowners than reliance on regulatory approaches. Because the environmental benefits obtained by using easements are either long-term or permanent protection, they are insulated from changing policies.

### Implementing an Easement Program

*Producers with lands currently under Conservation Reserve Program contracts should be offered the option of selling selected use rights, under long-term easements, to those lands currently under contract as a way of meeting compliance standards for bringing those lands back into production.*

About 15 million ha (36 million acres) of cropland have been enrolled in Conservation Reserve Program (CRP) contracts at a cost of more than $7 billion through 1991 for rental payments and cost-sharing to plant vegetative cover on croplands enrolled in CRP (U.S. Department of Agriculture, Economic Research Service, 1990). It is essential that the environmental benefits purchased through this program not be lost as contracts expire. Long-term easements that purchase rights to only those land uses that cause soil and water quality degradation, even under the best available farming systems, should be offered as a way to ease the transition of these lands to sustainable uses as CRP contracts expire. The lands eligible for such easements should be identified on the basis of the severity of soil and water quality damages expected if these lands were farmed using the best available farming systems. A program should be developed to review existing CRP contracts, as they near expiration, to identify tracts of land most appropriate for permanent protection and to solicit landowner interest in entering some form of long-term protection program.

*The various legal authorities for using easements contained in the 1990 Food, Agriculture, Conservation and Trade Act should be fully funded and implemented to expand public awareness of the concept, to gauge landowner attitudes to using easements, and to give the USDA and other agencies experience in using easements.*

Efforts should be undertaken at the federal level to expand the use of long-term easements, or similar mechanisms, for protecting soil and water quality. A major obstacle to using easements may be overcoming landowners' resistance to the concept of conveying partial interest in their lands to the public. Successful implementation of easements will require educational programs that focus on the reasons for landowner resistance and the levels of compensation that reflect the true costs to the landowners. Successful implementation of a system of long-term soil and water resource protection, such as easements, will also require agency commitment to the development of workable programs for promoting the availability of the programs, drafting easements, and implementing and managing the agreements. The recent success of the wetland reserve pilot program and the level of landowners' interest in selling wetland easements indicates that U.S. producers are interested in long-term easement programs.

## Need for Nonvoluntary Approaches

Although the opportunities to accelerate voluntary adoption of improved farming systems are great when such approaches also lead to increased profits, reliance on voluntary change or market-based incentives alone will not always be sufficient to achieve the improvements in soil and water quality increasingly demanded by the public.

The Chesapeake Bay Program is an example that shows that voluntary change alone has not been large enough or fast enough to meet environmental goals. By the year 2000, jurisdictions participating in the Chesapeake Bay program (Maryland, Virginia, the District of Columbia, and Pennsylvania) are to reduce their nutrient loadings to the bay by 40 percent. In 1990, a panel of producers, state agency staff, environmentalists, and academics assembled by the EPA Administrator William K. Reilly reported to EPA that current efforts would not be enough to meet the 40 percent goal. The panel concluded that voluntary incentives, at least as implemented in the past, had not been effective enough and that nutrient loadings were much larger than originally estimated (Nonpoint Source Evaluation Panel, 1990). The panel recommended that greater regulatory authority was needed to address agricultural as well as other sources of nutrient loadings to the bay watershed. The panel recommended that livestock operations, particularly large or intensive operations, or operations that were planning to expand should be targeted (Nonpoint Source Evaluation Panel, 1990).

Although comprehensive data on the production practices and management systems used by producers are not available, most of the data that were reported and discussed in Chapters 2 and 3 indicate that producers use a wide range of production practices and that there is wide variability in the degree to which they refine their management systems. These data suggest that a smaller set of problem farms may well be responsible for a substantial share of soil and water quality problems. If these producers fail to volunteer to participate in programs to improve their production practices and management systems, then voluntary programs may not improve soil and water quality enough to meet public demands.

### State and Local Legislation

The inherent limitations in programs to accelerate voluntary change have led to greater exploration of nonvoluntary approaches to accelerate adoption of improved farming systems. State and local governments have increasingly turned to more nonvoluntary approaches to changing farming systems in areas where soil and water quality damages are severe.

Ribaudo and Woo (1991) reviewed state water quality laws that affect agriculture and found that states were adopting a variety of approaches—including input control, land use controls, and economic incentives—to address water quality problems caused by agricultural production. Because of water quality concerns, 27 states have adopted laws that could affect farm management decisions (Figure 4-2).

### Evaluating the Role for Regulation

*Regulatory approaches based on clear planning or performance standards should receive greater attention to achieve more permanent protection in areas where soil and water quality degradation is severe and for problem farms that are unacceptably slow in implementing improved farming systems.*

Regulatory approaches will be needed to provide more permanent protection when commodity prices are high, damage to soil and water quality is severe, and voluntary change does not result in adequate improvements.

The arguments against regulatory approaches are well known and have been stated often. Mandating soil and water quality improvements can be expensive and ineffective if enforcement is inadequate or costly. Critics also claim that such regulations can potentially imperil private property rights. Furthermore, regulations might alter the relationships between the farming community and soil and water conservation agencies by turning the latter into "police officers" and the former into "lawbreakers" if regulations are not met. Others express concern that regulations cannot be written in a manner that provides the necessary flexibility to reflect the varying soil and water resources and farming systems found throughout the United States.

At the same time, there are advantages to grounding U.S. soil degradation and water pollution prevention efforts on a strong regulatory footing. Regulatory requirements can clearly state the objectives that a producer must meet and can be applied uniformly to all landowners and operators whose actions might degrade soil and water resources. If the producer meets certain standards, then compliance can be the basis for providing other benefits. Clearly defined planning or performance standards can provide the foundation on which other programs—including educational programs, programs that provide financial incentives, and cost-sharing programs—can be based.

Perhaps the most important benefit offered by using a regulatory approach is the promise of permanence. If landowners or operators are required to meet soil and water quality standards, these standards will apply in all circumstances regardless of changes in market prices, ownership of the land, production systems, the structure of the farm enterprise, or the goals of the producer.

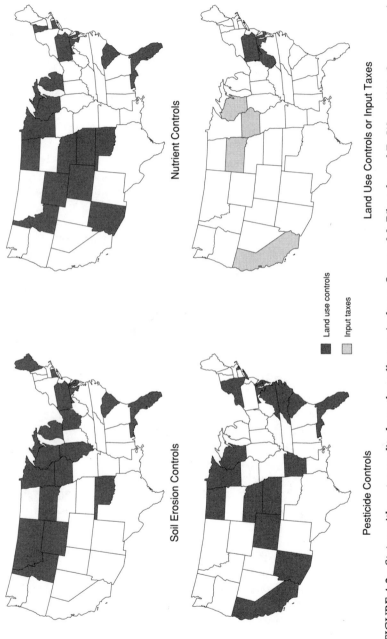

Nutrient Controls

Land Use Controls or Input Taxes

Soil Erosion Controls

Pesticide Controls

■ Land use controls
▨ Input taxes

FIGURE 4-2   States with water quality laws that affect agriculture. Source: M. Ribaudo and D. Woo. 1991. Summary of state water quality laws affecting agriculture. Pp. 50–54 in Agricultural Resources: Cropland, Water and Conservation Situation and Outlook. Report No. AR-23. Washington, D.C.: U.S. Department of Agriculture, Economic Research Service, Research and Technology Division.

The recognition of the value of including a regulatory approach in U.S. soil and water quality policies is not recent, as seen by the fact that the model state conservation law, on which all state enactments were originally based, included as a major component of the legal powers of the districts the power to implement land use regulations to protect soil and water quality. Although the history of the use of that authority by districts has been limited, there is growing interest in using regulations, as reflected in the innovative approaches being developed at the district level in many states. Likewise, the importance of regulatory methods for delivering and enforcing soil conservation policies has long been recognized and debated in the scholarly discussion of soil and water quality policy.

## Implications of the Structure of Agriculture for Regulation

The processing of many agricultural crops and most livestock is increasingly concentrated in a few firms (Barkema et al., 1991). Poultry is the most highly concentrated sector, and indications are that the hog industry is following the same processes of consolidation as the poultry industry (Barkema and Cook, 1993). It is probably not coincidental that the sectors of agriculture showing the most concentration are those that are not provided risk and income protection in federal agricultural price support and supply control programs.

In many cases, these firms contract directly with producers to provide crops and livestock. These contracts can specify, often in detail, the quality of the harvest required and, in some cases, the production practices that must be used. This vertical integration of contracted producers with processors has transferred substantial influence over the management of crop and livestock enterprises to processors.

The changing structure of agriculture suggests that regulation may increasingly be directed at processors and may seek to influence producers through their contracts with processors. Poultry processors, for example, could be required to specify requirements for the disposal and use of poultry manure in their contracts with poultry producers. If regulation required that whoever owned the poultry was to be responsible for appropriate manure disposal, then the liability for environmental damage would move from the individual poultry producer to the processor. Producer contracts would of necessity be altered to address the disposal problems associated with manure and dead poultry. Many individual producers could be affected by regulations directed at a handful of firms in agricultural sectors that are highly concentrated and integrated.

Furthermore, there would be incentives for processors to develop low-cost manure disposal alternatives. If the regulation covered the entire poultry sector, no one processor would be placed at an economic disadvantage to another. Although poultry prices might rise, depending on increased processor costs, consumers would be making choices based on prices that better represent the true social costs of poultry production. Similar regulation might be addressed toward other sectors of agriculture that evidence high rates of concentration.

## Clarifying Landowner Responsibilities and Rights

Bromley (1990) noted that landowners have enjoyed a wide range of actual and presumptive rights that have undergirded both environmental and agricultural policies. This arrangement automatically places the burden of proof—and of possible compensation—on the state when there is the need to (1) improve the environmental impacts of agriculture, (2) constrain agricultural output in the face of expensive surpluses, or (3) modulate swings in agricultural incomes. However, these arrangements of rights and responsibilities that have served well for many years are undergoing reevaluation.

*The legal responsibilities of landowners and land users to manage their lands in ways that do not degrade soil and water quality should be clarified in state and federal laws.*

Many landowners and land users manage their lands in ways that protect soil and water quality. Many others express their desire to improve their management to protect the environment. Clarifying society's expectations should encourage others to improve their management and help further public policies regarding the protection of soil and water resources.

The absence of a clear statement of the legal responsibilities as well as the rights of landowners and land users for managing their lands in ways that do not degrade soil and water quality has impeded efforts to protect soil and water quality. Basing publicly funded soil and water quality protection efforts on an articulated policy that establishes the legal responsibilities and rights of landowners and land users to protect soil and water quality offers the opportunity to provide a consistent and uniform basis for implementing soil and water quality protection efforts in a permanent manner.

Some argue that soil is simply a component of the land and thus is private property, meaning that efforts to limit the use and misuse of soil are constrained by constitutional limitations on exercise of the police power. Water quality, however, is more clearly a common good, the use and utility of which is subject to legitimate public policy concerns. Although landowners may own the soil on their farms, it is less arguable that they own the water that flows across it or under it, especially if ownership leads to degradation of its value for use by other individuals or the public. Soil also has some common-good characteristics since, as discussed in Chapter 2, soil quality is directly and indirectly linked to water quality and protecting soil quality is a fundamental step toward protecting water quality. The broadening of agricultural environmental issues from soil conservation to protecting soil and water quality brings with it a stronger public basis for codifying the responsibilities of landowners to protect soil and water quality.

The duty of landowners to protect soil and water quality, as reflected in the ideal that the landowner is ultimately responsible for preserving the sustainability of the land, is a powerful ideal that is reflected in many of the traditional approaches to soil and water conservation found in U.S. policy. In 1938, for example, Secretary of Agriculture Henry A. Wallace stated: "The social lesson of soil waste is that no man has the right to destroy soil even if he does own it in fee simple. The soil requires a duty of man which we have been slow to recognize" (Wallace, 1938: iii). And in 1943, the Iowa Supreme Court upheld a law requiring advance notice for terminating farm tenancies. The court stated:

> It is quite apparent that during recent years the old concept of duties and responsibilities of the owners and operators of farm land has undergone a change. Such persons, by controlling the food source of the nation, bear a certain responsibility to the general public. They possess a vital part of the national wealth, and legislation designed to stop waste and exploitation in the interest of the general public is within the sphere of the state's police power [8 N.W.2d 481 (Iowa 1943)].

This ideal of stewardship has been promoted in many ways, such as through education, financial incentives, ethical imperatives, and in some instances, legal mandates. Iowa, for example, enacted soil conservation legislation that states, in part:

> To conserve the fertility, general usefulness, and value of the soil and soil resources of this state, and to prevent the injurious effects of soil erosion, it is hereby made the duty of the owners of real property in this state to establish and maintain soil and water conservation practices, as required by the regulations of the commissioners of the respective soil conservation districts (Iowa Code, 1991:Section 467A.43).

The Iowa Supreme Court upheld the law's constitutionality, when it was challenged by a farmer required to implement soil conservation practices, noting: "[T]he state has a vital interest in protecting its soil as the greatest of its natural resources, and it is right to do so" [279 N.W.2d 276 (Iowa 1979)]. Several other states have enacted legislation that implicitly acknowledges the responsibilities of landowners and land users to protect soil and water quality (Figure 4-2).

This combination of incentives to encourage landowners to meet their stewardship responsibilities has resulted in significant progress, for example, as reflected through the work of the local soil and water conservation districts. However, the existence or acceptance of a duty to protect soil and water quality is not consistent among all landowners or in all soil and water conservation programs.

### Advantages of Defining Rights and Responsibilities

The national commitment to protect soil and water quality could be stated simply and directly in a manner that applies uniformly to all landowners and operators, regardless of their participation in federal farm programs. The concept of a duty to prevent soil and water quality degradation could be used as the basis for delivering and implementing other soil and water quality policies. These policies could include acceleration of voluntary adoption of improved farming systems, use of market-based incentives, reform of agricultural policies, implementation of nonvoluntary programs, and the administration of long-term easements.

The lack of a consistent definition of the legal responsibilities landowners have to protect soil and water quality as the foundation for soil and water quality programs has impeded the ability to build long-term comprehensive efforts in which publicly funded soil and water quality gains are made permanent. Basing soil and water quality protection efforts on an articulated policy that establishes landowners' responsibilities to manage their lands in ways that protect soil and water quality offers the United States the opportunity to provide a consistent and uniform basis for implementing soil and water quality protection efforts in a permanent manner.

The important value of establishing the responsibilities as well as right of landowners would be in the practical and psychological shift in the orientation of federal soil and water quality efforts. In codifying a landowner duty to protect soil and water quality, the burden of primary responsibility for protecting soil and water quality would shift from the government to the individual landowner. Rather than use programs that are based on inventing ways to educate, encourage, or coerce producers into protecting soil and water quality, the reorientation would establish that landowners and operators have a duty to protect soil and water quality from degradation. Rather than it being the government's responsibility to induce landowners to improve farming systems, it will be the duty of landowners to protect soil and water quality, with the government playing only a supporting role.

### Implementation

This clarification of landowners' and operators' responsibilities to protect soil and water quality could be moved forward in many ways:

- through education and voluntary compliance (essentially the history of the first 50 years of U.S. soil conservation programs);
- by integration of duties into existing federal farm programs as a condition for eligibility, as is now being done through Conservation Compliance, Sodbuster, and Swampbuster;
- by contractual agreement, as is the case with the CRP and the proposed use of long-term easements;
- as an imposed legal duty under state law, as is the case under the Iowa soil erosion control law, which makes it the duty of each landowner to protect his or her land from erosion by complying with the applicable county soil loss limits;
- as a function of private legal relationships, imposed either by the parties, such as through inclusion of such standards in the terms of a farm lease, or through the judicial imposition of stewardship under such common law concepts as the "covenant of good husbandry," which courts in many states attach to all farm lease relations; and/or
- a program to certify producers as stewards, analogous to the current programs requiring certification prior to using certain pesticides.

Articulation of landowners' responsibilities as well as rights to use their lands in ways that degrade soil or pollute water will allow producers who are committed to protecting soil and water quality to reaffirm their commitment to doing so and will offer a basis for public programs to change farming practices that are causing soil and water quality problems.

# PART TWO

# INTRODUCTION:
# Soil, Water, and
# Farming Systems

Farming systems are defined by the patterns in time and space in which producers grow their crops; the management decisions regarding the inputs and production practices used; the management skills, education, and objectives of the producer; the quality of the soil and water; and the nature of the landscapes and ecosystems within which production takes place. The production practices used to grow crops impinge on an agroecosystem made up of complex interactions among soil, water, biota, and the atmosphere. The interactions among the farming systems and the soil, water, biota, and atmosphere determine the effects those farming systems will have on soil and water quality.

Part One of this report recommended the most promising opportunities for manipulating these interactions to improve soil and water quality while still supporting the productive and economic production of food and fiber. Part Two of this report analyzes the individual pieces of these interactions.

Agricultural production profoundly affects the soil; the soil, in turn, mediates the effects of agricultural production on water quality. The functions soil performs in maintaining agricultural productivity and water quality and in regulating the global climate is discussed in Chapter 5. The soil is a living, dynamic system; conserving and enhancing the quality of U.S. soil resources is the first step toward improving the environmental performance of farming systems.

The effects of agricultural production on lakes, rivers, streams, and groundwater have become an important concern in agricultural and environmental policy. The transmission of nitrogen, phosphorus, pesticides, sediments, and salts from agricultural production to surface water and groundwater is an important source of water quality problems in the United States. Chapters 6 through 10 explore the complex fates of pollutants and the transport mechanisms that determine the amounts of these pollutants that are delivered to surface water and groundwater during agricultural production. The committee traces the links between nutrient, pesticide, and irrigation water use in agricultural production and the effects on water quality and pinpoints the most promising ways to improve both input management and water quality.

Livestock manures are important sources of the nitrogen and phosphorus used in agricultural production systems; they are also an important source of pollution if they are improperly managed. Chapter 11 explores the special problems managing animal wastes pose for livestock producers in their attempts to minimize the effects of animal waste on water quality. The chapter also both emphasizes the importance of improving the management of manures and identifies the barriers that may prevent management improvements.

Farming systems exist in landscapes made up of soils, slopes, streams, and lakes and adjacent ecosystems such as wetlands, forests, and riparian areas (the areas adjacent to rivers, lakes, and streams). The effects of farming systems on soil and water quality are strongly influenced by the landscape within which production takes place. Chapter 12 explores the interactions of farming systems and the landscape and suggests how these interactions can be managed by creating field and landscape buffer zones to mitigate the effects of agricultural production on soil and water quality.

The conclusions reached in the chapters in Part Two formed the basis of the recommendations put forth in Part One. A careful reading of the chapters in Part Two will provide a much firmer foundation for understanding the recommendations in Part One and will provide a solid background for those interested in understanding more fully the physical, chemical, and biological processes that determine how agricultural production affects soil and water quality.

# 5

# Monitoring and Managing
# Soil Quality

S oil, water, air, and plants are vital natural resources that help to produce food and fiber for humans. They also maintain the ecosystems on which all life on Earth ultimately depends. Soil serves as a medium for plant growth; a sink for heat, water, and chemicals; a filter for water; and a biological medium for the breakdown of wastes. Soil interacts intimately with water, air, and plants and acts as a damper to fluctuations in the environment. Soil mediates many of the ecological processes that control water and air quality and that promote plant growth.

Concern about the soil resource base needs to expand beyond soil productivity to include a broader concept of soil quality that encompasses all of the functions soils perform in natural and agricultural ecosystems. In the past, soil productivity and loss of soil productivity resulting from soil degradation have been the bases for concern about the world's soils. Equally important, however, are the functions soils perform in the regulation of water flow in watersheds, global emissions of greenhouse gases, attenuation of natural and artificial wastes, and regulation of air and water quality. These functions are impaired by soil degradation.

The ability of modern agricultural management systems to sustain the quality of soil, water, and air is being questioned. This chapter suggests methods that can be used to evaluate whether soil quality is being degraded, improved, or maintained under given management systems and methods of evaluating whether alternative management systems will sustain the quality of soil resources.

## DEFINING SOIL QUALITY

Soil quality is best defined in relation to the functions that soils perform in natural and agroecosystems. The quality of soil resources has historically been closely related to soil productivity (Bennett and Chapline, 1928; Lowdermilk, 1953; Hillel, 1991). Indeed, in many cases the terms *soil quality* and *soil productivity* have been nearly synonymous (Soil Science Society of America, 1984). More recently, however, there is growing recognition that the functions soils carry out in natural and agroecosystems go well beyond promoting the growth of plants. The need to broaden the concept of soil quality beyond traditional concerns for soil productivity have been highlighted at a series of recent conferences and symposia.

Johnson and colleagues (1992), in a paper presented at a Symposium on Soil Quality Standards hosted by the Soil Science Society of America in October 1990 suggested that soil quality should be defined in terms of the function soils play in the environment and defined soil function as "the potential utility of soils in landscapes resulting from the natural combination of soil chemical, physical, and biological attributes" (page 77). They recommended that policies to protect soil resources should protect the soil's capacity to serve several functions simultaneously including the production of food, fiber and fuel; nutrient and carbon storage; water filtration, purification, and storage; waste storage and degradation; and the maintenance of ecosystem stability and resiliency.

Larson and Pierce (1991) defined soil quality as "the capacity of a soil to function, both within its ecosystem boundaries (e.g., soil map unit boundaries) and with the environment external to that ecosystem (particularly relative to air and water quality)" (page 176). They proposed "fitness for use" as a simple operational definition of soil quality and stressed the need to explicitly address the function of soils as a medium for plant growth, in partitioning and regulating the flow of water in the environment, and as an environmental buffer. Parr and colleagues (1992), in a paper presented at a Workshop on Assessment and Monitoring of Soil Quality hosted by the Rodale Institute Research Center in July 1991, defined soil quality as "the capability of a soil to produce safe and nutritious crops in a sustained manner over the long term, and to enhance human and animal health, without impairing the natural resource base or harming the environment" (page 6). Parr and colleagues (1992) stressed the need to expand the notion of soil quality beyond soil productivity to include the role of the soil as an environmental filter affecting both air and water quality. They suggested that soil quality has important effects on the nutritional quality of the food produced in those soils but noted that these linkages are not well understood and research is needed to clarify the relationship between soil quality and the nutritional quality of food.

There is a growing recognition of the importance of the functions soils perform in the environment. The importance of those functions requires that scientists, policymakers, and producers adopt a broader definition of soil quality. Soil quality is best defined as the capacity of a soil to promote the growth of plants; protect watersheds by regulating the infiltration and partitioning of precipitation; and prevent water and air pollution by buffering potential pollutants such as agricultural chemicals, organic wastes, and industrial chemicals. The quality of a soil is determined by a combination of physical, chemical, and biological properties such as texture, water-holding capacity, porosity, organic matter content, and depth. Since these attributes differ among soils, soils differ in their quality. Some soils, because of their texture or depth, for example, are inherently more productive because they can store and make available larger amounts of water and nutrients to plants. Similarly, some soils, because of their organic matter content, are able to immobilize or degrade larger amounts of potential pollutants.

Soil management can either improve or degrade soil quality. Erosion, compaction, salinization, sodification, acidification, and pollution with toxic chemicals can and do degrade soil quality. Increasing soil protection by crop residues and plants; adding organic matter to the soil through crop rotations, manures, or crop residues; and careful management of fertilizers, pesticides, tillage equipment, and other elements of the farming system can improve soil quality.

## IMPORTANCE OF SOIL QUALITY

Soils have important direct and indirect impacts on agricultural productivity, water quality, and the global climate. Soils make it possible for plants to grow by mediating the biological, chemical, and physical processes that supply plants with nutrients, water, and other elements. Microorganisms in soils transform nutrients into forms that can be used by growing plants. Soils are the storehouses for water and nutrients. Plants draw on these stores as needed to produce roots, stems, leaves, and, eventually, food and fiber for human consumption. Soils—and the biological, chemical, and physical processes they make possible—are a fundamental resource on which the productivities of agricultural and natural ecosystems depend.

The soil, which interacts with landscape features and plant cover, is a key element in regulating and partitioning water flow through the environment (Jury et al., 1991). Rainfall in terrestrial ecosystems falls on the soil surface where it either infiltrates the soil or moves across the soil surface into streams or lakes. The condition of the soil surface determines whether rainfall infiltrates or runs off. If it enters the soil it may be stored and later taken up by plants, it may move into groundwaters or move laterally through the earth, appearing later in springs. This partitioning of rainfall determines whether a rainstorm results in a replenishing rain or a damaging flood. The movement of water through soils to streams, lakes, and groundwater is an essential component of the hydrological cycle.

The biological, chemical, and physical processes that occur in soils buffer environmental changes in air quality, water quality, and global climate (Lal and Pierce, 1991). The soil matrix is the major incubation chamber for the decomposition of organic wastes, for example, pesticides, sewage, and solid wastes. Depending on how they are managed, soils can be important sources or sinks of carbon dioxide and other gases, also known as greenhouse gases, that contribute to the so-called greenhouse effect. Soils store, degrade, or immobilize nitrates, phosphorus, pesticides, and other substances that can become air or water pollutants.

Soil degradation through erosion, compaction, loss of biological activity, acidification, salinization, or other processes can reduce soil quality. These processes reduce soil quality by changing the soil attributes, such as nutrient status, organic and labile carbon content (organic carbon is the total amount of carbon held in the organic matter in the soil; labile carbon is that fraction of organic carbon that is most readily decomposable by soil microorganisms), texture, available water-holding capacity (the amount of water that can be held in the soil and made available to plants), structure, maximum rooting depth, and pH (a measure of the acidity or alkalinity). Some changes in these soil attributes can be reversed by external inputs. Nutrient losses, for example, can be replaced by adding fertilizers. Other changes such as loss of the soil depth available for rooting because of soil erosion or degradation of soil structure because of subsoil compaction are much more difficult to reverse.

### Soil Quality and Agricultural Productivity

Damage to agricultural productivity has historically been the major concern regarding soil degradation. Agricultural technology has, in some cases, improved the quality of soils. In other cases, improved technology has masked much of the yield loss that could be attributed to declining soil quality, except on those soils that are vulnerable to rapid and irreversible degradation.

### Effect of Soil Degradation on Productivity

Four major studies predicted that yield losses resulting from soil erosion would be less than 10 percent over the next 100 years (Crosson and Stout, 1983; Hagen and Dyke, 1980; Pierce et al., 1984; Putnam et al., 1988). Such projections of low-yield losses, coupled with increasing concern over off-site water quality damages from agricultural production, have begun to shift the emphasis of federal policy to the off-site damages caused by erosion.

On-site losses of soil productivity from current degradative forces, however, have been underestimated. The projections for low levels of erosion-induced losses in agricultural productivity largely result from the hypothesis that almost two-thirds of U.S. croplands will suffer little or no yield loss over the next 100 years (Pierce, 1991). Productivity losses on the remaining one-third of the lands may be serious (Pierce et al., 1984), but the losses are masked by the larger area of soils that are less vulnerable to erosion (Pierce, 1991).

More important, estimates of productivity losses resulting from erosion have not accounted for damages caused by gully and ephemeral erosion, sedimentation (Pierce, 1991), or reduced water availability because of decreased infiltration of precipitation. Those studies also assumed that the optimum nutrient status is maintained on the eroding lands through application of fertilizers, manures, or other sources of plant nutrients. Replacing these nutrients comes at a cost. Larson and colleagues (1983) estimated that in 1982 the amount of nitrogen, phosphorus, and potassium from U.S. croplands lost in eroded sediments was 9,494, 1,704, and 57,920 metric tons, respectively (10,465, 1,878, and 63,846 tons, respectively). The value of the nitrogen, phosphorus, and potassium lost was estimated at $677 million, $17 million, and $381 million, respectively.

In addition, estimates of the effects of soil degradation on productivity have focused on the yield losses expected from erosion-induced damage to croplands. The nation's croplands are also being damaged by compaction, salinization, acidification, and other forces. These damages will add to the yield losses resulting from erosion. More important, erosion accelerates the processes of compaction, salinization, and acidification. The reverse is also true. Yield losses will be greater than those projected in the past if all degradation processes and their interactions are considered.

Walker and Young (1986) have suggested that the use of absolute crop yield reductions as the measure of productivity losses masks more

Even though this cropland has been tilled, ephemeral rills are still evident. During heavy rains, water will collect in these small channels and increase the severity of runoff. Credit: U.S. Department of Agriculture.

subtle but important productivity losses. The analyses concluded that losses in potential yields will occur sooner and will be of greater magnitude than losses in absolute yields resulting from reduced soil quality. New, high-yielding crop varieties often require increased inputs of nutrients and more stable water regimes in order to produce maximum yield. Loss of soils' ability to hold and store nutrients and water can significantly restrain achievement of the full yield potentials of new agricultural technologies. New technologies may allow yields to increase or stay the same, even in the face of soil degradation, but these yields may mask important losses in the productive potential that could have been realized if soil quality had not been reduced. The true loss of productivity because of soil mismanagement or degradation is this loss in productive potential (Walker and Young, 1986).

*Effect of Soil Degradation on Costs of Production*

Crosson and colleagues (1985) indicated that it is the cost of erosion, not predicted yield losses, that is really of interest. They suggested that farmers can substitute fertilizers, tillage, and other inputs for losses in soil productivity caused by soil erosion and that, from a production standpoint, increases in costs to reduce erosion are no different than higher input costs to compensate for erosion. Similarly, it is the cost of compensating for reduced soil quality resulting from degradation by compaction, acidification, salinization, loss of biological activity, and erosion that is most important when assessing the effects of soil degradation on soil productivity.

Estimating the effect of soil degradation from erosion on the costs of production has proved difficult. Larson and colleagues (1983) suggested that soil degradation results in both replaceable and irreplaceable losses in soil productivity. A replaceable loss, for example, may be nutrients lost in eroded soil; an irreplaceable loss may be the loss in water-holding capacity resulting from decreased soil depth. Similarly, Walker and Young (1986) and Young (1984) distinguished between reparable and residual loss of yields resulting from soil erosion. Reparable yield losses were those that could be compensated for by substitution of other inputs such as fertilizer. Residual yield losses were those that remain even after substitution of other inputs and represent the cost to the yield of losing irreplaceable elements of soil quality such as soil depth. A total assessment of the costs of erosion would have to account for the costs of both the substituted inputs and the residual yield losses.

Few data are available to estimate the effects of soil degradation from compaction, salinization, acidification, loss of biological activity, and other processes of soil degradation on production costs. Estimates of the extent or cost of compaction nationwide are not available. Eradat Oskoui and Voorhees (1990) extrapolated data from studies on yield losses resulting from subsoil compaction in Minnesota. They suggested that the value of the lost corn yield (based on a corn price of $0.06/kg [$2/bushel]) in Minnesota, Wisconsin, Iowa, Illinois, Indiana, and Ohio could be $100 million annually. In years with high levels of water stress, when root growth is limited because of too much or too little water, yield losses would be higher. The U.S. Department of Agriculture (USDA), Soil Conservation Service (1989a) estimated that the productivity of 9 percent of the nation's croplands and pasturelands, including more than one-fifth of the irrigated lands, was being lowered by salinization or sodification. No data are available to suggest the extent or the cost of soil degradation resulting from the loss of biological activity or acidification.

*Sustaining Soil Quality Is Essential to Improving Agricultural Productivity*

Given the multiple processes of soil degradation and the probable underestimation of the full cost of erosion on the cost of production, it can be concluded that soil degradation may have significant effects on the ability of the United States to sustain a productive agricultural system. The costs of reversing multiple causes of soil degradation to maintain yields may be large enough to affect the costs of production, even if absolute yields are not affected. To date, improvements in agricultural technologies have kept the costs of compensation for losses in soil quality low enough or increases in yields large enough to offset the costs of soil degradation on most croplands.

## Soil Management

Finally, although attention has understandably been focused on soil degradation, soil management to improve soil quality holds the promise of producing gains in productivity. Current research suggests that soil management to improve infiltration, aeration, and biological activity can lead to significant gains in crop yields (Allmaras et al., 1991; Edwards, 1991). Yield gains from improved soil quality can be large on croplands that have suffered historic degradation from erosion. Soil management to improve soil quality is an opportunity to simultaneously improve profitability and environmental performance.

## Soil Quality and Water Quality

Soil quality losses increase environmental as well as production costs. Indeed, investigators have argued that the costs of off-site damages from soil erosion are greater than the costs imposed by decreased productivity (Clark et al., 1985; Crosson and Stout, 1983). Soil degradation causes both direct and indirect degradation of water quality.

## Direct Effects

Soil degradation from erosion leads directly to water quality degradation through the delivery of sediments and agricultural chemicals to surface water. Clark and colleagues (1985), using admittedly imperfect methods, estimated that the cost of sediment delivery on recreation, water storage facilities, navigation, flooding, water conveyance facilities, and water treatment facilities, among other damages, at $2.2 billion (1980 dollars) annually. Soil degradation resulting from compaction, salinization, acidification,

Soil degradation leads directly to water pollution by sediments and attached agricultural chemicals from eroded fields. Soil degradation indirectly causes water pollution by increasing the erosive power of runoff and by reducing the soil's ability to hold or immobilize nutrients and pesticides. Credit: U.S. Department of Agriculture.

or loss of biological activity can increase the vulnerability of soils to erosion and exacerbate the water quality problems associated with sedimentation.

### Indirect Effects

The indirect effects of soil quality degradation may be as important as the direct damages resulting from sediment delivery, but they are often overlooked. Soil degradation impairs the capacity of soils to regulate water flow through watersheds. The physical structure, texture, and condition of the soil surface determine the portion of precipitation that runs off or infiltrates soils. In the process, the volume, energy, and timing of seasonal stream flows and recharge to groundwater are determined. Soil erosion and compaction degrade the capacities of watersheds to capture and store precipitation. Stream flow regimes are altered: seasonal patterns of flow are exaggerated, increasing the frequency, severity, and unpredictability of high-flow periods and extending the duration of low-flow periods. The increased energy of runoff water causes stream channels to erode, adding to sediment loads and degrading aquatic habitat for fish and other wildlife. Channel erosion was estimated to contribute from 25 to 60 percent of the sediment load in rivers in Iowa, Illinois, and Mississippi (see Chapter 6).

Soil degradation that leads to the loss of a soil's capacity to buffer nutrients, pesticides, and other inputs accelerates the degradation of surface water or groundwater quality. Erosion not only results in the direct transport of sediment, nutrients, and pesticides to surface waters but also reduces the nutrient storage capacity of soils. A reduced nutrient storage capacity may lead to less efficient use of applied nutrients by crop plants and a greater potential for loss of nutrients to surface water and groundwater (Power, 1990). The pesticides held by soil organic matter or clay may become more mobile in the soil environment as erosion reduces organic matter levels and changes the soil's texture (Wagenet and Rao, 1990). Reduced biological activity can slow the rate at which pesticides are degraded, increasing the likelihood that the pesticides will be transported out of the soil to surface water or groundwater (Sims, 1990). Compaction in combination with other soil degradation processes can reduce the health of crop root systems, leading to less efficient nutrient use and increasing the pool of residual nutrients that can be lost to surface water or groundwater (Dolan et al., 1992; Parish, 1971).

### Soil Quality and Water Quality Are Linked

Soil degradation results in both direct and indirect degradation of surface water and groundwater quality. Protecting or improving soil quality is a fundamental step toward improving the environmental performance of agricultural ecosystems. Changes in farming systems that attempt to address the loss of nutrients, pesticides, salts, or other pollutants will not be as effective unless soil quality is also protected or improved.

Soil quality improvement alone, however, will not be sufficient to address all water quality problems unless other elements of the agricultural system are addressed. Soil quality improvement alone, for example, will not solve the problem of nitrate contamination of surface water and groundwater if excessive nitrogen is applied to the cropping system. If nitrogen applications are excessive, changes in soil quality may change the proportion of nitrates delivered to surface waters rather than to groundwaters, but total nitrate losses may remain the same.

### Soil Quality and the Global Climate

Recently, the role of the soil resource as a global climate regulator has received more attention as a result of heightened concern over human-induced climate changes. Depending on how it is managed, soil is a source (or sink) of carbon and nitrogen. Lal and Pierce (1991), for example, estimated that if 1 percent of the organic carbon stored in the most widely occurring types of tropical soils is mineralized annually, 128 billion metric tons (130 billion tons) of carbon will be released into the atmosphere. Lal and Pierce (1991) point out that this quantity compares with annual carbon emissions of an estimated 325 million metric tons (330 million tons) from burning of fossil fuels and 1,659 million metric tons (1,686 million tons) from deforestation (Brown et al., 1990). Little is known, however, about the contribution of soil-related processes to greenhouse gas emissions under different systems of soil and crop management. What is known suggests that the soil resource may play an important role in regulating greenhouse gas concentrations.

### Soil Quality as a Long-Term Goal of Soil Management

The ways that humans use soils affect soil quality. Soil erosion can strip away fertile topsoils and leave the soil less hospitable to plants. Heavy farm machinery can compact the soil and impede its capacity to accept and store water. Loss of organic matter because of erosion or poor cropping practices can seriously impede the soil's ability to filter out potential pollutants.

In the past, soil erosion was used as a convenient proxy for all of the processes of soil degradation, and efforts to control erosion have dominated programs and policies to protect soil resources. Soil erosion has been and continues to be the single most important process that degrades soil quality. In the long term, however, all processes of soil degradation—compaction, salinization, acidification, loss of biological activity, pollution, and erosion—need to be considered when making soil management decisions. In the past, soil productivity was the primary value attached to soils and crop yield reductions were the primary measure used to assess the significance of soil degradation. In the long term, however, soil management goals need to be broadened to include the roles that soils play in regulating water flow through watersheds and buffering environmental changes. Conservation of soil quality should become the goal of long-term soil management policies and programs.

The relative importance of the three components of soil quality and the relative importance of the processes of soil degradation vary from area to area. Soil quality varies dramatically from soil to soil. Some soils are shallow and restrict plant growth. Others have impermeable layers beneath the surface that limit the soil's capacity to store water and restrict plant root growth. Still others are so acidic or basic that the biological activity needed to recycle wastes is seriously impaired. Certain soils are more vulnerable to the loss of one or the other components of soil quality and vary in their resistance to different soil degradation processes. The value that society places on the three components of soil quality also vary from place to place. In some cases, management to protect a soil's capacity to accept and degrade wastes may take precedence over conservation of soil's productivity. Similarly, on some soils compaction is a more important problem than erosion.

Setting soil quality as the long-term goal of soil management has implications for national-level assessments of soil resources, for the design of programs to conserve soil resources, and for analyses of sustainable farming systems.

### National Assessments of Soil Resources

Since the 1930s, investigators have periodically made national-level assessments of the amounts of erosion and its consequences. The 1938 yearbook of agriculture (U.S. Department of Agriculture, 1938) reported, based on a minimum of quantitative data, that of the total U.S. land area (770 billion ha [1,903 million acres]), 114 million ha (282 million acres) was ruined or severely damaged and 314 million ha (775 million acres) was moderately damaged. In the 1950s and 1960s, investigators made periodic estimates of the amounts of erosion on the basis of reconnaissance surveys. In 1977, the U.S. Congress passed the Soil, Water and Related Resources Conservation Act, which called for an assessment every 5 years of the status of U.S. natural resources and their ability to provide the long-term resource needs of the United States. In response to the Soil, Water and Related Resources Conservation Act, the Soil Conservation Service of the USDA established, on croplands throughout the United States, primary sampling units where information was gathered. Although investigators gathered many kinds of information, the emphasis was on soil erosion.

The 1977, 1982, and 1987 National Resources Inventories were by far the most extensive and quantitative inventories of soil resources in the United States. These inventories and assessments, however, were limited by their focus on quantifying rates of erosion and other processes of soil degradation rather than assembling and assessing the information needed to monitor the changes in soil attributes that can be related to changes in soil quality. The 1977, 1982, and 1987 National Resources Inventories, for example, did not include direct measurements of the changes in soil attributes caused by erosion. Rather, the inventories concentrated on comparing estimated erosion amounts with values such as $T$, usually defined as the maximum amount of erosion that can be tolerated; below this level of erosion, crop yields can be maintained economically and indefinitely (Wischmeier and Smith, 1978). The estimated erosion was calculated by the universal soil loss equation and the wind erosion equation.

Although widely used, the accuracies of the equations need improvement and the scientific validity of $T$ has been questioned (Johnson, 1987). National-level assessments need to be redirected to include quantifiable measures of soil attributes if soil quality changes are to be estimated. In addition, soil resource assessments need to be broadened to include all soil degradation processes. Monitoring of the processes of soil degradation such as erosion is an important component of such assessments, but monitoring must be strengthened by the collection and interpretation of data that can be related quantitatively to changes in the soil resource itself. A system that enables more direct quantification of actual changes in soil attributes will allow policies and programs to be directed more closely to alleviating actual degradation of soil quality.

### Soil Quality and Soil Conservation

The concept of soil quality should be the principle guiding the recommendations for use of conservation practices and the targeting of programs and resources. Soil quality can be defined as the ability of a

TABLE 5-1    Reference and Measured Values of Minimum Data Set
for a Hypothetical Typic Hapludoll from North-Central United States

| Horizon and Characteristic | Reference Value | Measured Value |
|---|---|---|
| Surface horizon | | |
| Phosphorus (mg/kg) | 30 | 15 |
| Potassium (mg/kg) | 300 | 300 |
| Organic carbon (percent) | | |
| Total | 2 | 1.5 |
| Labile | 0.2 | 0.15 |
| Bulk density (mg/m$^3$) | 1.3 | 1.5 |
| pH | 6.0 | 5.5 |
| Electrical conductivity (S/m) | 0.10 | 1.0 |
| Texture (percent clay) | 30 | 32 |
| Subsoil horizon | | |
| Texture (percent clay) | 35 | 35 |
| Depth of root zone (m) | 1.0 | 0.95 |
| Bulk density (mg/m$^3$) | 1.5 | 1.5 |
| pH | 5.5 | 5.5 |
| Electrical conductivity (S/m) | 0.10 | 0.10 |

soil to perform its three primary functions: to function as a primary input
to crop production, to partition and regulate water flow, and to act as an
environmental filter. Many attributes (or properties) of a soil contribute to
soil quality, and the attributes are highly interrelated. Thus, no single
attribute can be used as an index of soil quality. However, a few key
attributes can be selected as indicators. Because many soil attributes are
interrelated, the indicator attributes can often be used to estimate other
attributes. The indicator attributes can then be used in simple models to
predict a soil's ability to perform its three primary functions.

The utility of the concept of soil quality in guiding soil management can
be seen in a simple example. Table 5-1 lists the changes in specific soil
attributes for a Typic Hapludoll soil from the north-central United States.
The changes in these soil attributes suggest the process of degradation that
needs to be addressed, the corrective management practices that are
needed, and whether further investigation is needed to improve soil
quality. In this example, the available phosphorus is low, suggesting the
need for improved nutrient management. The organic carbon level in the
soil has declined, indicating that additional organic matter is needed.
Management of residues needs to be improved or the crop sequence
needs to be changed to include more closely grown crops, cover crops,
legumes, or other sources of organic matter. Alternatively, tillage intensity
could be reduced to slow the rate of organic matter decline.

The bulk density of the surface soil has increased, perhaps as a consequence of the lowered organic carbon content, the use of intensive tillage practices, or the use of heavy harvest machinery (compaction). The increase in the surface clay content (enriched from the subsoil) and the decrease in the rooting depth suggest that erosion may be serious. The decrease in surface soil pH may have resulted from the use of acid-forming fertilizer or natural leaching and may signal the need for lime. On this soil, an analysis of changes in soil attributes suggests that conservation practices need to focus on the maintenance of organic carbon, phosphorus, and pH and on erosion control. Reductions in erosion rates alone may not be sufficient to reduce evident compaction or the declining organic matter levels. Changes in these soil attributes, however, are directly linked not only to the maintenance of soil productivity but also to the regulation of water flow through the environment and the capacity of the soil to buffer environmental changes.

Such analyses of changes in soil attributes would be enriched if data on management variables such as cropping sequence, residue levels, tillage practices, and nutrient management were available. The combination of data on changes in soil attributes with management variables would allow for a more useful analysis of the kinds of conservation practices needed to protect soil quality.

### Soil Quality and Sustainability

The concept of sustainable agricultural systems, whether for croplands, rangelands, or forestlands, is gaining acceptance as a framework that can be used to guide research and bridge the apparent conflicts between agricultural production and environmental goals. Investigators have provided several definitions and descriptions of sustainability, but the critical analysis of sustainable production systems has been constrained by the lack of systematic and scientifically sound criteria against which to compare alternative production systems.

The soil quality criteria proposed for inclusion in a minimum data set can serve as the first step toward the development of systematic criteria of sustainability. These criteria will be discussed in the next section. Current research and historical data should allow researchers and managers to predict the impact of a particular farming system on soil quality. An accepted set of criteria for soil quality would permit comparative analyses of farming systems and would help to systematize the debate and research that attempt to define and implement the concept of sustainability.

The comparison of alternative farming systems with a set of soil quality criteria would be only the first step in a more systematic investigation of sustainability. The impacts of farming systems on air and water quality would also need to be evaluated, and economic analyses of those alternative farming systems would be required to complete the picture. Because of the soil's role in integrated impacts on both air and water quality, development of systematic soil quality criteria is an important first step.

## IMPORTANCE OF MONITORING CHANGES IN SOIL QUALITY

A system that measures changes in soil quality is needed if conservation of soil quality is to become the long-term goal for management of the soil resource. A system that monitors changes in soil quality could be used for three major purposes. First, such a system can be used to track national trends in soil quality by incorporating measures of soil quality indicators into national resource surveys and assessments. Second, such a system can improve the management of soil conservation programs by aiding in setting tolerable soil erosion standards, targeting lands that need conservation measures, and identifying lands most suitable for inclusion in long-term easement programs. Finally, a system of soil quality indicators can aid in the analysis of the sustainability of farming systems by providing a set of criteria against which farming systems can be compared.

The need for systems to monitor changes in soil quality has received increasing attention recently. Several authors have called for soil quality monitoring as a basic component of national policies to protect soil resources (Haberen, 1992; Hortensius and Nortcliff, 1991; Johnson et al., 1992; Larson and Pierce, 1991, 1994; Parr et al., 1992; Pierce and Larson, 1993; Young, 1991). Larson and Pierce (1991) compared a system that measures the quality of a soil to a medical clinic that assesses human health. A routine health assessment includes measurements of key indicators of health such as temperature, blood pressure, heart beat, and a few simple blood characteristics. These are considered indicators of possible problems. If the assessment finds abnormalities in any of the key indicators, more detailed information will be requested. Likewise, key soil quality indicators are needed so that investigators can monitor changes in soil quality. Over time, changes in soil quality indicators will provide the information needed to assess the effects of current farming systems and land use on soil quality, develop new farming systems that improve soil quality, and guide the development of national policies to protect soil and water quality.

Soils, however, are difficult to inventory and assess. Soils vary greatly, with variations often occurring at distances of only a few meters. Gross differences in soil surfaces can be seen or felt and usually reflect differences in organic matter content, mineralogy, or texture. The soil characteristics below the usual depth of cultivation, however, are often not carefully observed and characterized except by soil specialists. It is often difficult and laborious to obtain samples from the deeper horizons of the soil.

A system to monitor changes in soil quality will require the following:

- identification of the soil attributes that can serve as indicators of change in soil quality,
- standard field and laboratory methodologies that can be used to measure changes in indicators of soil quality,
- a coordinated monitoring program that can quantify changes in soil quality indicators, and
- a coordinated research program designed to support, test, and confirm models that can be used to predict the impact of management practices on soil quality.

### Indicators of Soil Quality

The quality of a soil is a composite of its physical, chemical, and biological properties. Indicators of soil quality are needed that relate to all three functions soils perform in natural and agroecosystems: (1) promote plant growth, (2) protect watersheds by partitioning and regulating precipitation, and (3) prevent air and water pollution by buffering agricultural chemicals, organic wastes, industrial chemicals, and other potential pollutants.

It will be impossible and unnecessary to monitor changes in all of the soil attributes that relate to these three soil functions. Monitoring of a select set of soil attributes that can serve as indicators of change in soil quality is possible and can yield useful information on trends in soil quality. Many soil attributes could serve as indicators of soil quality. Soil attributes are often highly correlated, which makes interpretation of the significance of changes in selected indicators of soil quality difficult (Larson and Pierce, 1991). A change in soil organic matter, for example, has a direct effect on soil quality, but it also changes other measurable indicators of soil quality such as structure or bulk density. A system made up of soil quality indicators that are independent of one another would be ideal, but such a system is not possible because of the interrelated nature of the soil system.

In addition, the measured value of a selected soil quality indicator will have a different interpretation depending on the soil or region from which the sample was obtained. Critical bulk density values, for example, vary with the texture of tne soil. The correlation between soil quality indicators and their soil or region specificity means that change in any set of soil quality indicators must be evaluated as a group. Interpretation of changes in one indicator without relating the change to other indicators may lead to misleading results. Interpretation of changes in soil quality indicators will also vary with soil taxa (classification group), probably at the suborder level.

### Minimum Data Set

A great deal is known about the relationship of specific soil attributes to soil quality, and several authors have recently recommended various soil attributes as indicators of soil quality. Larson and Pierce (1991) recommended a combination of physical, chemical, and biological attributes as a minimum data set of soil quality indicators including nutrient availability, organic carbon, texture, water-holding capacity, structure, rooting depth, and pH. Griffith et al. (1992) reported that the Forest Service of USDA was using soil quality standards including amount of soil cover, soil porosity, and organic matter content to protect long-term soil productivity on National Forest System lands.

Olson (1992) suggested that surface soil properties such as erosion phase, aggregation, organic carbon content, texture, and amount of coarse fragments coupled with subsoil properties including mechanical strength, aeration porosity, residual porosity, bulk density, permeability and rooting depth could be used to quantify and monitor changes in soil quality. They further suggested that soil quality thresholds could be set for each indicator depending on the effect of a change in that indicator on soil productivity. Hornsby and Brown (1992) reviewed the soil properties most important for determining the fate and transport of pesticides and suggested organic matter content, ion exchange capacity, type and amount of clay minerals, metal oxide content, pore size distribution, soil-water content, temperature, pH, and bioactivity as important parameters.

Alexander and McLaughlin (1992) suggested that changes in soil structure were particularly important indicators of change in soil quality on forests and rangelands and suggested the use of bulk density and cone penetrometer reading to monitor changes in structure. Granatstein and Bezdick (1992) stressed the need to integrate a combination of soil tests into a meaningful index that correlates with productivity, environmental, and health goals. They suggested that indicators of improved soil quality included increases in infiltration, macropores, aeration, biological activity, water-holding capacity, aggregate stability, and soil organic matter. Decreases in bulk density, runoff, erosion, nutrient losses, soil resistance, diseases, and production costs were also suggested as indicators of improving soil quality.

Physical and chemical indicators of soil quality were suggested by Arshad and Coen (1992) including soil depth to a restricting layer, available water-holding capacity, bulk density, penetration resistance, hydraulic conductivity, aggregate stability, organic matter, nutrient availability, pH, electrical conductivity, and exchangeable sodium. Visser and Parkinson (1992) noted that indicators of biological activity are less well developed than physical and chemical properties, and suggested that ecosystem processes such as carbon cycling, nitrogen cycling, nutrient leaching from soils, and soil enzymes may be the most useful indicators of soil microbial activity. Stork and Eggleton (1992) suggested that measures of changes in soil invertebrate populations including the abundance, biomass, and density of keystone species or of selected orders and classes of invertebrates along with species richness of dominant groups of soil invertebrates could serve as useful indicators of soil biological activity.

Finally, Reagnold and colleagues (1993) compared the effect of different farming systems on soil quality by measuring differences in texture, structure, bulk density, penetration resistance, percent carbon, respiration rates, mineralizable nitrogen, ratio of mineralizable nitrogen to carbon, topsoil thickness, cation exchange capacity, total nitrogen and phosphorus, exchangeable phosphorus, sulfur, calcium, magnesium, potassium, and pH.

There are many soil properties that may serve as indicators of soil quality, as shown by the diverse list of indicators suggested by the authors cited above. The minimum data set, however, need only include those indicators that are most generally applicable to soils in varying climates and landscapes. Additional indicators could be added to the minimum data set to address properties that are particularly important in certain types of soils or regions. Table 5-2 presents a list of indicators that may be most useful for a minimum data set. The indicators suggested in the table are those that have been commonly recommended or used by several authors and should serve as a useful starting point for the development of a system to monitor changes in soil quality. A brief discussion of each suggested indicator follows.

*Nutrient Availability*

Nutrient availability is an important soil attribute for plant productivity and water quality and is significantly altered by soil management practices. Nutrient availability can be estimated by extracting nutrients

TABLE 5-2   Indicators of Change in Soil Quality and Their
Relationship to Soil Functions

| Soil Quality Indicator | Soil Functions | | |
|---|---|---|---|
| | Promote Plant Growth | Regulate Water Flow | Buffer Environmental Changes |
| Nutrient availability | Direct | Indirect | Direct |
| Organic carbon | Indirect | Indirect | Direct |
| Labile carbon | Indirect | Direct | Direct |
| Texture | Direct | Direct | Direct |
| Water-holding capacity | Direct | Direct | Indirect |
| Soil structure | Direct | Direct | Indirect |
| Maximum rooting depth | Direct | Indirect | Indirect |
| Salinity | Direct | Direct | Indirect |
| Acidity/alkalinity | Direct | Direct | Indirect |

from the different components in the soil with chemicals and measuring
the nutrient content in the extract. Nitrogen, phosphorus, and potassium are the major nutrients in the soil that are measured by extraction.

*Organic Carbon*

Soil organic carbon or soil organic matter is perhaps the single most
important indicator of soil quality and productivity. Depletion of soil
organic carbon is followed by depletion of plant nutrients, deterioration
of soil structure, diminished soil workability (Frye, 1987), and lower
water-holding capacity of the soil. The amount of organic carbon in the
soil affects permeability, water retention, and hydraulic conductivity,
which all determine the way rainfall is portioned and potential pollutants transported. It also alters the efficacies and fates of applied
pesticides. Depletion of soil organic carbon and erosion are interrelated,
since a decrease in organic carbon increases the susceptibility of a soil to
erosion, thereby increasing the rate of depletion of soil organic carbon.

Because of its importance and its susceptibility to change by soil
erosion, organic carbon should be included in the minimum data set and
monitored periodically. Total organic carbon in the soil can be affected
by management and has been shown to be directly related to the
amount of organic matter added to the soil in crop residues, manures, or
other sources (Larson and Stewart, 1992). The total organic matter in the
soil may change slowly, however; even changes restricted to the few
millimeters of surface soil can have substantial effects on infiltration,

aeration, and erosion (Bruce et al., 1988). Organic residues and soil organic matter at the interface between the soil and atmosphere are extremely influential in partitioning water and influencing surface soil structure. Tillage and residue management can stratify the organic carbon content at various levels on or within the soil.

## Labile Carbon

Although total organic carbon provides important information, it is the labile carbon fraction that is most active in the soil. The amount of labile carbon is most directly related to important biological processes in the soil including rates of mineralization of nutrients, the generation of soil structure, and the attenuation of potential pollutants. Simple chemical procedures to assess the labile carbon fraction are available.

## Texture

The soil's texture (particle-size distribution) in the surface soil layer may be altered as a result of the selective removal of fine particles during the erosion process, as a result of mixing subsoil into the surface layer during tillage (as cumulative erosion reduces the thickness of the surface soil layer) and as a result of deposition of eroded sediments on the soil surface. Changing the surface soil texture can have important effects on crop productivity (Frye, 1987; Lal, 1987), for example, by reducing the amount of nutrients or water the soil can hold or by restricting the growth of plant roots. Texture also influences the partitioning of rainfall and the flow of water and potential pollutants through the soil.

## Water-Holding Capacity

An important attribute of a soil is its ability to store and release available water to plants. The importance of water available to plants and its measurement were discussed by Ritchie (1981). Plant-available water capacities are a required input for nearly all crop simulation models. The plant-available water capacities should be determined to the depth of rooting, and temporal changes in plant-available water capacities—those that are either natural or induced by management or erosion—should be determined in the surface layers.

Water-holding capacity is also directly related to the effect of changes in soil quality on water quality. Water-holding capacity is related to the rate at which water enters and leaves the soil. The rate and direction of water flow through soils is an important factor determining the effect of farming practices on water quality. Models for estimating water retention curves from particle-size distributions, organic matter, and bulk density are available and were recently reviewed by Rawls et al. (1992). Management of the soil can have significant effects on water-holding capacity by changing the depth and texture of surface layers, the structure and compactness of surface and subsurface layers, and by affecting the rate of infiltration of precipitation.

*Structure*

The term *soil structure*, as broadly defined by Kay (1989), has three components. The first is structural form, which refers to the geometry of the soil pore space (porosity, pore size distribution, and pore continuity). The second is aggregate stability, which refers to the size distribution and resistance of aggregates to degradation. The third is structural resiliency, which refers to the ability of the soil structure to re-form once it has been degraded. Measurements of structural form include bulk density, macroporosity (pores >60 $\mu$m), and saturated hydraulic conductivity.

Determination of compact soil layers that impede root growth are important for determination of effective soil rooting volume. Either bulk density or penetration resistance measurements (interpreted with respect to water contents) can be used to identify root-impeding layers. Critical limits of bulk density for soils of different textures were given by Pierce and colleagues (1983). Compaction by wheeled traffic has direct and sometimes irreversible effects on soil structure. Texture, organic matter, and labile carbon are also related to structure, and soil management that results in changes in these soil attributes will also affect soil structure.

*Rooting Depth*

Soil thickness has been related to crop productivity, particularly in mine reclamation studies (Power et al., 1981). Soils in which the rooting depth is limited by the presence of a physical or chemical constraint are generally less productive. As limiting layers are moved closer to the soil surface as erosion removes the topsoil, crop productivity generally declines. Rooting depth varies by crop species, and the limits for various species were given by Taylor and Terrell (1982). Maximum rooting depth should be determined at the time of physiological maturity of the crop species under study. Management of the soil can have important effects on rooting depth. Erosion reduces rooting depth by removing layers of surface soil, and compaction reduces rooting depth by creating layers in the soil that are impenetrable by crop roots.

*Acidity and Alkalinity*

Soil pH is a measure from which many general interpretations about the chemical properties of a soil can be made. The acidity, neutrality, or alkalinity of a soil suggests the solubilities of various compounds in the soil, the relative bonding of ions to exchange sites, and the activities of microorganisms (McLean, 1982). A pH of less than 4 indicates the presence of free acids, generally from oxidation of sulfides; a pH of less than 5.5 indicates the likely occurrence of exchangeable aluminum; and a pH from 7.8 to 8.2 indicates the presence of calcium carbonate (Thomas, 1967). Acidity and alkalinity can be readily managed, in many soils, by careful management of fertilizer and lime applications.

*Pedotransfer Functions*

Prediction of the direction in which soil attributes are changing can often be made without direct measurement of the specific attribute. For example, Larson and Stewart (1992) showed that a simple regression equation could predict organic matter changes in several U.S. soils on the basis of the amount of crop residue that was added to the soil. Important hydraulic properties of soils including water retention, hydraulic conductivity, and water-holding capacity can be estimated from data on texture and organic matter content (Gupta and Larson, 1979a; Rawls et al., 1992). Larson and Pierce (1991) have suggested the development of pedotransfer functions that can be used to evaluate changes in soil quality from a minimum data set of indicators.

Pedotransfer functions could dramatically increase the utility of a minimum data set of soil quality indicators. Estimations of changes in many important soil attributes could be simulated from measures of relatively few indicators of soil quality. Larson and Pierce (1991) suggest that a review of the literature would uncover many already developed pedotransfer functions that could be used to simulate changes in soil quality. Table 5-3 provides some of the pedotransfer functions described by Larson and Pierce (1991). The development and validation of pedotransfer functions for use in simulation of soil quality should be an urgent research priority.

*Quantifying Soil Quality*

The preceding section demonstrates that a great deal of information and understanding is available to select soil attributes. The selection of soil attributes for use in a minimum data set, however, depends not only

TABLE 5-3   Some Pedotransfer Functions

| Estimate | Relationship | Reference |
|---|---|---|
| **Chemical** | | |
| Phosphate sorption capacity | $PSC = 0.4 \, (Al_{ox} + Fe_{ox})$ | Breeusma et al., 1986 |
| Cation-exchange capacity | $CEC = a \, OC + b \, C$ | Breeusma et al., 1986 |
| Change in organic matter | $C = a + b \, OR$ | Larson and Stewart, 1992 |
| **Physical** | | |
| Bulk density | $D_b = b_0 + b_1 \, OC + b_2 \, \%si + b_3 \, M$ | Bouma, 1989 |
| Bulk density | Random packing model using particle size distribution | Gupta and Larson, 1979 |
| Bulk density | $D_b = f(OC, cl)$ | Manrique and Jones, 1991 |
| Water retention | $WR = b_0 \, \%sa + b_2 \, \%si + b_3 \, \%cl + b_4 \, \%OC$ | Gupta and Larson, 1979 |
| Water retention | $WR = b_0 + b_1 \, C + b_2 \, Sy$ | Bouma, 1989 |
| Random roughness from moldboard plowing | $RR = f(\text{soil morphology})$ | Allmaras et al., 1967 |
| Porosity increase | $P = f(Mr, IP, cl, si, OC)$ | Allmaras et al., 1967 |
| **Hydraulic** | | |
| Hydraulic conductivity | $K^s = f(\text{texture})$ | Childs and Collis-George, 1950; Marshall, 1958; Millington and Quirk, 1961 |
| Seal conductivity | $SC = f(\text{texture})$ | Gupta et al., 1991 |
| Saturated hydraulic conductivity | $D_s = f(\text{soil morphology})$ | McKeague et al., 1982 |
| **Productivity** | | |
| Soil productivity | $PI = f(D_b, AWHC, pH, EC, ARE)$ | Pierce et al., 1983; Kiniry et al., 1983 |
| Rooting depth | $RD = f(D_b, WHC, pH)$ | Pierce et al., 1983 |

NOTE: Variables other than italicized coefficients are defined as follows: PSC, phosphate sorption capacity; $Al_{ox}$, oxalate extractable aluminum; $Fe_{ox}$, oxalate extractable iron; CEC, cation-exchange capacity; OC, organic carbon; C, change in organic carbon; OR, organic residue; $D_b$, bulk density; si, silt; M, median sand fraction; WR, water retention; sa, sand; cl, clay; Sy, $1/D_b$; RR, random roughness; P, porosity; Mr, moisture ratio; IP, initial porosity; $K^s$, hydraulic conductivity; SC, seal conductivity; $D_s$, saturated hydraulic conductivity; PI, productivity index; AWHC, available water-holding capacity; EC, electrical conductivity; ARE, available rooting environment; RD, rooting depth; WHC, water-holding capacity.

SOURCE: W. E. Larson and F. J. Pierce. 1991. Conservation and enhancement of soil quality. Pp. 175–203 in Evaluation for Sustainable Land Management in the Developing World. Volume 2: Technical Papers. Bangkok, Thailand: International Board for Soil Research and Management. Reprinted with permission from © International Board for Soil Research and Management.

on our understanding of the relation of those attributes to soil quality, but also on the utility of the attributes for use in sampling programs and in pedotransfer functions to provide more quantitative estimates of change in soil quality. Sampling can be expensive. It is important that the attributes selected for a minimum data set be as few as possible. Testing and empirical evaluation of proposed indicators will be required to identify those most suited for use based on the ease and accuracy with which they can be sampled, the degree of spatial and temporal variability of the attribute, their utility as parameters in pedotransfer functions, and their applicability to a wide range of soils, climates, and landscapes. The following discussion summarizes some of the experience with soil attributes in quantitative assessments of soil quality.

*Indicators of Productivity*

Pierce and colleagues (1983) used a simple model to estimate the potential soil productivity loss over time on the soils in the Corn Belt. The model was expressed as follows:

$$PI = \sum_{i=1}^{r} (A_i \cdot C_i \cdot D_i \cdot WF)$$

where *PI* is the productivity index, $A_i$ is sufficiency of available water-holding capacity, $C_i$ is sufficiency of bulk density adjusted for permeability, $D_i$ is sufficiency of pH, *WF* is a weighing factor based on root distribution, and *r* is the number of horizons in the maximum rooting depth. In the study by Pierce and colleagues (1983), $A_i$, $C_i$, and $D_i$ and were taken from the Soil Conservation Service's (U.S. Department of Agriculture) SOILS-5 data base for the soil mapping unit at each of the primary sampling units in the 1982 National Resources Inventory. The sufficiency curves for the soil attributes as used by Pierce and colleagues (1983) were modified from those presented by Kiniry and colleagues (1983). Erosion was simulated by removing the erosion amount (depth) given in the National Resources Inventory from the surface for a soil mapping unit and then adding an equal depth to the base of the 100-cm (40-inch) profile by using the attributes and values for that horizon. In this way the productivity index was computed initially and after 25, 50, and 100 years of erosion.

The study by Pierce and colleagues (1983) illustrates the usefulness of making predictions with simple models on the basis of soil quality attributes or indicators. Although the study by Pierce and colleagues

(1983) used estimated values for soil attributes given in the SOILS-5 data base, the estimations could have been enhanced if actual measured values at the primary sampling unit had been available.

### Indicators of Water Regulation

Soil quality and the changes in soil quality that occur with soil management can be expected to affect natural resource models. For example, consider the Water Erosion Prediction Project (WEPP) model (Laflen et al., 1991a,b), in which soil quality is assumed to affect both the water infiltration and the soil erosion portions of the model.

Infiltration in the WEPP model is quantified with the Green-Ampt infiltration model (a simple mathematical equation used to estimate how much and how quickly water soaks into the soil) (Laflen et al., 1991a,b). The soil capillary potential is assumed to be proportional to bulk density and soil texture (sand, clay, and porosity). The saturated hydraulic conductivity is determined by the amount of coarse fragments, the amount of soil cover, whether the soil is frozen, and whether crusting occurs. Thus, for example, an increase in the amount of clay decreases hydraulic conductivity and a decrease in bulk density increases water infiltration. A baseline bulk density is assumed to be proportional to the amount of sand, clay, and organic matter plus the cation-exchange capacity.

Erosion in the WEPP model is predicted by interrill and rill soil erodibility terms. For croplands, the interrill erodibility term is assumed to be a function of the soil's texture and the magnesium and aluminum concentrations. Rill erodibility is estimated from the soil organic matter, cation-exchange capacity, sodium absorption ratio, aluminum concentration, and soil texture (very fine sand, clay, and sand). The critical shear (the amount of water-induced shear that initiates sediment movement) is also quantified from soil properties such as the amounts of clay and sand, the specific surface area of the soil, and a sodium absorption ratio.

On rangelands, interrill erodibility is estimated from the amount of sand, silt, and organic matter and estimates of the soil's water-holding capacity. Rill erodibility is estimated from the amount of clay, organic matter, bulk density, and root mass in the soil. Critical shear is assumed to be proportional to the amount of sand, organic matter, and bulk density.

In general, improvements in such soil quality attributes as organic matter and bulk density can be expected to increase infiltration, reduce runoff, and decrease soil erosion (with the exception of interrill erodibility).

This unprotected soil is cut by numerous channels caused by water runoff. The more shallow channels, rills, will be filled in by tillage; however, the deeper channels, ephemeral rills, will remain and enhance the processes of soil erosion. Credit: U.S. Department of Agriculture.

*Indicators of Buffering Capacity*

Soil quality attributes are needed to make such estimations as the desirability of soils for use in waste management. For example, in Minnesota soils, texture, pH, total organic carbon content, and cation-exchange capacity are used as indicators of the suitability of applying processed sewage sludge to the land and the application amount. Reliable measures or estimates are required for each mapping unit in the proposed area of application. In this example, texture is used as an indicator of the susceptibility of soils to leaching of chemicals, pH is an indicator of the solubility of heavy metals and their ease of uptake by plants, and total organic carbon content and cation-exchange capacity are indicators of the capacity of the soil to absorb chemicals.

### Temporal and Spatial Variabilities

The desired frequency of measurement of the minimum data set depends on the particular use of the data as well as on the particular attribute and soil and climatic conditions. For national-level or area assessments, the frequency of measurement may be less than that if the minimum data set is to be used to guide management systems.

#### Temporal Variability

The frequency of measurement also varies between indicators, since the rates of change of the various indicators differ. Depending on the indicator selected, rates of change can vary from less than 1 year to more than 1,000 years. Some indicators selected for inclusion in the minimum data set may need to be measured more frequently than others. The frequency of measurement may also depend on both the climate and the management system used. In temperate regions considerable periods of time may be needed to measure differences in organic matter content. On the other hand, under a slash-and-burn system in the tropics, significant changes in organic matter can occur rapidly. Alarming changes in bulk density may occur during one pass of a heavy vehicle.

Arnold and colleagues (1990) categorized the fluctuations and trends apparent in soils into three groups: (1) nonsystematic or random changes; (2) regular, periodic, or cyclical changes; and (3) trend changes. Nonsystematic changes are short-term changes brought about by daily weather fluctuations or episodic human or natural disturbances. These kinds of changes are difficult to predict. Periodic or cyclical changes may be brought about by annual fluctuations in weather, crop growth periods, or soil moisture content. Monitoring of soil quality should be primarily directed toward the detection of trend changes. Trend changes show a definite tendency in a general direction over time. Such changes may include an increase or decrease in soil organic matter content, for example, and an increase or decrease in soil nutrient status. Longer-term changes might be brought about by the slow processes of soil development.

Monitoring changes in soil quality for assessment of the sustainability of current management systems and land uses requires a focus on trend changes that are measurable over a 1- to 10-year period, for example, soil water content at the permanent wilting point of plants, soil acidity, soil cation-exchange capacity, exchangeable cation content, and the ion composition of soil extracts (Larson and Pierce, 1991). The detected changes must be real, but they must change rapidly enough so that human intervention can correct problems before serious and perhaps irreversible loss of soil quality occurs. The separation of trend changes in soil quality from periodic or random changes will be a major challenge.

*Spatial Variability*

Soils and landscapes vary spatially, sometimes dramatically. This means that the choice of the basic unit for soil quality assessments is important. Three unit sizes can be considered (Larson and Pierce, 1991): (1) the local landscape unit (Van Diepen et al., 1991), (2) the soil mapping unit (Lamp, 1986), or (3) the pedon (the smallest unit or volume of soil that represents all the horizons of a soil profile) (Lamp, 1986). The local landscape unit represents a combination of topographic, climatic, and management units that typify a particular region. The soil mapping unit covers a number of hectares with soil variability that is recognized as part of the mapping unit description. The pedon may represent only a few square meters.

To date most land evaluation efforts have focused on the landscape unit. The Food and Agriculture Organization of the United Nations, for example, has suggested 25 land qualities—such as radiation, temperature, nutrient availability, rooting conditions, flood hazard, soil workability, and soil degradation hazard—as factors that can be used in the evaluation of rain-fed agricultural systems (Food and Agriculture Organization of the United Nations, 1983). In the United States, much effort has gone into the delineation and description of soil mapping units, which appear to be useful units of study. In many cases, mapping of these units has been completed, and soil surveys represent a wealth of information about the soil attributes characteristic of different mapping units. Spatial variation, however, can be substantial at the mapping unit scale because of natural soil development processes or human-induced variability. Bulk density, for example, can vary over short distances because of alternating rows and wheel tracks in a row crop field. Because of spatial variability, sampling at the pedon level, that is, within a few square meters may be needed in some cases to estimate variability of soil quality indicators.

## EXTENT OF DEGRADATION OF U.S. SOILS

A decline in soil quality results from soil degradation. Soil degradation is an outcome of human activities that deplete soil and the interaction of these activities with natural environments. The three principal types of soil degradation are physical, chemical, and biological. Each type is made up of different processes, as illustrated in Figure 5-1.

### Physical Degradation

Physical degradation leads to a deterioration of soil properties that can have a serious impact on water infiltration and plant growth. Wind and water erosion are generally the dominant physical degradation processes, but compaction is also a widespread concern in places where heavy machinery is commonly used. Hardsetting of a cultivated soil is also a process of compaction, but it results from wetting structurally weak or unstable soil rather than from the application of an external load. Laterization is the desiccation and hardening of exposed plinthitic material (material consisting of clay and quartz with other diluents; it is rich in sesquioxides, poor in humus, and highly weathered), but lateritic soils are rare in the United States.

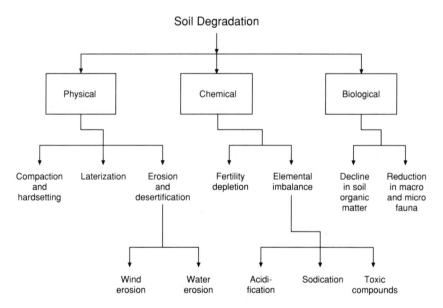

FIGURE 5-1 Processes of soil degradation. Source: R. Lal and B. A. Stewart. 1990. Soil degradation: A global threat. Advances in Soil Science 11:13–17. Reprinted with permission from © Springer-Verlag New York.

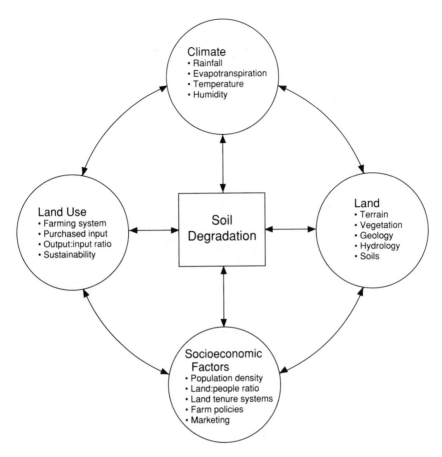

FIGURE 5-2 Interactions of factors that cause soil degradation. Source: R. Lal and B. A. Stewart. 1990a. Need for action: Research and development priorities. Advances in Soil Science 11:331–336. Reprinted with permission from © Springer-Verlag New York.

Soil degradation is a complex phenomenon driven by strong interactions among socioeconomic and biophysical factors (Figure 5-2). Soil degradation is fueled worldwide by increasing human populations, fragile economies, and misguided farm policies. There is also often a conflict between short-term benefits and long-term consequences. An example in the United States is the difficulty of developing sustainable agricultural systems.

The physical and chemical attributes of a soil can be protected from degradation by a number of general types of management practices.

The Badlands of South Dakota is an example of the more extreme long-term effects of erosion by water and wind. The area is primarily rocky spires, buttes, and gorges embedded with the petrified remains of prehistoric camels, three-toed horses, and saber-toothed tigers. Credit: The South Dakota Department of Tourism.

These include crop rotation, fertilizer and lime applications, tillage, residue management, strip cropping, and mechanical practices. One or several of these practices used in various combinations is usually needed. The choice of practices will depend on the soil, landscape, crop sequence, climate, and social conditions.

### Erosion

Soil erosion is a natural phenomenon that has occurred since Earth was formed. Erosion by water and wind has helped shape the land-scapes that people know today. Quantitative studies of the amount of erosion that occurred during periods of geologic and historic time show that the rate is highly variable in space and time. This variability can be caused by external factors, such as changes in climate and vegetation, or by internal factors that result in episodic erosion. Usually, however, erosion rates under most current farming systems are much greater than they were before farming began and are greater than those in uncultivated areas. According to a review by Franzmeier (1990), erosion rates in the central United States before European settlement varied from 0.02 to 11 metric tons/ha/year (0.009 to 5 tons/acre/year), but postsettlement rates have varied from 7 to 86 metric tons/ha/year (3 to 38 tons/acre/ year).

Erosion represents the major agent of soil degradation worldwide, although the amounts of erosion and the damage that it causes are difficult to quantitate (Dudal, 1982; Lal, 1990). Although damage from erosion was recognized as a serious threat to agriculture in U.S. colonial times, rapid expansion of agriculture to new lands west of the Allegheny Mountains lessened the need for careful husbandry of soil resources. Sporadic reports of the effects of soil erosion in the United States occurred in the literature in the early part of the twentieth century (McDonald, 1941), but no concerted national effort for control occurred until the 1930s when the Soil Erosion Service of the USDA was founded. The Soil Erosion Service was quickly reorganized into the Soil Conservation Service and was then expanded.

Damage from water erosion on croplands, although widespread, generally increases as the slope (steepness and length) of the land increases. Other factors that affect the rate of erosion are rainfall amount and intensity, the nature of the soil, the cropping and tillage practices used, and mechanical farming practices such as terracing and contouring.

### Erosion Estimates

Many estimates of the amounts of erosion and the consequences of erosion have been made since the formation of the Soil Conservation Service. The early estimates were made on the basis of a minimum of quantitative data (U.S. Department of Agriculture, 1938). The most comprehensive analysis of the amounts of wind and water erosion was contained in the 1982 National Resources Inventory published in the Second Resource Conservation Act Appraisal (U.S. Department of Agriculture, Soil Conservation Service, 1989a), which was in response to the Resource Conservation Act passed by Congress in 1977.

The 1977 National Resources Inventory concentrated on the amounts of erosion from water and comparison of the amounts of erosion with the established soil loss tolerance values. Erosion was estimated by statistically establishing primary sampling units, identifying the appropriate coefficients for the universal soil loss equation, and solving the equation to estimate the amount of soil eroded.

The calculated erosion amounts represent movement of soil material from the sampled point. The computed amounts of erosion tell nothing about the eventual deposition of the eroded sediment. The sediment may be moved a few meters to other places in the field or to nearby riparian areas or wetlands or it may be deposited in streams. The National Resources Inventory estimates of 1982 were made for sheet and rill erosion by water as well as wind erosion and did not include ephemeral, gully, or stream bank erosion, all of which are significant sources of sediment.

In the Soil Conservation Service's second appraisal of the Resource Conservation Act (U.S. Department of Agriculture, Soil Conservation Service, 1989a), the average amount of soil lost through sheet and rill erosion on all cropland (170 million ha, or 421 million acres) was 9.8 metric tons/ha/year (4.4 tons/acre/year) and the average amount lost through wind erosion was 6.7 metric tons/ha/year (3.0 tons/acre/year). Areas within the United States with particularly serious erosion include the Palouse (Washington, Oregon, Idaho), southeastern Idaho, the Texas Blackland Prairie, the southern Mississippi Valley, the Corn Belt, and Aroostock County, Maine. Nationwide in 1982, 40 percent of cropland was eroding at rates higher than the soil loss tolerance level, and 20 percent was eroding at rates higher than twice the soil loss tolerance level. Nearly 9 percent of pasturelands and 18 percent of rangelands were eroding excessively. About 43 million ha (106 million acres) of cropland was considered highly erodible on the basis of the results of the 1982 National Resources Inventory.

Wind erosion is a severe problem in the western half of the United States and in certain sandy areas in the eastern part of the country. The National Resources Inventory estimated that wind erosion for all croplands was 6.7 metric tons/ha/year (3.0 tons/acre/year). However, in Texas, average wind erosion amounts were 29.3 metric tons/ha/year (13.1 tons/acre/year); average wind erosion rates in Colorado, Nevada, and Montana were 20.8 (9.3), 20.6 (9.2), and 18.6 metric tons/ha/year (8.3 tons/acre/year), respectively. Wind erosion is greatest in the Great Plains and Mountain States.

*Effect of Erosion on Soil Quality*

Sheet and rill erosion significantly reduced the productivity indices of Pierce and colleagues (1983) on most of the 75 soils selected from the north-central United States after 25 and 50 cm (10 and 20 inches) of soil

Unprotected soil is vulnerable to wind erosion. Here sandy soil is blown over and between rows of crops. Credit: U.S. Department of Agriculture.

was removed from the surface by erosion. Fifty percent of the soils exhibited a reduction in the productivity index of more than 0.1; 32 percent of the soil exhibited a reduction of more than 0.2; and 16 percent exhibited a reduction of more than 0.3 when 50 cm (20 inches) of the soil was eroded (Larson et al., 1985). In the productivity index model, the productivity indices range from 0 to 1.0, with 1.0 being the most productive soil.

The lack of direct measurements of soil attributes that can be linked to changes in soil quality make assessments of soil quality degradation caused by erosion difficult. Some analyses are available for some sections of the United States and are suggestive of the effects of current erosion rates on soil quality.

Larson and colleagues (1972) added a variety of organic residues at various rates to a Typic Hapludoll in Iowa for 11 years while cropping the soil to corn (*Zea mays* L.) by moldboard plowing. At the end of the 11 years, the organic carbon content varied linearly with the amount of residues added. Rasmussen and Collins (1991) have shown a similar

The effects of sheet erosion are subtle and difficult to recognize. The sheeting action of the runoff has deposited thin layers of eroded soil in the foreground and in the background (note the darker bands across the rows of crops). Even though difficult to recognize, sheet erosion can be quite damaging. Credit: U.S. Department of Agriculture.

linear relationship for the Palouse region of Oregon and Washington. From these linear equations, they calculated the annual amount of residue that was required to be returned to the soil to maintain the organic content at the level present at the start of the experiment. The linear relationships developed by Larson and colleagues (1972) and Rasmussen and Collins (1991) were used to estimate the changes in organic matter content presented in Table 5-4. They also calculated the average amount of organic carbon in the residues produced from corn (*Zea mays* L.) in major land resource area 107 (Iowa) and wheat (*Triticum aestivum* L.) in major land resource area 9 (Oregon, Washington), assuming a harvest ratio of 1.0 for corn and 1.5 for wheat (*Iridium aestivum* L.). Because changes in the amount of organic carbon in the soil are difficult to measure, the ratio between the amount of organic carbon in crop residues and the amount needed to maintain organic matter in the soil is a useful index of whether organic carbon is increasing or decreasing in a given major land resource area. For the four areas described in Table 5-4, the amount of residues returned to

TABLE 5-4  Organic Carbon Additions Necessary to Maintain Soil
Organic Carbon at Present Levels at Several Locations

| Location | Crop Rotation[a] | Amount of Organic Carbon (kg/ha/year) | | Harvest Ratio |
| | | To Maintain Soil Organic Carbon | Produced from Crops | |
| --- | --- | --- | --- | --- |
| Shenandoah, Iowa | C-C | 3,272 | 3,020 | 0.9 |
| Pendleton, Oregon | W-F | 2,288 | 1,764 | 0.8 |
| Pullman, Washington | W-F | 1,933 | 1,764 | 0.9 |
| Pullman, Washington | W-W | 780 | 3,528 | 3.9 |

[a]C, corn; W, wheat; F, fallow.

SOURCE: Adapted from W.E. Larson and B.A. Stewart. 1992. Thresholds for soil removal for maintaining cropland productivity. Pp. 6–14 in Proceedings of the Soil Quality Standards Symposium, San Antonio, Texas, October 23, 1990. Washington, D.C.: U.S. Department of Agriculture, Forest Service.

the soil either exceeded or nearly equaled the amount required to maintain organic matter levels in the soil, a key indicator of trends in soil quality.

Erosion, however, removes organic carbon along with sediments. Table 5-5 provides estimates of the organic carbon needed in crop residues to replace organic carbon lost from soil as a result of different average erosion rates and different slope classes. The data in Table 5-5 indicate that at any erosion rate exceeding 5 metric tons/ha/year (2.2 tons/acre/year), the amount of residues returned to the soil from corn production is not enough to prevent declines in organic carbon content. On lands with steeper slopes, the amount of residuesrequired far exceeds that produced by corn or other grain crops. Hence, on lands with steeper slopes where erosion is severe, the organic carbon content will decline to low levels.

Since the organic carbon content is an important indicator of soil quality, the analyses presented above suggest that current rates of erosion may have significant effects on long-term soil quality.

Soil erosion influences most of the soil attributes that determine soil quality. Eroded sediments usually contain higher amounts of plant nutrients than do bulk soils, thus degrading the soil of the important attributes of nitrogen, phosphorus, potassium, and total organic carbon (Barrows and Kilmer, 1963; Young et al., 1985). Erosion can also bring soil horizons closer to the surface of the soil profile. These horizons usually have low pHs, low available water-holding capacities, and high bulk densities and can thus influence soil quality. Using

TABLE 5-5 Amounts of Organic Carbon Needed Annually in Residues to Maintain Soil Organic Carbon on Lands with Different Slopes and Erosion Levels[a]

| Area (1,000 hectares) | Slope (percent) | Average Erosion (metric tons/ha/year)[b] | Organic Carbon (kg/ha) | |
|---|---|---|---|---|
| | | | In sediment[c] | Needed in Residue |
| 853 | 0-2 | 5 | 135 | 1,900 |
| 1,157 | 2-6 | 18 | 486 | 6,840 |
| 819 | 6-12 | 61 | 1,647 | 23,180 |
| 376 | 12-20 | 114 | 3,078 | 43,320 |

[a]Major land resource area 107 (Iowa and Missouri Deep Loess hills).

[b]From U.S. Department of Agriculture, Soil Conservation Service. 1982. Basic Statistics 1977 National Resources Inventory. Statistical Bulletin No. 686. Washington, D.C.: U.S. Department of Agriculture.

[c]Enrichment ratio of 1.5; organic carbon in soil = 1.8 percent.

SOURCE: Adapted from W. E. Larson and B. A. Stewart. 1992. Thresholds for soil removal for maintaining cropland productivity. Pp. 6–14 in Proceedings of the Soil Quality Standards Symposium. Watershed and Air Management Report No. WO-WSA-2. Washington, D.C.: U.S. Department of Agriculture, Soil Conservation Service.

the productivity index model of Pierce and colleagues (1983), Larson and colleagues (1985) calculated which of the four soil attributes in the subsoil—available water-holding capacity, bulk density, pH, or rooting depth—would cause the greatest decline in soil productivity on 75 major soils of the Corn Belt, assuming erosion removed 50 cm (20 inches) of soil from the surface. Of the 75 soils, the productivity index decreased significantly (for example, the productivity index was less than 0.1) in 37 of the soils. Thirteen of the soils showed a significant degradation in the available water-holding capacity in the subsoil, 4 were degraded because of increased bulk density, 7 were degraded because of decreased rooting depth, and 13 were degraded because of both bulk density and decreased rooting depth.

Rijsberman and Wolman (1985) reported that nutrients and total organic carbon, in addition to available water-holding capacity, pH, and bulk density, were attributes readily degraded by erosion and were important in maintaining soil productivity. Maintenance of total organic carbon was important in preventing the formation of soil surface crusts.

The specific soil quality attributes degraded by erosion depend on the soil characteristics, climate, and amount of erosion. In most cases, erosion reduces the quality of more than one attribute.

## Compaction

One aspect of soil degradation that is of increasing concern is soil compaction caused by the wheeled traffic involved in normal farming operations.

### Surface Soil Compaction

Most of the farm machinery in use today is sufficiently heavy to cause compaction in the surface 10 to 20 cm (4 to 8 inches) of soil. Although tillage operations after the use of heavy machinery are often sufficient to alleviate compaction, the increasing use of no-till and ridge-till farming systems can result in areas within a field that remain relatively dense in the surface 10 cm (4 inches), despite annual freezing (Voorhees, 1983). Under these conditions, the soil water runoff control benefits associated with reduced tillage can be lost (Lindstrom and Voorhees, 1980; Lindstrom et al., 1981). These studies showed that interrow wheeled traffic during spring planting operations may negate the beneficial tillage management effects and compact the soil to the point of significantly decreasing infiltration rates in those interrows. Young and Voorhees (1982) reported that about 34 percent of the total runoff and 49 percent of total soil loss from a bare field can originate from the 22 percent of the field surface that is used as wheel tracks during planting operations. A concentration of plant residues in the wheel-trafficked interrows may be a practical solution.

Plant growth response is another aspect of surface layer soil compaction (less than 30 cm [12 inches] deep) that relates to soil degradation. Several researchers have reported a wide range of growth responses to surface layer compaction by a variety of crops (Draycott et al., 1970; Fausey and Dylla, 1984; Johnson et al., 1990; Van-Loon et al., 1985; Voorhees et al., 1990). The theory is that the crop yield response to surface compaction should follow a parabolic relationship (Soane et al., 1982), inferring that there is an optimum degree of compactness for maximum crop yield. There is evidence for this inference (Voorhees, 1987), and efforts are under way to develop the technology needed to assess the economic extent of soil compaction in the Corn Belt (Eradat Oskoui and Voorhees, 1990). Even though surface soil compaction may be economically important because it decreases crop yields, it is potentially manageable because surface layer compaction can be alleviated by normal tillage equipment. In systems that do not require annual tillage, the detrimental effects of surface compaction on either runoff and erosion or crop yield can be minimized with the application of interrow plant residues and proper management of wheeled traffic.

### Subsoil Compaction

Soil compaction deeper than the normal tillage depth (subsoil compaction) is a much more serious consequence of modern agricultural production and should be considered an important factor in soil degradation. The reasons are threefold: (1) subsoil compaction persists longer than surface soil compaction, (2) the trends of increasing farm and farm machinery sizes tend to worsen the potential difficulties with subsoil compaction, and (3) it is difficult and costly to remedy subsoil compaction by mechanical means.

The previously cited research reporting the persistence of surface soil compaction, despite annual freezing and thawing cycles, forewarns of even more persistence of compaction at deeper depths. More than one freeze-thaw cycle may be required to ameliorate compacted soil. As the soil depth increases, the number of freeze-thaw cycles decreases. In northern latitudes, the front extends to a deeper depth than in warmer climates, but the volume of soil subjected to more than one freeze-thaw cycle is greatly diminished.

Lowery and Schuler (1991) reported that a 11.3-metric ton (12.5-ton) axle load causes a significant and persistent increase in penetrometer (an instrument used to measure soil compaction) resistance in the subsoil of silt loam and silty clay loam soils in Wisconsin for 4 years. In Sweden, increased vane shear resistance was measured in the subsoil 7 years after application of a 9-metric ton (10-ton) axle load (Hakansson, 1985). Higher bulk density was still measurable 9 years after compacting the subsoil of a clay loam in Minnesota (Blake et al., 1976). In all of those studies, the subsoil went through at least one freeze-thaw cycle each winter.

Current harvest equipment in the Corn Belt ranges from about 9 metric tons per axle (10 tons per axle) for an empty six-row combine to 36 metric tons per axle (40 tons per axle) for a loaded grain cart. Many hard-surfaced public highways have axle load limits ranging from 5 to 8 metric tons (6 to 9 tons). The subsoil compacting effect of the increased axle weights of current farm machinery can be partially offset by increasing the surface areas of the tires that carry the load. However, with current machinery design, there are practical limits to this approach. Prototype models of new tire track designs that reduce subsoil compaction have not been proven in the field. Meanwhile, the mechanical forces applied to soils will likely continue to increase, as will the potential for increasing subsoil compaction.

*Alleviation of Subsoil Compaction*

Since tillage is often shown to be an effective way of alleviating surface compaction, subsoil tillage should alleviate subsoil compaction. However, deep tillage experiments in Iowa and Illinois (Larson et al., 1960) and more recently in Minnesota (Johnson et al., 1992) show that while deep tillage effectively loosens the soil, it does not automatically lead to increased crop yields. Data from the research in Minnesota showed that deep tillage (about 55 cm [22 inches]) followed by a relatively dry growing season results in a 1,176-kg/ha (15-bushel/acre) decrease in corn yield. Deep tillage is an expensive operation, so a considerable increase in crop yield is needed to pay for the operation. The inconsistent effects of deep tillage on crop yield, coupled with the slow rate at which natural forces ameliorate a compacted subsoil, emphasize the potential degrading effect that wheeled traffic-induced soil compaction can have on productivity.

*Corn Yield Response to Subsoil Compaction*

A series of field experiments recently conducted across the Corn Belt of the United States and southern Canada showed that wheeled traffic with axle loads typical for harvest operations can cause soil to be excessively compacted to depths of 60 cm (24 inches). The subsoil compaction can persist for a number of years, despite annual freezing and thawing, and crop yields may be decreased for a number of years after a one-time application of heavy wheeled traffic. Data from the Minnesota site illustrate the response. In the fall of 1981 a Webster clay loam in southern Minnesota was trafficked with a load of 18 metric tons per axle (20 ton/axle). The surface 20 cm (8 inches) was intensively tilled to alleviate surface compaction. All subsequent wheeled traffic on the plots was limited to an axle load of less than 4.5 metric tons (5 tons). Corn yields were then measured for the next 9 years. Corn yields were significantly reduced by 30 and 13 percent in the first and second years, respectively, after heavy wheeled traffic. The yields were reduced by 7 and 3 percent in the third and fourth years, respectively, but the reductions were not statistically significant. There were no significant yield responses the fifth, sixth, or eighth years, but yields were significantly decreased by 15 percent in year 7 (a relatively dry year) and year 9 (a relatively wet year). Ignoring the yield data for the first year, which may have been a combination of surface soil and subsoil compaction effects, the average yield reduction over the 8 years was 6 percent. Yield responses at other sites across the Corn Belt were similar or even more negatively affected by subsoil compaction.

There are two facts that must be considered in extrapolating these data to whole-field situations. First, the entire plot surface was tracked with wheeled traffic four times at the beginning of the experiment. Producers do not do this during normal farming operations; thus it could be argued that the experimental yield responses overestimate the real farm situation. However, the wheel tracks from a six-row combine alone cover about 27 percent of the field. A typical grain cart for a combine of that size also tracks about 27 percent of the field if it is pulled beside the combine for on-the-go unloading. Then, the tractor pulling the grain cart must be considered. When these three types of wheeled traffic are considered in total, a major portion of a field may be covered with heavy wheeled traffic or a significant portion of the field may be covered with wheeled tracks more than once. Thus, the actual situation in the field may not be so different from the experimental conditions outlined above.

The second difference between plots and whole fields is that the producer puts heavy wheeled traffic on the field every harvest season, whereas it was applied only once on the experimental plots. Since natural forces are relatively ineffective in ameliorating the subsoil compaction in 1 year, it can be argued that subsoil compaction in a real farming situation may be long lasting, if not permanent.

If one is willing to accept the assumption that the experimental data are somewhat typical for a real farming situation, and conservatively extrapolating the long-term 6 percent average plot yield reduction in Minnesota to 30 percent of a given field and 50 percent of the corn acreage in the states of Minnesota, Wisconsin, Iowa, Illinois, Indiana, and Ohio, the annual monetary loss caused by subsoil compaction is estimated at about $100 million (assuming corn prices at $63/metric ton [$2/bushel]). It could be more in high-stress years, when root growth is limited because of either too much or too little water.

## Chemical Degradation

Chemical degradation processes can lead to a rapid decline in soil quality. Nutrient depletion, acidification, and salinization are common soil degradation processes in the United States that have had a serious impact on crop production. Chemical degradation is also caused by the buildup of toxic chemicals resulting from human activities.

### Salinization

Investigators normally distinguish between saline and sodic soils. Saline soils suffer from an excess of salinity caused by a range of ions. When

TABLE 5-6   Extent of Salinity and Associated Problems by Land Use in California

| Primary Land Use | Millions of Hectares | | | |
| --- | --- | --- | --- | --- |
| | Nonfederal Land Area | Saline or Sodic Soils[a] | High Water Table[b] | Water Quality |
| Irrigated cropland | 4.01 | 1.18 | 1.09 | 1.38 |
| Dry cropland | 0.73 | 0.00 | 0.04 | 0.04 |
| Grazed land | 7.94 | 0.32 | 0.16 | 0.16 |
| Timberland | 3.60 | 0.00 | 0.00 | 0.04 |
| Wildlife land | 0.49 | 0.08 | 0.04 | 0.08 |
| Urban | 2.02 | 0.04 | 0.04 | 0.08 |
| Other | 3.64 | 0.08 | 0.04 | 0.12 |
| Total | 22.51 | 1.70 | 1.41 | 2.02 |

[a]Areas with electrical conductivity of 4 dS/m (about 2.500 mg/L) or greater and/or exchangeable sodium values of more than 15 percent.

[b]High water table indicates a depth of 1.5 m or less or at a depth that affects the growth of commonly grown crops. Includes parameters such as salinity or boron toxicity.

SOURCE: U.S. Department of Agriculture, Soil Conservation Service. 1983. California's Soil Salinity. Davis, Calif.: U.S. Department of Agriculture, Soil Conservation Service.

sodium is the prevalent cation, the soils are generally classified as sodic.

Salinity problems are not restricted to irrigated areas. In fact, huge dryland areas suffer from salinity and/or sodicity problems. In reverse, all irrigated areas in arid (and semiarid) regions are subject to salinization if adequate drainage is not provided.

Investigators have attempted to inventory the extent of salinity problems and to establish trends. The data base for the United States is weak at best; many of the figures are only estimates. Large-scale mapping projects in Europe and other continents provide a more reliable data base. According to Szabolcs (1989), the total area of salt-affected soils in the world approaches 1 billion ha (2.5 billion acres).

Postel (1989) has made a different estimate, but a large part of the difference is that Postel's estimate included only irrigated lands, while Szabolcs' estimate included nonirrigated lands.

Another estimate comes from the Soil Conservation Service for California (Backlund and Hoppes, 1984). Backlund and Hoppes reported that the area of the San Joaquin Valley with salinity problems would increase to 1.46 million ha (3.6 million acres) by the year 2000 (Tables 5-6 and 5-7).

Although good statistics are hard to find, it is the consensus of specialists that, worldwide, the salinity problem continues to increase substantially. In the United States, contamination of irrigation drainage

TABLE 5-7   Salinity and Drainage Problems by Major Irrigated Areas (approximate area)

| Location | Millions of Hectares | | | |
| --- | --- | --- | --- | --- |
| | Irrigated Area | Saline or Sodic Soil[a] | High Water Table[b] | Water Quality |
| San Joaquin Valley | 2.27 | 0.89 | 0.61 | 0.93 |
| Sacramento Valley | 0.85 | 0.08 | 0.16 | 0.12 |
| Imperial Valley | 0.20 | 0.08 | 0.20 | 0.20 |
| Other areas | 0.77 | 0.12 | 0.12 | 0.12 |
| Total | 4.09 | 1.17 | 1.09 | 1.37 |

[a]Areas with electrical conductivity of 4 dS/m (about 2.500 mg/L) or greater and/or exchangeable sodium values of more than 15 percent.

[b]High water table indicates a depth of 1.5 m or less or at a depth that affects the growth of commonly grown crops. Includes parameters such as salinity or boron toxicity.

SOURCE: U.S. Department of Agriculture, Soil Conservation Service. 1983. California's Soil Salinity. Davis, Calif.: U.S. Department of Agriculture, Soil Conservation Service.

water with toxic trace elements has been added to concerns about salinity. This newly identified component has led to greater emphasis on the off-site effects of irrigation as opposed to the on-site effects of salinization. Thus, sustainability and evaluation of the impacts of salinity caused by irrigation have taken on an entirely new aspect (van Schilfgaarde, 1990). It is too early to give reliable estimates of the areas affected, but it is not too early to recognize this potentially serious problem (National Research Council, 1989b).

*Acidification*

Acidity is an important attribute of a soil because it influences many of the chemical and biological reactions that occur in the soil. Through these reactions, the pH influences plant growth. Soil pH influences microbial populations and activities and thus is important in buffering environmental reactions. Soils become more acidic when bases (for example, calcium, magnesium, potassium, and sodium) are leached from the soil and replaced by hydrogen ions on the exchange complex. In humid regions soils usually become more acidic with time, even if they are uncultivated. Many cultivated soils in the eastern, southeastern, and midwestern United States were too acidic for optimum plant growth when they were first cultivated from the native forests and prairies. Without the addition of lime, they have become more acidic under cultivation as a result of leaching and the addition of nitrogenous fertilizers.

The pH of a soil is a reflection of the nature of the cations on the exchange complex. Soils with pHs of less than 7.0 (acidic soils) have hydrogen in exchangeable form, whereas those with pHs of greater than 7.0 (alkaline soils) have exchange complexes that are dominated by bases, usually calcium and magnesium. Soils with pHs of less than about 5.5 may have significant amounts of exchangeable aluminum. Soils with free calcium and magnesium carbonates usually have pHs of greater than 8.2.

In the soil's native state in the humid regions of the United States, the lowest pH is often below the tilled layer, whereas in arid areas, the highest pH is below the tilled layer. Corrective management practices such as application of lime to raise the pHs of the acid soils or lower the pHs of sodic soils is common. It is more difficult to alter the pH below the tilled layer.

Some poorly drained soils contain significant concentrations of pyrite (iron disulfide), which oxidizes to sulfuric acid when it is drained of water, creating unusually low soil pHs. These soils usually have pHs below 3.5.

Acidic soils may limit plant growth because they have insufficient calcium or magnesium or toxic concentrations of exchangeable aluminum or because they decrease the availability of certain essential nutrients. As soils become more acidic, the microbial populations tend to shift from bacteria to fungi, changing the decomposition rates of soil organic matter and organic residues. Acidic soil conditions often cause a reduction in the amount of nitrogen fixed by legumes. Sodic soils may limit plant growth by having toxic concentrations of exchangeable sodium or sodium concentrations that keep the soil dispersed and that maintain a poor soil structure.

Application of nitrogenous fertilizers can lower pHs in both surface soils and subsoils (Pierre et al., 1970). The use of large amounts of nitrogenous fertilizers has accentuated the lowering of pHs on croplands in humid regions. Movement of acidity to depths below the tilled layer is of particular concern because of the difficulty in modifying the acidity in lower soil layers.

A pH near neutrality (pH 6.5 to 7.5) is usually considered best for plant growth. Many soils in the humid regions of the United States are acidic, with pHs ranging from 7.0 to 5.0 or lower. Soils in arid regions may have pHs greater than 8.5, which usually indicates excessive amounts of exchangeable sodium. The optimum soil pH for plants varies. The optimum pH ranges for selected field crops, for example, are: corn, 5.5 to 7.5; soybeans, 6.0 to 7.0; wheat, 5.5 to 7.5; oats, 5.0 to 7.5; sorghum, 5.5 to 7.5; alfalfa, 6.2 to 7.8; and sweet clover, 6.5 to 7.5.

The acidity of a soil may be reduced (pH increased) by the addition of basic materials. Application of ground limestone (calcium and magnesium carbonates) is most common. The amount of ground limestone needed to raise the pH to acceptable levels depends on the initial pH of the soil, the desired pH for crop growth, the texture of the soil, and the soil's clay properties. The amount of lime required to raise an 18-cm (7-inch) layer of a silt loam soil from pH 5.5 to 6.5 in the northern and central United States is about 5.2 metric tons/ha (2.3 tons/acre); the value is 3.6 metric tons/ha (1.6 tons/acre) in the southern United States (Kellogg, 1957).

Approximately 30 million metric tons (33 million tons) of pulverized limestone was applied to soils in the United States in 1980 at a cost of $180 million. Assuming that the lime was applied at 1.8 metric tons/ha (2 tons/acre), it would be applied to 7 million ha (16.5 million acres) of the approximately 162 million ha (400 million acres) of cropland. This is probably a small fraction of what is needed for maximum crop production. Pierre and colleagues (1970) concluded that it requires about 300 kg (660 lb) of calcium carbonate to neutralize the acidity produced in about 1 metric ton (1 ton) of ammonium nitrate fertilizer.

Figure 5-3 shows the percentage of U.S. soils with pHs of 6 or less. Soils of the northeast, southeast, and northwest have higher percentages of acidic soils. The proportion of states east of the Mississippi River with soils that have pHs of 6.0 or less range from 13 percent in Wisconsin to 75 percent in New Hampshire and Vermont. Management of acidic subsoils with high amounts of toxic exchangeable aluminum, which restricts root growth, is a major problem in the Southeast. Relatively few soils in the Great Plains have pHs of less than 6.0. The soils west of the Cascade Mountains in the Pacific Northwest are usually acidic.

## Biological Degradation

Biological degradation includes reductions in organic matter content, declines in the amount of carbon from biomass, and decreases in the activity and diversity of soil fauna. Biological degradation is perhaps the most serious form of soil degradation because it affects the life of the soil and because organic matter significantly affects the physical and chemical properties of soils. Biological degradation can also be caused by indiscriminate and excessive use of chemicals and soil pollutants. Biological degradation is generally more serious in the tropics and subtropics than it is in temperate zones because of the prevailing high soil and air temperatures. Tillage also stimulates biological degradation because it increases the exposure of organic matter to decomposition processes.

FIGURE 5-3   U.S. pH soil test summary as percentage of soils testing 6.0 or less in 1989. Source: Potash and Phosphate Institute. 1990. Soil test summaries: Phosphorus, potassium, and pH. Better Crops with Plant Food 74(2):16–18. Reprinted with permission from © Potash & Phosphate Institute.

### Organic Matter Content

Although losses of mineral and organic soil particles through erosion are relatively well studied and documented, changes in the biological properties of soils induced by agricultural activities are less well known. Biological degradation of soil can be analyzed by looking at changes in total living soil biomass or by quantifying changes in specific biological populations or functions. Biological activities are associated with organic matter decomposition, nutrient cycling, the genesis of soil structure, degradation of pollutants, and disease suppression (Sims, 1990). Degradation of these activities through erosion, compaction, organic matter depletion, or toxic inputs results in subtle but significant changes in cropping system performance.

### Carbon from Biomass

Cultivation has long been known to cause marked reductions in the total organic carbon content of between 20 and 50 percent (Paul and Clark, 1989). More recent work has shown even more dramatic reductions in labile or microbial carbon associated with cultivation (Bowman et al., 1990; Follett and Schimel, 1989; Schimel et al., 1985). These reductions should lead to decreases in various soil biological activities such as nitrogen mineralization, genesis of soil structure, and specific soil enzyme activities (Sims, 1990).

A study at Pendleton, Oregon, found that intensive cultivation and fallow decreased the total carbon content and microbial biomass in the soil, whereas increasing returns of crop residues, manure, and grass increased the organic carbon content and biomass (Granatstein, 1991).

### Soil Fauna Activity and Diversity

The effects of toxic compounds such as pesticides and other organic compounds on soil biology are not as well studied as cultivation effects. The effects of a wide range of pesticides on microbial growth, biomass, and activity have been tested, but mostly in short-term laboratory studies (Sims, 1990). Most of these studies have found that, at the pesticide concentrations found under field conditions, pesticides have little effect on microbial parameters. Certain compounds (in particular, fungicides) have been found to reduce total microbial biomass, soil fauna populations, or both. Organic compounds other than pesticides, such as petroleum products, have been found to have much more marked effects than pesticides on soil biology (Sims, 1990), but these compounds are rarely encountered in agricultural soils.

### Effects of Biological Degradation

The effects of biological degradation should be more important in cropping systems that rely heavily on biological nutrient cycling processes than systems that rely on chemical fertilizers for fertility. Similarly, systems that rely on natural biological pest suppression rather than pesticides for pest control are more sensitive to biological degradation. Since understanding of specific nutrient cycling and biological pest suppression mechanisms is limited for conventional, chemical-based cropping systems and low-input systems, the extent of the effects of biological degradation on cropping system performance are not known.

# 6

# Nitrogen in the Soil-Crop System

Nitrogen is ubiquitous in the environment. It is one of the most important plant nutrients and forms some of the most mobile compounds in the soil-crop system. Nitrogen is continually cycled among plants, soil organisms, soil organic matter, water, and the atmosphere (Figure 6-1). Nitrogen enters the soil from many different sources and leaves the root zone of the soil in many different ways. This flux of nitrogen into, out of, and within the soil takes place through complex biochemical transformations.

The mounting concerns related to agriculture's role in nitrogen delivery into the environment are reflected in several detailed reviews (Follett and Schimel, 1989; Follet et al., 1991; Hallberg, 1987, 1989b; Keeney, 1986a,b; Power and Schepers, 1989). A brief review of the nitrogen cycle and nitrogen budget or mass balance considerations is necessary to understand the options for management improvements in farming systems to mitigate the environmental impacts of nitrogen.

## THE NITROGEN CYCLE

The nitrogen cycle is critical to crop growth. The balance between inputs and outputs and the various transformations in the nitrogen cycle determine how much nitrogen is available for plant growth and how much may be lost to the atmosphere, surface water, or groundwater.

Nitrogen is an important component of soil organic matter, which is made up of decaying plant and animal tissue and the complex organic

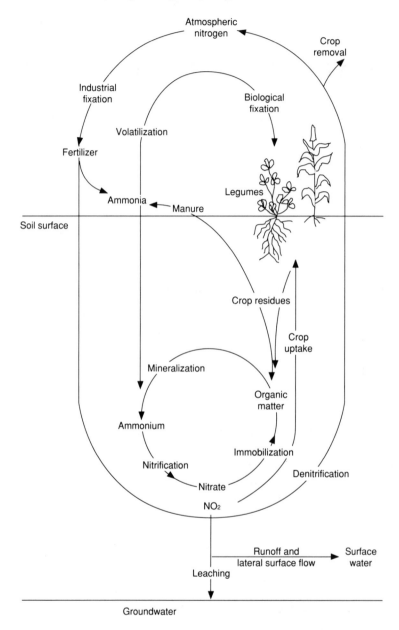

FIGURE 6-1 The nitrogen cycle. Source: Pennsylvania State University, College of Agriculture. 1989. Groundwater and Agriculture in Pennsylvania. Circular 341. College Station: Pennsylvania State University. Reprinted with permission from © The Pennsylvania State University.

compounds that form the soil humus. At any one time, most of the nitrogen held in the soil is stored in soil organic matter.

## Mineralization

Mineralization processes transform the nitrogen in soil organic matter to ammonium ions ($NH_4$), releasing them into the soil. Ammonium is relatively immobile in the soil, being strongly adsorbed to clay minerals and organic matter. Ammonium may be delivered to surface water, attached to sediment or suspended matter, or in solution. It is readily converted into nitrate, through nitrification, at appropriate soil temperatures (above about 9°C [48°F]). Ammonium can create water quality problems for fish and aquatic life under certain temperature and dissolved oxygen conditions.

## Nitrification

Nitrification processes transform ammonium ions, which are produced by mineralization or added to the soil, to nitrite ($NO_2$) and to nitrate ($NO_3$), which is easily absorbed by plant roots. Nitrification is typically mediated by soil bacteria and can take place rapidly with adequate soil moisture and temperature under oxidizing conditions in the soil. Except for some atmospheric processing, nitrification in the soil is the sole natural source of nitrate in the environment. Nitrate is soluble and mobile in water and is the form of nitrogen most commonly related to water quality problems. Nitrates that are not absorbed by plants or microorganisms or otherwise immobilized may readily move with percolating water and may leach through the soil to groundwater. Nitrates in the groundwater can move through springs and seeps or shallow flow systems to pollute surface waters, or they can leach into deeper aquifers.

## Immobilization

Immobilization includes various processes through which ammonium ions and nitrates are converted to organic nitrogen (referred to as organic-N) and immobilized or bound up in the soil. Ammonium and nitrate ions can be taken up by plants or microorganisms in the soil, transforming the nitrogen into organic matter. Mineralized nitrogen can rapidly recycle through transformations to ammonium and nitrate and then back into the organic-N pool. This occurs primarily through the action of microbes.

## Denitrification

Denitrification, another biological transformation, converts nitrate into nitrite and then to gaseous nitrogen ($N_2$) and nitrous oxide ($N_2O$). This is the major pathway that returns nitrogen from the soil environment to the atmosphere. Such losses are of environmental concern because these gases are among those that contribute to the so-called greenhouse effect and may affect the protective layer of ozone in the stratosphere.

## Interactive Processes

Mineralization, nitrification, immobilization, and denitrification are interactive processes through which a nitrogen molecule may move many times. The processes are affected by oxidizing and reducing conditions and the availability of oxygen and organic carbon in the soil. These processes go on simultaneously; they can coexist in close proximity and vary temporally in the same setting. In the small pores within aggregates in the soil profile, oxygen may be depleted and reducing conditions may become dominant, resulting in denitrification. Yet, on the exteriors of aggregates, around macropores, oxygen may be available and nitrification occurs. Seasonally, in a setting where the soil is normally dominated by air-filled pores and oxidizing conditions, the soil may become saturated with water during recharge events, and reducing conditions and denitrification may dominate temporarily. It is the balance between these processes and their seasonal timing that determines how much nitrogen is available for crops and how much nitrogen may be lost from the soil to groundwater and surface water or the atmosphere.

## NITROGEN MASS BALANCE

A molecule of nitrogen may enter the soil system as organic-N from crop residues or other plant or microbial biomass, from animal manures or organic wastes (for example, sewage sludge or food processing residues), and through the action of leguminous plants such as alfalfa that take nitrogen from the atmosphere and incorporate it into the plant's tissue (nitrogen fixation). The nitrogen in commercial fertilizer is directly added to soil systems in many forms, but the dominant forms are ammonium, nitrate, and urea. Some nitrogen, primarily as nitrate and ammonium, is also added with precipitation.

Nitrogen is taken up by crops and can be removed from the soil

TABLE 6-1   Nitrogen (N) Inputs, Outputs, and Balances in the
United States under the Low, Medium, and High Scenarios

| Input or Output | Metric Tons of N (Percent of Total Inputs)[a] | | |
| --- | --- | --- | --- |
| | Low Scenario | Medium Scenario | High Scenario |
| Input | | | |
| Fertilizer-N | 9,390,000 (47) | 9,390,000 (45) | 9,390,000 (42) |
| Manure-N | 1,730,000 (9) | 1,730,000 (8) | 1,730,000 (8) |
| Legume-N | 6,120,000 (30) | 6,870,000 (33) | 8,560,000 (38) |
| Crop residues | 2,890,000 (14) | 2,890,000 (14) | 2,890,000 (13) |
| Total input | 20,100,000 (100) | 20,900,000 (100) | 22,600,000 (100) |
| Output | | | |
| Harvested crops | 10,600,000 (53) | 10,600,000 (51) | 10,600,000 (47) |
| Crop residues | 2,890,000 (14) | 2,890,000 (14) | 2,890,000 (13) |
| Total output | 13,500,000 (67) | 13,500,000 (64) | 13,500,000 (60) |
| Balance | 6,670,000 (33) | 7,420,000 (36) | 9,110,000 (40) |

NOTE: See the Appendix for a full discussion of the methods used to estimate nitrogen
inputs and outputs.

[a]Input, output, or balance as a percent of the total mass of inputs.

system with the harvested portion of the crop (for example, grain) or can
be left in the soil system as root mass or crop residues. Nitrogen can be
lost to the atmosphere through denitrification or the volatilization of
ammonia from the fertilizers and manures applied to the soil surface. It
can also move through or over the soil with water to pollute surface
water or groundwater.

Even under native prairies and forests, some nitrogen loss occurs
through leaching, denitrification, erosion, and biomass. Biomass nitrogen can be lost because of a limited harvest, lost from senescing
vegetation, or carried away by wind or smoke when the biomass is
burned. Nutrient gains and losses in natural ecosystems are roughly in
balance, however; and nitrogen losses from natural ecosystems into
water are significantly lower than losses from agricultural ecosystems.
Numerous studies on various scales have shown from 3- to 60-fold
greater nitrate concentrations in surface water and groundwater in
agricultural areas compared with those in forested or grassland areas
(Hallberg, 1987, 1989b; Keeney, 1986a,b; McArthur et al., 1985; Omernik, 1976). Continued growth of plants in natural ecosystems depends
on the cycling of nutrients between biomass and organic and inorganic
stores (Miller and Larson, 1990).

Table 6-1 estimates the major, manageable, national nitrogen inputs
and outputs for harvested croplands in 1987. Inputs of nitrogen include
nitrogen applied to croplands as synthetic fertilizers, nitrogen in crop

residues voided in manures, and nitrogen supplied by legumes (alfalfa and soybeans). Outputs include nitrogen in harvested crops and crop residues. (See the Appendix for a full discussion of the methods used to estimate nitrogen inputs and outputs.) Only the manure that is collectible and that can be applied to croplands was considered. Some of the nitrogen in collectible manures is lost through volatilization, runoff, leaching, or other processes before it can be applied to croplands. The amount of nitrogen lost depends on the methods used to collect, store, and apply manures. In Table 6-1, only that portion of total nitrogen voided in manures that was estimated to be economically collectible and recoverable for use on croplands was used as the nitrogen inputs from manure.

Estimates of the rate of nitrogen fixation by alfalfa and soybeans vary widely. Estimates of rates of fixation by alfalfa range from 70 to 600 kg/ha/yr (62 to 532 lb/acre/yr) and from 15 to 310 kg/ha/yr (13 to 275 lb/acre/yr) for soybeans (Appendix Table A-4). Such large ranges in reported values are related, in part, to differences in soil nitrogen availability, climate, and crop variety. In addition, the amount of nitrogen fixed by alfalfa depends on the density and age of the stand. Estimates are further complicated because the fixed nitrogen is not immediately available for use by crop plants and some of the reduced need for nitrogen by crops following legumes is related to rotation effects other than the nitrogen supplied by fixation. Because of these difficulties, nitrogen replacement values are usually used to estimate the effect of legumes on the need for supplemental nitrogen by succeeding crops. The nitrogen replacement values include both the rotation effects and the influence of fixed nitrogen when determining the need for supplemental nitrogen.

Because of the wide range of estimates of nitrogen fixation by legumes (alfalfa and soybeans), the committee used three fixation-nitrogen replacement value estimates (low, medium, and high scenarios) to calculate nitrogen inputs. The nitrogen fixation rates and replacement values under the three scenarios are given in Table 6-2. The nitrogen replacement value, as used here, is the difference between the nitrogen input (fixed and accumulated nitrogen) and the nitrogen removed with the harvested legume crop (see Appendix Table A-5.)

Estimates of nitrogen outputs in harvested crops and crop residues are also reported in Table 6-1. The difference between nitrogen inputs and outputs is reported as nitrogen balances. A more detailed analysis of nitrogen inputs and outputs from agricultural lands helps to identify opportunities for reducing nitrogen losses from farming systems.

TABLE 6-2   Nitrogen Accumulation and Nitrogen Replacement Value Estimated for Alfalfa and Soybeans

| Legume | Scenario | Nitrogen Accumulation (kg/ha) | |
| --- | --- | --- | --- |
| | | Total Nitrogen Input | Nitrogen Replacement Value[a] |
| Alfalfa | Low | 230 | 45 |
| | Medium | 250 | 65 |
| | High | 380 | 195 |
| Soybeans | Low | 175 | 10 |
| | Medium | 200 | 35 |
| | High | 220 | 55 |

NOTE: See the Appendix for a full discussion of the methods used to estimate nitrogen accumulation and replacement values.

[a]The nitrogen replacement value includes the amount of fixed nitrogen available to a succeeding crop and the reduced need for supplemental nitrogen that may be a result of rotation effects.

## Nitrogen Inputs

The nitrogen delivered in rainfall; obtained from fertilizers; mineralized from soil organic-N, crop residues, manure, or legumes; or even delivered in irrigation water contributes to the nitrogen budget of a particular agricultural field. All of these nitrogen sources are subject to the transformations of the nitrogen cycle and all can contribute to environmental nitrogen losses. The importance of any particular source depends on the type of agricultural enterprise, its geographic location and climate, and the soil's microclimate. This variation is evident in Tables 6-3 and 6-4, which report state- and national-level nitrogen mass balances.

### Nitrogen in Fertilizers

The nitrogen in fertilizers is the single largest source of nitrogen applied to most croplands. In 1987, 9.39 million metric tons (10.4 million tons) of nitrogen was applied nationwide in the form of synthetic fertilizers. For the low, medium, and high scenarios, the amount of synthetic fertilizer applied represents 47, 45, and 42 percent of nitrogen inputs, respectively. The importance of synthetic fertilizers as a nitrogen source (fertilizer-N) varies widely around the United States, depending on the crop and the region where that crop is grown. Three of the four major commodity crops—corn, wheat, and cotton—use 61 percent of U.S. fertilizer-N. Corn, which covers about 21 percent of U.S. cropland,

TABLE 6-3 State and National Nitrogen Inputs and Outputs (metric tons)

| State | Inputs | | | | | Outputs | | | |
|---|---|---|---|---|---|---|---|---|---|
| | Fertilizer-N | Recoverable Manure-N | Legume-N Fixation | Crop Residues | Total | Harvested Crop | Crop Residues | Total | Balance |
| | | | | | *Low Scenario* | | | | |
| Alabama | 111,000 | 25,300 | 44,300 | 9,420 | 190,000 | 60,800 | 9,420 | 70,200 | 120,000 |
| Alaska | 1,850 | 0 | 0 | 61 | 1,910 | 620 | 61 | 681 | 1,230 |
| Arizona | 73,500 | 22,200 | 12,800 | 3,980 | 112,000 | 35,500 | 3,980 | 39,500 | 73,000 |
| Arkansas | 188,000 | 40,300 | 229,000 | 54,600 | 512,000 | 220,000 | 54,600 | 275,000 | 237,000 |
| California | 482,000 | 106,000 | 88,400 | 32,800 | 709,000 | 246,000 | 32,800 | 279,000 | 431,000 |
| Colorado | 126,000 | 61,300 | 66,600 | 48,000 | 302,000 | 175,000 | 48,000 | 223,000 | 78,500 |
| Connecticut | 6,450 | 4,880 | 1,900 | 81 | 13,300 | 3,800 | 81 | 3,880 | 9,430 |
| Delaware | 15,200 | 5,850 | 16,300 | 4,380 | 41,700 | 14,000 | 4,370 | 18,400 | 23,400 |
| Florida | 258,000 | 15,700 | 7,900 | 3,090 | 285,000 | 25,100 | 3,090 | 28,200 | 257,000 |
| Georgia | 175,000 | 31,400 | 56,500 | 20,000 | 283,000 | 105,000 | 20,000 | 125,000 | 158,000 |
| Hawaii | 14,900 | 1,030 | 10 | 0 | 15,900 | 110 | 0 | 110 | 15,800 |
| Idaho | 187,000 | 26,600 | 85,800 | 37,900 | 337,000 | 186,000 | 37,900 | 224,000 | 113,000 |
| Illinois | 805,000 | 37,200 | 676,000 | 380,000 | 1,900,000 | 1,120,000 | 380,000 | 1,500,000 | 402,000 |
| Indiana | 462,000 | 32,900 | 349,000 | 200,000 | 1,040,000 | 595,000 | 200,000 | 795,000 | 249,000 |
| Iowa | 780,000 | 87,400 | 688,000 | 394,000 | 1,950,000 | 1,200,000 | 394,000 | 1,590,000 | 356,000 |
| Kansas | 438,000 | 114,000 | 202,000 | 163,000 | 917,000 | 547,000 | 163,000 | 710,000 | 207,000 |
| Kentucky | 165,000 | 24,500 | 103,000 | 35,300 | 327,000 | 163,000 | 35,300 | 199,000 | 129,000 |
| Louisiana | 138,000 | 5,670 | 111,000 | 26,100 | 280,000 | 109,000 | 26,100 | 135,000 | 145,000 |
| Maine | 11,600 | 7,780 | 2,200 | 1,230 | 22,800 | 11,700 | 1,230 | 12,900 | 9,880 |
| Maryland | 38,800 | 21,100 | 36,500 | 12,600 | 109,000 | 47,100 | 12,600 | 59,700 | 49,300 |
| Massachusetts | 9,860 | 4,390 | 3,000 | 161 | 17,400 | 5,300 | 161 | 5,460 | 12,000 |
| Michigan | 220,000 | 37,900 | 163,000 | 64,000 | 485,000 | 238,000 | 64,000 | 302,000 | 183,000 |
| Minnesota | 525,000 | 79,400 | 450,000 | 220,000 | 1,270,000 | 737,000 | 220,000 | 957,000 | 317,000 |
| Mississippi | 142,000 | 13,400 | 145,000 | 23,800 | 324,000 | 115,000 | 23,800 | 138,000 | 186,000 |
| Missouri | 322,000 | 42,400 | 386,000 | 113,000 | 863,000 | 487,000 | 113,000 | 600,000 | 263,000 |

| | | | | | | | | |
|---|---|---|---|---|---|---|---|---|
| Montana | 91,100 | 12,000 | 114,000 | 52,400 | 270,000 | 212,000 | 52,400 | 264,000 | 5,200 |
| Nebraska | 608,000 | 93,000 | 273,000 | 236,000 | 1,210,000 | 662,000 | 236,000 | 898,000 | 312,000 |
| Nevada | 4,880 | 2,220 | 22,500 | 521 | 30,100 | 29,900 | 521 | 30,400 | (300) |
| New Hampshire | 2,040 | 1,980 | 1,700 | 24 | 5,740 | 3,210 | 24 | 3,230 | 2,510 |
| New Jersey | 20,800 | 3,170 | 11,700 | 3,040 | 38,700 | 14,900 | 3,040 | 17,900 | 20,800 |
| New Mexico | 24,000 | 14,500 | 18,300 | 5,410 | 62,200 | 36,800 | 5,410 | 42,200 | 20,000 |
| New York | 93,900 | 78,200 | 84,300 | 18,700 | 275,000 | 138,000 | 18,700 | 156,000 | 119,000 |
| North Carolina | 176,000 | 33,000 | 94,000 | 30,900 | 334,000 | 122,000 | 30,900 | 153,000 | 181,000 |
| North Dakota | 302,000 | 13,700 | 162,000 | 108,000 | 586,000 | 357,000 | 108,000 | 465,000 | 121,000 |
| Ohio | 356,000 | 40,100 | 327,000 | 134,000 | 856,000 | 468,000 | 134,000 | 602,000 | 253,000 |
| Oklahoma | 246,000 | 30,500 | 44,300 | 34,100 | 355,000 | 162,000 | 34,100 | 196,000 | 159,000 |
| Oregon | 114,000 | 13,500 | 34,100 | 16,600 | 178,000 | 93,900 | 16,700 | 111,000 | 67,600 |
| Pennsylvania | 47,500 | 79,200 | 90,900 | 29,700 | 247,000 | 156,000 | 29,700 | 186,000 | 61,200 |
| Rhode Island | 1,490 | 0 | 200 | 8 | 1,700 | 330 | 8 | 338 | 1,360 |
| South Carolina | 65,100 | 5,680 | 43,200 | 11,300 | 125,000 | 45,600 | 11,300 | 56,900 | 68,400 |
| South Dakota | 164,000 | 35,300 | 278,000 | 94,900 | 572,000 | 360,000 | 94,900 | 455,000 | 117,000 |
| Tennessee | 140,000 | 20,600 | 96,600 | 23,500 | 281,000 | 118,000 | 23,600 | 141,000 | 140,000 |
| Texas | 674,000 | 153,000 | 29,700 | 94,800 | 951,000 | 340,000 | 95,800 | 436,000 | 515,000 |
| Utah | 27,000 | 11,300 | 44,900 | 4,480 | 87,700 | 57,100 | 4,480 | 61,500 | 26,100 |
| Vermont | 4,480 | 17,800 | 9,500 | 256 | 32,000 | 17,400 | 256 | 17,700 | 14,400 |
| Virginia | 71,300 | 25,400 | 47,800 | 11,000 | 156,000 | 73,500 | 11,000 | 84,500 | 71,000 |
| Washington | 185,000 | 25,300 | 40,600 | 39,200 | 290,000 | 158,000 | 39,200 | 197,000 | 93,200 |
| West Virginia | 11,900 | 5,610 | 7,800 | 948 | 26,300 | 14,800 | 948 | 15,700 | 10,500 |
| Wisconsin | 243,000 | 161,000 | 271,000 | 84,000 | 759,000 | 441,000 | 84,000 | 524,000 | 235,000 |
| Wyoming | 20,900 | 9,570 | 50,500 | 5,180 | 86,200 | 53,700 | 5,180 | 58,900 | 27,300 |
| United States | 9,390,000 | 1,730,000 | 6,120,000 | 2,890,000 | 20,100,000 | 10,600,000 | 2,890,000 | 13,500,000 | 6,670,000 |

*Medium Scenario*

| | | | | | | | | |
|---|---|---|---|---|---|---|---|---|
| Alabama | 111,000 | 25,300 | 50,500 | 9,420 | 196,000 | 60,800 | 9,420 | 70,200 | 126,000 |
| Alaska | 1,850 | 0 | 0 | 61 | 1,910 | 620 | 61 | 681 | 1,230 |
| Arizona | 73,500 | 22,200 | 13,900 | 3,980 | 114,000 | 35,500 | 3,980 | 39,500 | 74,100 |
| Arkansas | 188,000 | 40,300 | 261,000 | 54,600 | 544,000 | 220,000 | 54,600 | 275,000 | 269,000 |

*(continued)*

TABLE 6-3 (Continued)

| State | Inputs | | | | | Outputs | | | Balance |
|---|---|---|---|---|---|---|---|---|---|
| | Fertilizer-N | Recoverable Manure-N | Legume-N Fixation | Crop Residues | Total | Harvested Crop | Crop Residues | Total | |
| California | 482,000 | 106,000 | 96,100 | 32,800 | 717,000 | 246,000 | 32,800 | 278,000 | 438,000 |
| Colorado | 126,000 | 61,300 | 72,400 | 48,000 | 308,000 | 175,000 | 48,000 | 223,000 | 84,300 |
| Connecticut | 6,450 | 4,880 | 2,100 | 81 | 13,500 | 3,800 | 81 | 3,880 | 9,630 |
| Delaware | 15,200 | 5,850 | 18,600 | 4,380 | 44,000 | 14,000 | 4,370 | 18,400 | 25,700 |
| Florida | 258,000 | 15,700 | 8,900 | 3,090 | 286,000 | 25,100 | 3,090 | 28,200 | 258,000 |
| Georgia | 175,000 | 31,400 | 64,400 | 20,000 | 291,000 | 105,000 | 20,000 | 125,000 | 166,000 |
| Hawaii | 14,900 | 1,030 | 15 | 0 | 15,900 | 110 | 0 | 110 | 15,800 |
| Idaho | 187,000 | 26,600 | 93,200 | 37,900 | 345,000 | 186,000 | 37,900 | 224,000 | 121,000 |
| Illinois | 805,000 | 37,200 | 770,000 | 380,000 | 1,990,000 | 1,120,000 | 380,000 | 1,500,000 | 496,000 |
| Indiana | 462,000 | 32,900 | 397,000 | 200,000 | 1,090,000 | 595,000 | 200,000 | 795,000 | 297,000 |
| Iowa | 780,000 | 87,400 | 780,000 | 394,000 | 2,040,000 | 1,200,000 | 394,000 | 1,590,000 | 450,000 |
| Kansas | 438,000 | 114,000 | 227,000 | 163,000 | 942,000 | 547,000 | 163,000 | 710,000 | 232,000 |
| Kentucky | 165,000 | 24,500 | 116,000 | 35,300 | 340,000 | 163,000 | 35,300 | 199,000 | 142,000 |
| Louisiana | 138,000 | 5,670 | 126,000 | 26,100 | 296,000 | 109,000 | 26,100 | 135,000 | 160,000 |
| Maine | 11,600 | 7,780 | 2,400 | 1,230 | 23,000 | 11,700 | 1,230 | 12,900 | 10,100 |
| Maryland | 38,800 | 21,100 | 41,300 | 12,600 | 114,000 | 47,100 | 12,600 | 59,700 | 54,100 |
| Massachusetts | 9,860 | 4,390 | 3,300 | 161 | 17,700 | 5,300 | 161 | 5,460 | 12,300 |
| Michigan | 220,000 | 37,900 | 181,000 | 64,000 | 503,000 | 238,000 | 64,000 | 302,000 | 201,000 |
| Minnesota | 525,000 | 79,400 | 507,000 | 220,000 | 1,330,000 | 737,000 | 220,000 | 957,000 | 373,000 |
| Mississippi | 142,000 | 13,400 | 165,000 | 23,800 | 344,000 | 115,000 | 23,800 | 138,000 | 206,000 |
| Missouri | 322,000 | 42,400 | 439,000 | 113,000 | 916,000 | 487,000 | 113,000 | 600,000 | 315,000 |
| Montana | 91,100 | 12,000 | 124,000 | 52,400 | 280,000 | 212,000 | 52,400 | 264,000 | 15,200 |
| Nebraska | 608,000 | 93,000 | 306,000 | 236,000 | 1,240,000 | 662,000 | 236,000 | 898,000 | 345,000 |
| Nevada | 4,880 | 2,220 | 24,400 | 521 | 32,000 | 29,900 | 521 | 30,400 | 1,600 |
| New Hampshire | 2,040 | 1,980 | 1,800 | 24 | 5,840 | 3,210 | 24 | 3,230 | 2,610 |
| New Jersey | 20,800 | 3,170 | 13,200 | 3,040 | 40,200 | 14,900 | 3,040 | 17,900 | 22,300 |

| | | | | | | | | | |
|---|---|---|---|---|---|---|---|---|---|
| New Mexico | 24,000 | 14,500 | 19,900 | 5,410 | 63,800 | 36,800 | 5,410 | 42,200 | 21,600 |
| New York | 93,900 | 78,200 | 91,700 | 18,700 | 283,000 | 138,000 | 18,700 | 156,000 | 126,000 |
| North Carolina | 176,000 | 33,000 | 107,000 | 30,900 | 347,000 | 122,000 | 30,900 | 153,000 | 194,000 |
| North Dakota | 302,000 | 13,700 | 178,000 | 108,000 | 602,000 | 357,000 | 108,000 | 465,000 | 137,000 |
| Ohio | 356,000 | 40,100 | 369,000 | 134,000 | 899,000 | 468,000 | 134,000 | 602,000 | 296,000 |
| Oklahoma | 246,000 | 30,500 | 49,000 | 34,100 | 360,000 | 162,000 | 34,100 | 196,000 | 164,000 |
| Oregon | 114,000 | 13,500 | 37,500 | 16,600 | 181,000 | 93,900 | 16,700 | 111,000 | 70,500 |
| Pennsylvania | 47,500 | 79,200 | 99,500 | 29,700 | 256,000 | 156,000 | 29,700 | 186,000 | 69,800 |
| Rhode Island | 1,490 | 0 | 230 | 8 | 1,730 | 330 | 8 | 338 | 1,390 |
| South Carolina | 65,100 | 5,680 | 49,400 | 11,300 | 131,000 | 45,600 | 11,300 | 56,900 | 74,600 |
| South Dakota | 164,000 | 35,300 | 307,000 | 94,900 | 601,000 | 360,000 | 94,900 | 455,000 | 146,000 |
| Tennessee | 140,000 | 20,600 | 110,000 | 23,500 | 294,000 | 118,000 | 23,600 | 141,000 | 153,000 |
| Texas | 674,000 | 153,000 | 33,000 | 94,800 | 955,000 | 341,000 | 95,800 | 436,000 | 518,000 |
| Utah | 27,000 | 11,300 | 48,800 | 4,480 | 91,600 | 57,100 | 4,480 | 61,600 | 30,000 |
| Vermont | 4,480 | 17,800 | 10,300 | 256 | 32,800 | 17,400 | 256 | 17,700 | 15,200 |
| Virginia | 71,300 | 25,400 | 53,800 | 11,000 | 162,000 | 73,500 | 11,000 | 84,500 | 77,000 |
| Washington | 185,000 | 25,300 | 44,200 | 39,200 | 294,000 | 158,000 | 39,200 | 197,000 | 96,800 |
| West Virginia | 11,900 | 5,610 | 8,600 | 948 | 27,100 | 14,800 | 948 | 15,700 | 11,300 |
| Wisconsin | 243,000 | 161,000 | 296,000 | 84,000 | 784,000 | 441,000 | 84,000 | 525,000 | 259,000 |
| Wyoming | 20,900 | 9,570 | 54,800 | 5,180 | 90,500 | 53,700 | 5,180 | 58,900 | 31,600 |
| United States | 9,390,000 | 1,730,000 | 6,870,000 | 2,890,000 | 20,900,000 | 10,600,000 | 2,890,000 | 13,500,000 | 7,420,000 |
| *High Scenario* | | | | | | | | | |
| Alabama | 111,000 | 25,300 | 57,000 | 9,420 | 203,000 | 60,800 | 9,420 | 70,200 | 133,000 |
| Alaska | 1,850 | 0 | 0 | 61 | 1,910 | 620 | 61 | 681 | 1,230 |
| Arizona | 73,500 | 22,200 | 21,000 | 3,980 | 121,000 | 35,500 | 3,980 | 39,500 | 81,200 |
| Arkansas | 188,000 | 40,300 | 289,000 | 54,600 | 572,000 | 220,000 | 54,600 | 275,000 | 297,000 |
| California | 482,000 | 106,000 | 146,000 | 32,800 | 767,000 | 246,000 | 32,800 | 279,000 | 488,000 |
| Colorado | 126,000 | 61,300 | 110,000 | 48,000 | 345,000 | 175,000 | 48,000 | 223,000 | 122,000 |
| Connecticut | 6,450 | 4,880 | 3,200 | 81 | 14,600 | 3,800 | 81 | 3,880 | 10,700 |
| Delaware | 15,200 | 5,850 | 20,800 | 4,380 | 46,200 | 14,000 | 4,370 | 18,400 | 27,900 |

(continued)

TABLE 6-3 (Continued)

| State | Inputs | | | | | Outputs | | | Balance |
|---|---|---|---|---|---|---|---|---|---|
| | Fertilizer-N | Recoverable Manure-N | Legume-N Fixation | Crop Residues | Total | Harvested Crop | Crop Residues | Total | |
| Florida | 258,000 | 15,700 | 10,500 | 3,090 | 287,000 | 25,100 | 3,090 | 28,200 | 259,000 |
| Georgia | 175,000 | 31,400 | 72,000 | 20,000 | 298,000 | 105,000 | 20,000 | 125,000 | 174,000 |
| Hawaii | 14,900 | 1,030 | 20 | 0 | 16,000 | 110 | 0 | 110 | 15,800 |
| Idaho | 187,000 | 26,600 | 142,000 | 37,900 | 393,000 | 186,000 | 37,900 | 224,000 | 169,000 |
| Illinois | 805,000 | 37,200 | 871,000 | 380,000 | 2,090,000 | 1,120,000 | 380,000 | 1,500,000 | 597,000 |
| Indiana | 462,000 | 32,900 | 453,000 | 200,000 | 1,150,000 | 595,000 | 200,000 | 795,000 | 354,000 |
| Iowa | 780,000 | 87,400 | 917,000 | 394,000 | 2,180,000 | 1,200,000 | 394,000 | 1,590,000 | 586,000 |
| Kansas | 438,000 | 114,000 | 282,000 | 163,000 | 997,000 | 547,000 | 163,000 | 710,000 | 286,000 |
| Kentucky | 165,000 | 24,500 | 141,000 | 35,300 | 366,000 | 163,000 | 35,300 | 199,000 | 167,000 |
| Louisiana | 138,000 | 5,670 | 139,000 | 26,100 | 309,000 | 109,000 | 26,100 | 135,000 | 173,000 |
| Maine | 11,600 | 7,780 | 3,700 | 1,230 | 24,300 | 11,700 | 1,230 | 12,900 | 11,400 |
| Maryland | 38,800 | 21,100 | 48,900 | 12,600 | 121,000 | 47,100 | 12,600 | 59,700 | 61,700 |
| Massachusetts | 9,860 | 4,390 | 5,000 | 161 | 19,400 | 5,300 | 161 | 5,460 | 14,000 |
| Michigan | 220,000 | 37,900 | 241,000 | 64,000 | 563,000 | 238,000 | 64,000 | 302,000 | 261,000 |
| Minnesota | 525,000 | 79,600 | 621,000 | 220,000 | 1,450,000 | 737,000 | 220,000 | 957,000 | 488,000 |
| Mississippi | 142,000 | 13,400 | 183,000 | 23,800 | 362,000 | 115,000 | 23,800 | 138,000 | 224,000 |
| Missouri | 322,000 | 42,400 | 502,000 | 113,000 | 980,000 | 487,000 | 113,000 | 600,000 | 379,000 |
| Montana | 91,100 | 12,000 | 188,000 | 52,400 | 344,000 | 212,000 | 52,400 | 264,000 | 79,600 |
| Nebraska | 608,000 | 93,000 | 387,000 | 236,000 | 1,320,000 | 662,000 | 236,000 | 898,000 | 426,000 |
| Nevada | 4,890 | 2,220 | 37,100 | 521 | 44,700 | 29,900 | 521 | 30,400 | 14,300 |

| | | | | | | | | |
|---|---|---|---|---|---|---|---|---|
| New Hampshire | 2,040 | 1,980 | 2,760 | 24 | 6,800 | 3,210 | 24 | 3,230 | 3,570 |
| New Jersey | 20,800 | 3,170 | 16,400 | 3,040 | 43,400 | 14,900 | 3,040 | 17,900 | 25,500 |
| New Mexico | 24,000 | 14,500 | 30,300 | 5,410 | 74,200 | 36,800 | 5,410 | 42,200 | 32,000 |
| New York | 93,900 | 78,200 | 139,000 | 18,700 | 329,000 | 138,000 | 18,700 | 156,000 | 173,000 |
| North Carolina | 176,000 | 33,000 | 120,000 | 30,900 | 360,000 | 122,000 | 30,900 | 153,000 | 207,000 |
| North Dakota | 302,000 | 13,700 | 254,000 | 108,000 | 678,000 | 357,000 | 108,000 | 465,000 | 212,000 |
| Ohio | 356,000 | 40,100 | 434,000 | 134,000 | 964,000 | 468,000 | 134,000 | 602,000 | 362,000 |
| Oklahoma | 246,000 | 30,500 | 66,700 | 34,100 | 377,000 | 162,000 | 34,100 | 196,000 | 182,000 |
| Oregon | 114,000 | 13,500 | 56,300 | 16,600 | 200,000 | 93,900 | 16,700 | 111,000 | 89,800 |
| Pennsylvania | 47,500 | 79,200 | 145,000 | 29,700 | 302,000 | 156,000 | 29,700 | 186,000 | 115,000 |
| Rhode Island | 1,490 | 0 | 350 | 8 | 1,850 | 330 | 8 | 338 | 1,510 |
| South Carolina | 65,100 | 5,680 | 54,700 | 11,300 | 137,000 | 45,600 | 11,300 | 56,900 | 79,9000 |
| South Dakota | 164,000 | 35,300 | 423,000 | 94,900 | 717,000 | 360,000 | 94,900 | 455,000 | 262,000 |
| Tennessee | 140,000 | 20,600 | 126,000 | 23,500 | 311,000 | 118,000 | 23,600 | 141,000 | 169,000 |
| Texas | 674,000 | 153,000 | 44,200 | 94,800 | 966,000 | 341,000 | 95,800 | 437,000 | 529,000 |
| Utah | 27,000 | 11,300 | 74,200 | 4,480 | 117,000 | 57,100 | 4,480 | 61,600 | 554,000 |
| Vermont | 4,480 | 17,800 | 15,600 | 256 | 38,100 | 17,400 | 256 | 17,700 | 20,500 |
| Virginia | 71,300 | 25,400 | 66,100 | 11,000 | 174,000 | 73,500 | 11,000 | 84,500 | 89,300 |
| Washington | 185,000 | 25,300 | 67,100 | 39,200 | 317,000 | 158,000 | 39,200 | 197,000 | 120,000 |
| West Virginia | 11,900 | 5,610 | 12,800 | 948 | 31,300 | 14,800 | 948 | 15,800 | 15,500 |
| Wisconsin | 243,000 | 161,000 | 440,000 | 84,000 | 928,000 | 441,000 | 84,000 | 525,000 | 403,000 |
| Wyoming | 20,900 | 9,570 | 83,400 | 5,180 | 119,000 | 53,700 | 5,180 | 58,900 | 60,200 |
| United States | 9,390,000 | 1,730,000 | 8,560,000 | 2,890,000 | 22,600,000 | 10,600,000 | 2,890,000 | 13,500,000 | 9,110,000 |

NOTE: See the Appendix for a full discussion of the methods used to estimate nitrogen inputs and outputs.

TABLE 6-4 State and National Nitrogen Contributions to Total Inputs and Outputs

| | Inputs | | | | Outputs | | |
|---|---|---|---|---|---|---|---|
| | Percentage of Total Input Mass | | | | | | |
| State | Fertilizer-N | Recoverable Manure-N | Legume-N Fixation | Crop Residues | Harvested Crop | Crop Residues | Balance |
| | | | *Low Scenario* | | | | |
| Alabama | 58 | 13 | 23 | 5 | 32 | 5 | 63 |
| Alaska | 97 | 0 | 0 | 3 | 32 | 3 | 64 |
| Arizona | 65 | 20 | 11 | 4 | 32 | 4 | 65 |
| Arkansas | 37 | 8 | 45 | 11 | 43 | 11 | 46 |
| California | 68 | 15 | 12 | 5 | 35 | 5 | 61 |
| Colorado | 42 | 20 | 22 | 16 | 58 | 16 | 26 |
| Connecticut | 48 | 37 | 14 | 1 | 29 | 1 | 71 |
| Delaware | 36 | 14 | 39 | 10 | 34 | 10 | 56 |
| Florida | 91 | 6 | 3 | 1 | 9 | 1 | 90 |
| Georgia | 62 | 11 | 20 | 7 | 37 | 7 | 56 |
| Hawaii | 93 | 6 | 0 | 0 | 1 | 0 | 99 |
| Idaho | 55 | 8 | 25 | 11 | 55 | 11 | 34 |
| Illinois | 42 | 2 | 36 | 20 | 59 | 20 | 21 |
| Indiana | 44 | 3 | 33 | 19 | 57 | 19 | 24 |
| Iowa | 40 | 4 | 35 | 20 | 61 | 20 | 18 |
| Kansas | 48 | 12 | 22 | 18 | 60 | 18 | 23 |
| Kentucky | 50 | 7 | 31 | 11 | 50 | 11 | 39 |
| Louisiana | 49 | 2 | 39 | 9 | 39 | 9 | 52 |
| Maine | 51 | 34 | 10 | 5 | 51 | 5 | 43 |
| Maryland | 36 | 19 | 33 | 12 | 43 | 12 | 69 |
| Massachusetts | 57 | 25 | 17 | 1 | 30 | 1 | 69 |
| Michigan | 45 | 8 | 34 | 13 | 49 | 13 | 38 |
| Minnesota | 41 | 6 | 35 | 17 | 58 | 17 | 25 |

| | | | | | | | |
|---|---|---|---|---|---|---|---|
| Mississippi | 44 | 4 | 45 | 7 | 35 | 7 | 57 |
| Missouri | 37 | 5 | 45 | 13 | 56 | 13 | 30 |
| Montana | 34 | 4 | 42 | 19 | 79 | 19 | 2 |
| Nebraska | 50 | 8 | 23 | 20 | 55 | 20 | 26 |
| Nevada | 16 | 7 | 75 | 2 | 99 | 2 | (1) |
| New Hampshire | 36 | 34 | 30 | 0 | 56 | 0 | 44 |
| New Jersey | 54 | 8 | 30 | 8 | 38 | 8 | 54 |
| New Mexico | 39 | 23 | 29 | 9 | 59 | 9 | 32 |
| New York | 34 | 28 | 31 | 7 | 50 | 7 | 43 |
| North Carolina | 53 | 10 | 28 | 9 | 36 | 9 | 54 |
| North Dakota | 52 | 2 | 28 | 18 | 61 | 18 | 21 |
| Ohio | 42 | 5 | 38 | 16 | 55 | 16 | 30 |
| Oklahoma | 69 | 9 | 12 | 10 | 46 | 10 | 45 |
| Oregon | 64 | 8 | 19 | 9 | 53 | 9 | 38 |
| Pennsylvania | 19 | 32 | 37 | 12 | 63 | 12 | 25 |
| Rhode Island | 88 | 0 | 12 | 0 | 19 | 0 | 80 |
| South Carolina | 52 | 5 | 34 | 9 | 36 | 9 | 55 |
| South Dakota | 29 | 6 | 49 | 17 | 63 | 17 | 20 |
| Tennessee | 50 | 7 | 34 | 8 | 42 | 8 | 50 |
| Texas | 71 | 16 | 3 | 10 | 36 | 10 | 54 |
| Utah | 31 | 13 | 51 | 5 | 65 | 5 | 30 |
| Vermont | 14 | 56 | 30 | 1 | 54 | 1 | 45 |
| Virginia | 46 | 16 | 31 | 7 | 47 | 7 | 46 |
| Washington | 64 | 9 | 14 | 14 | 54 | 14 | 32 |
| West Virginia | 45 | 21 | 30 | 4 | 58 | 4 | 40 |
| Wisconsin | 32 | 21 | 36 | 11 | 56 | 11 | 31 |
| Wyoming | 24 | 11 | 59 | 6 | 62 | 6 | 32 |
| United States | 47 | 9 | 30 | 14 | 53 | 14 | 33 |

(continued)

TABLE 6-4 (*Continued*)

| State | Percentage of Total Input Mass | | | | | | |
|---|---|---|---|---|---|---|---|
| | Inputs | | | | Outputs | | |
| | Fertilizer-N | Recoverable Manure-N | Legume-N Fixation | Crop Residues | Harvested Crop | Crop Residues | Balance |
| | | | *Medium Scenario* | | | | |
| Alabama | 57 | 13 | 26 | 5 | 31 | 5 | 64 |
| Alaska | 97 | 0 | 0 | 3 | 32 | 3 | 64 |
| Arizona | 65 | 20 | 12 | 4 | 31 | 4 | 65 |
| Arkansas | 35 | 7 | 48 | 10 | 40 | 10 | 50 |
| California | 67 | 15 | 13 | 5 | 34 | 5 | 61 |
| Colorado | 41 | 20 | 24 | 16 | 57 | 16 | 27 |
| Connecticut | 48 | 36 | 16 | 1 | 28 | 1 | 71 |
| Delaware | 35 | 13 | 42 | 10 | 32 | 10 | 58 |
| Florida | 90 | 5 | 3 | 1 | 9 | 1 | 90 |
| Georgia | 60 | 11 | 22 | 7 | 36 | 7 | 57 |
| Hawaii | 93 | 6 | 0 | 0 | 1 | 0 | 99 |
| Idaho | 54 | 8 | 27 | 11 | 54 | 11 | 35 |
| Illinois | 40 | 2 | 39 | 19 | 56 | 19 | 25 |
| Indiana | 42 | 3 | 36 | 18 | 54 | 18 | 27 |
| Iowa | 38 | 4 | 38 | 19 | 59 | 19 | 22 |
| Kansas | 46 | 12 | 24 | 17 | 58 | 17 | 25 |
| Kentucky | 48 | 7 | 34 | 10 | 48 | 10 | 42 |
| Louisiana | 47 | 2 | 43 | 9 | 37 | 9 | 54 |
| Maine | 50 | 34 | 10 | 5 | 51 | 5 | 44 |
| Maryland | 34 | 19 | 36 | 11 | 41 | 11 | 48 |
| Massachusetts | 56 | 25 | 19 | 1 | 30 | 1 | 40 |
| Michigan | 44 | 8 | 36 | 13 | 47 | 13 | 40 |
| Minnesota | 39 | 6 | 38 | 17 | 55 | 17 | 28 |

| | | | | | | | |
|---|---|---|---|---|---|---|---|
| Mississippi | 41 | 4 | 48 | 7 | 33 | 7 | 60 |
| Missouri | 35 | 5 | 48 | 12 | 53 | 12 | 34 |
| Montana | 33 | 4 | 44 | 19 | 76 | 19 | 5 |
| Nebraska | 49 | 7 | 25 | 19 | 53 | 19 | 28 |
| Nevada | 15 | 7 | 76 | 2 | 93 | 2 | 5 |
| New Hampshire | 35 | 34 | 31 | 0 | 55 | 0 | 45 |
| New Jersey | 52 | 8 | 33 | 8 | 37 | 8 | 55 |
| New Mexico | 38 | 23 | 31 | 8 | 58 | 8 | 34 |
| New York | 33 | 28 | 32 | 7 | 49 | 7 | 45 |
| North Carolina | 51 | 10 | 31 | 9 | 35 | 9 | 56 |
| North Dakota | 50 | 2 | 30 | 18 | 59 | 18 | 23 |
| Ohio | 40 | 4 | 41 | 15 | 52 | 15 | 33 |
| Oklahoma | 68 | 8 | 14 | 9 | 45 | 9 | 46 |
| Oregon | 63 | 7 | 20 | 9 | 52 | 9 | 39 |
| Pennsylvania | 19 | 31 | 39 | 12 | 61 | 12 | 27 |
| Rhode Island | 86 | 0 | 13 | 0 | 19 | 0 | 80 |
| South Carolina | 50 | 4 | 38 | 9 | 35 | 9 | 57 |
| South Dakota | 27 | 6 | 51 | 16 | 60 | 16 | 24 |
| Tennessee | 48 | 7 | 37 | 8 | 40 | 8 | 52 |
| Texas | 71 | 16 | 3 | 10 | 36 | 10 | 54 |
| Utah | 29 | 12 | 53 | 5 | 62 | 5 | 33 |
| Vermont | 14 | 54 | 31 | 1 | 53 | 1 | 46 |
| Virginia | 44 | 16 | 33 | 7 | 46 | 7 | 48 |
| Washington | 63 | 9 | 15 | 13 | 54 | 13 | 33 |
| West Virginia | 44 | 21 | 32 | 4 | 55 | 4 | 42 |
| Wisconsin | 31 | 21 | 38 | 11 | 56 | 11 | 33 |
| Wyoming | 23 | 11 | 61 | 6 | 59 | 6 | 35 |
| United States | 45 | 8 | 33 | 14 | 51 | 14 | 36 |

(continued)

TABLE 6-4  (Continued)

| State | Inputs | | | | Outputs | | |
|---|---|---|---|---|---|---|---|
| | Fertilizer-N | Recoverable Manure-N | Legume-N Fixation | Crop Residues | Harvested Crop | Crop Residues | Balance |
| | | | *High Scenario* | | | | |
| Alabama | 55 | 12 | 28 | 5 | 30 | 5 | 65 |
| Alaska | 97 | 0 | 0 | 3 | 32 | 3 | 64 |
| Arizona | 61 | 18 | 17 | 3 | 29 | 3 | 67 |
| Arkansas | 33 | 7 | 51 | 10 | 38 | 10 | 52 |
| California | 63 | 14 | 19 | 4 | 32 | 4 | 64 |
| Colorado | 37 | 18 | 32 | 14 | 51 | 14 | 35 |
| Connecticut | 44 | 33 | 22 | 1 | 26 | 1 | 73 |
| Delaware | 33 | 13 | 45 | 9 | 30 | 9 | 60 |
| Florida | 90 | 5 | 4 | 1 | 9 | 1 | 90 |
| Georgia | 59 | 11 | 24 | 7 | 35 | 7 | 58 |
| Hawaii | 93 | 6 | 0 | 0 | 1 | 0 | 99 |
| Idaho | 48 | 7 | 36 | 10 | 47 | 10 | 43 |
| Illinois | 38 | 2 | 42 | 18 | 53 | 18 | 29 |
| Indiana | 40 | 3 | 39 | 17 | 52 | 17 | 31 |
| Iowa | 36 | 4 | 42 | 18 | 55 | 18 | 27 |
| Kansas | 44 | 11 | 28 | 16 | 55 | 16 | 29 |
| Kentucky | 45 | 7 | 39 | 10 | 45 | 10 | 46 |
| Louisiana | 45 | 2 | 45 | 8 | 35 | 8 | 56 |
| Maine | 48 | 32 | 15 | 5 | 48 | 5 | 47 |
| Maryland | 32 | 17 | 40 | 10 | 39 | 10 | 51 |
| Massachusetts | 51 | 23 | 26 | 1 | 27 | 1 | 72 |
| Michigan | 39 | 7 | 43 | 11 | 42 | 11 | 46 |
| Minnesota | 36 | 5 | 43 | 15 | 51 | 15 | 34 |

| | | | | | | | |
|---|---|---|---|---|---|---|---|
| Mississippi | 39 | 4 | 51 | 7 | 32 | 7 | 62 |
| Missouri | 33 | 4 | 51 | 12 | 50 | 12 | 39 |
| Montana | 26 | 3 | 55 | 15 | 62 | 15 | 23 |
| Nebraska | 46 | 7 | 29 | 18 | 50 | 18 | 32 |
| Nevada | 11 | 5 | 83 | 1 | 67 | 1 | 32 |
| New Hampshire | 30 | 29 | 41 | 0 | 47 | 0 | 52 |
| New Jersey | 48 | 7 | 38 | 7 | 34 | 7 | 59 |
| New Mexico | 32 | 20 | 41 | 7 | 50 | 7 | 43 |
| New York | 29 | 24 | 42 | 6 | 42 | 6 | 53 |
| North Carolina | 49 | 9 | 33 | 9 | 34 | 9 | 58 |
| North Dakota | 45 | 2 | 37 | 16 | 53 | 16 | 31 |
| Ohio | 37 | 4 | 45 | 14 | 49 | 14 | 38 |
| Oklahoma | 65 | 8 | 18 | 9 | 43 | 9 | 48 |
| Oregon | 57 | 7 | 28 | 8 | 47 | 8 | 45 |
| Pennsylvania | 16 | 26 | 48 | 10 | 52 | 10 | 38 |
| Rhode Island | 81 | 0 | 19 | 0 | 18 | 0 | 82 |
| South Carolina | 48 | 4 | 40 | 8 | 33 | 8 | 58 |
| South Dakota | 23 | 5 | 59 | 13 | 50 | 13 | 36 |
| Tennessee | 45 | 7 | 41 | 8 | 38 | 8 | 55 |
| Texas | 70 | 16 | 5 | 10 | 35 | 10 | 55 |
| Utah | 23 | 10 | 63 | 4 | 49 | 4 | 47 |
| Vermont | 12 | 47 | 41 | 1 | 46 | 1 | 54 |
| Virginia | 41 | 15 | 38 | 6 | 42 | 6 | 51 |
| Washington | 58 | 8 | 21 | 12 | 50 | 12 | 38 |
| West Virginia | 38 | 18 | 41 | 3 | 47 | 3 | 50 |
| Wisconsin | 26 | 17 | 47 | 9 | 47 | 9 | 43 |
| Wyoming | 18 | 8 | 70 | 4 | 45 | 4 | 51 |
| United States | 42 | 8 | 38 | 13 | 47 | 13 | 40 |

NOTE: See the Appendix for a full discussion of the methods used to estimate nitrogen inputs and outputs.

TABLE 6-5    Nitrogen and Phosphorus Fertilizer Use: Top Ten States

| Rank | State | Percent Nitrogen Use | Rank | State | Percent Phosphorus Use |
|------|-------|---------------------|------|-------|------------------------|
| 1. | Illinois | 9 | 1. | Illinois | 9 |
| 2. | Iowa | 9 | 2. | Iowa | 7 |
| 3. | Texas | 8 | 3. | Texas | 6 |
| 4. | Nebraska | 7 | 4. | Minnesota | 6 |
| 5. | Minnesota | 5 | 5. | Indiana | 5 |
| 6. | California | 5 | 6. | Missouri | 4 |
| 7. | Kansas | 5 | 7. | California | 4 |
| 8. | Indiana | 4 | 8. | Ohio | 4 |
| 9. | Missouri | 4 | 9. | Kansas | 4 |
| 10. | Oklahoma | | 10. | Nebraska | 4 |
| | Subtotal | 59 | | Subtotal | 53 |

SOURCE: H. Vroomen. 1989. Fertilizer Use and Price Statistics: 1960–88. Statistical Bulletin 780. Washington, D.C.: U.S. Department of Agriculture, Economic Research Service, Resources and Technology Division.

is by far the major nitrogen user in the United States, accounting for about 41 percent of the fertilizer-N applied (Vroomen, 1989).

Rates of application of fertilizer-N also vary by crop and region. Of the major commodity crops, little or no nitrogen is applied to soybean crops; but in 1988, an average of 153 kg of nitrogen/ha (137 lb/acre) was applied to corn crops nationwide. Corn crops receive the highest amounts of fertilizer-N, which have increased nationally from about 67 kg/ha (60 lb/acre) in 1965 to about 157 kg/ha (140 lb/acre) in 1985. Rates have declined slightly since 1985. The rates of fertilizer-N application to crops such as sorghum and potatoes are also significant, but these crops cover more limited areas (Vroomen, 1989). Geographically, fertilizer-N use parallels cropping patterns; 10 states—predominantly grain-producing states—account for nearly 60 percent of fertilizer-N use (Table 6-5).

*Nitrogen Fixed by Legumes*

The symbiotic bacteria associated with leguminous crops such as alfalfa and soybeans can fix and add substantial amounts of nitrogen to the soil. The amount of nitrogen fixed by alfalfa and soybeans under the low, medium, and high scenarios is approximately 6.1 million metric tons (6.6 million tons), 6.9 million metric tons (7.5 million tons), and 8.6 million metric tons (9.5 million tons), respectively. These estimates represent 30, 33, and 38 percent of nitrogen inputs, respectively (depending on the rate of fixation and the nitrogen replacement values used for alfalfa and soybeans). Alfalfa has been reported to fix as little as

70 kg of nitrogen/ha (62 lb/acre) and as much as 600 kg of nitrogen/ha (532 lb/acre). Soybeans have been found to fix as little as 15 kg of nitrogen/ha (13 lb/acre) and as much as 310 kg of nitrogen/ha (275 lb/acre) (Appendix Table A-4). Some of that fixed nitrogen is removed when the crop is harvested, but some remains in the soil and is available for subsequent crops.

Estimates of the amount of nitrogen actually fixed by particular legumes are problematic because there are no unequivocal methods for measurement (see Appendix). Crop rotation with legumes, however, consistently produces a yield benefit to succeeding crops with reduced inputs of nitrogen. The contribution of legumes to the national nitrogen balance is very important (Tables 6-3 and 6-4). To minimize environmental losses of nitrogen and to optimize crop yields, an estimate of the legume contribution to nitrogen in the farming system must be considered.

### Nitrogen in Animal Manure

The importance of the nitrogen in manure (manure-N) in the mass balance varies from region to region (Tables 6-3 and 6-4). When livestock is a component of the farming system, the contribution of manure-N to the mass balance can be significant.

Economically recoverable manure-N represents 9, 8, and 8 percent of total nitrogen inputs in the low, medium, and high scenarios, respectively. The mass of economically recoverable manure-N, however, is relatively low compared with the total mass of manure-N. Nationally, only 34 percent of the total nitrogen voided in manures is estimated to be economically recoverable for use elsewhere. The portion of manures that are economically recoverable can be increased through better management.

The amount of manure-N actually applied to croplands depends on the kind of manure and, particularly, the way that the producer handles the manure. Application rates vary dramatically from farm to farm, and manure is often applied by using manure-spreading equipment that makes careful calibrations of the nitrogen application rate difficult. In Lancaster County, Pennsylvania, for example, Schepers and Fox (1989) found that manure was applied to fields at rates ranging from 29 to 101 metric tons/ha (13 to 45 tons/acre), even though producers thought they were applying 45 metric tons/ha (20 tons/acre). Different animal manures contain different proportions of nitrogen, and the nitrogen occurs in various forms. A large portion of the nitrogen in manures may be found in the organic form and is not immediately available for crops when it is applied. This nitrogen becomes available over time as it is mineralized and can contribute nitrogen over several crop seasons.

These and other special problems in managing nutrients in manures are discussed at greater length in Chapter 11.

### Nitrogen in Crop Residue

Crop residue is the mass of plant matter that remains in the field after harvest (such as corn stover). The harvested portion of crops remove nutrients from the system, but most of the crop residues remain in the soil system and effectively enter the organic-N storage component. Although crop residues from a previous year may be factored as an input, the crop residues of the current crop year must be considered an output, and for a given field this often results in a relative balance. Hence, in routine management and nutrient-yield response evaluations, residues are often ignored as inputs and are implicitly factored into the soil-mineralization contribution.

### Other Nitrogen Inputs

Synthetic fertilizers, legumes, and manures are the most important sources of nitrogen inputs to soil-crop systems. Nitrogen is, however, added to soil-crop systems in rainfall and irrigation water and through mineralization from soil organic matter. In certain farming systems and at certain times, these other inputs can be important. Because of their variability and the difficulty in estimating the amount of nitrogen obtained from these sources on a state or national basis, they were not used to estimate the mass balances given in Tables 6-1, 6-3, and 6-4. There are other inputs sources, such as nitrogen in dry deposition, crop seed, foliar absorption, and nonsymbiotic fixation of nitrogen. These are minor or secondary inputs that are not typically manageable and are seldom measured. These sources have been implicitly included in nutrient-yield response evaluations and are explicitly ignored in most studies.

### Nitrogen in Rainfall

The amount of nitrogen found in rainfall varies from storm to storm and region to region. The total inorganic nitrogen deposited in rainfall ranged from 3.9 to 12.4 kg/ha/year (3.5 to 11.1 lb/acre/year) in studies done in Indiana, Iowa, Minnesota, Wisconsin, and Nebraska (Tabatabai et al., 1981); and annual averages across the eastern United States in National Atmospheric Deposition Program Monitoring range from 3 to 7 kg/ha (3 to 6 lb/acre) (Schepers and Fox, 1989). These sources provide low amounts of nitrogen compared with the nitrogen inputs from fertilizers, manure, and legumes in intensively managed croplands. Hence, they are not typically considered in cropland nitrogen mass balances (Oberle and Keeney, 1990). They can be, however, an important source of nitrogen in rangelands and natural ecosystems (Schepers and Fox, 1989). Nitrogen inputs from precipitation are generally low, and they are often assumed to be about equivalent to the annual nitrogen losses through runoff and erosion (Meisinger, 1984).

*Nitrogen in Irrigation Water*

The amount of nitrogen in irrigation water is often quite low and is not normally considered in nitrogen mass balances. However, in areas where irrigated and fertilized crop production has been practiced for some time, the nitrogen in the form of nitrate (nitrate-N) in the groundwater used to irrigate crops has become a significant nitrogen source. In the Central Platte River Valley in Nebraska, nitrate contamination of the shallow groundwater has been increasing at a rate of 0.4 to 1.0 mg/liter/year (0.4 to 1.0 ppm/year) (Exner, 1985; Exner and Spalding, 1976, 1990; Spalding et al., 1978). The contamination is related to the nitrogen output losses from intensive nitrogen fertilization in irrigated corn production. In many areas the nitrate-N concentrations in the groundwater have increased from 2 mg/liter (2 ppm) to between 10 and more than 20 mg/liter (10 to >20 ppm) (Exner and Spalding, 1990). With increased nitrate-N concentrations, irrigation water can become an important source of nitrogen. For example, 30 cm (12 inches) of irrigation water with a nitrate-N concentration of 20 mg/liter (20 ppm) would result in an application of 60 kg of nitrogen to each hectare (54 lb/acre) irrigated.

Schepers and colleagues (1986) noted that in the Central Platte River Valley, the groundwater used to irrigate corn contributed an average of 46 kg of nitrogen/ha (41 lb/acre), or 31 percent of the nitrogen applied as fertilizer. The groundwater used to irrigate potatoes in Wisconsin contributed an average of 57 kg/ha (51 lb/acre), or 25 percent of the nitrogen added as fertilizer (Saffigna and Keeney, 1977). Surface waters used as sources of irrigation water usually contain much lower concentrations of nitrogen (Schepers and Fox, 1989). In some natural resource districts in Nebraska, the nitrate-N in the irrigation water must now be accounted for and is used to reduce the amount of fertilizer-N applied (Central Platte Natural Resources District, 1992; Schepers et al., 1991).

*Soil Nitrogen and Mineralization*

Mineralization is a relatively slow process that is dependent on temperature and moisture; only 2 to 3 percent of the organic nitrogen stored in soil is mineralized annually (Buckman and Brady, 1969; Oberle and Keeney, 1990). This 2 to 3 percent, however, is the basis for natural ecosystem nutrient cycling and, depending on the amount of organic matter in the soil, can supply a significant portion of the nitrogen needed by crops each year.

Despite the relatively slow rate of mineralization, this process can be an important factor in determining the year-to-year variability in the amount of nitrogen available to crop plants. The 2 to 3 percent mineralization rate is an average and the moisture and temperature regimes that are optimal for plant growth are also optimal for nitrogen mineralization and nitrification that make the nitrogen stored in soil organic matter available to plants. In years when conditions are optimal, more nitrogen may be released; this natural interaction is an important contributor of nitrogen in climatically optimal years that produce bumper crops. However, the mineralization of nitrogen from the soil, related to inherent soil fertility, has been implicitly included in nutrient-yield response and management evaluations for different soils. New tools are needed to measure the actual nitrogen available from mineralization and other residuals to account for and take advantage of annual variability.

## Nitrogen Outputs

The primary desired output is nitrogen taken up in harvested crops and crop residues. Nitrogen is lost to the atmosphere by volatilization and denitrification and is washed away in runoff in solution, attached to eroded particulates or organic matter. Nitrogen is also leached as nitrate to locations deeper in the soil or to groundwater. Other minor outputs can include gaseous losses such as $N_2O$ evolution during nitrification; decomposition of nitrous acid, or losses directly from maturing or senescent crops (Bremner et al., 1981; Meisinger and Randall, 1991; Nelson, 1982). Some nutrients are taken up by weeds or immobilized by microbes, but these nutrients primarily enter the organic-N storage pool. These minor outputs are secondary factors and have typically been implicitly included in nutrient-crop yield response models.

*Nitrogen in Crops and Residues*

The nitrogen found in harvested crops represents the greatest and most important output of nitrogen from croplands. For 1987, the nitrogen harvested in crops and residues was estimated at more than 13 million metric tons (14 million tons) (Table 6-1). The amount of nitrogen harvested in crops and residues was estimated to be 67, 64, and 60 percent of total inputs under the low, medium, and high scenarios, respectively. The balance of total nitrogen inputs not accounted for in crops or residues was 6.7 million, 7.4 million, and 9.1 million metric tons (7.4 million, 8.1 million, and 10 million tons) under the low, medium, and high scenarios, respectively. These balances represent 33, 36, and 40 percent of total nitrogen inputs, respectively.

## Nitrogen Balance

The national nitrogen balance summarized in Tables 6-1, 6-3, and 6-4, is a partial cropland budget (see Appendix for details). The balance term in Table 6-1 is simply the residual of the major cropland inputs minus the major output of nitrogen taken up in crop production. The balance term, or residual in this treatment, is an estimate of the amount of nitrogen available that (1) may go into storage or (2) may potentially be lost into the environment. The magnitude of the balance and the relative magnitude of the inputs provide insights into the opportunities to improve the environmental and financial performance of farming systems.

Nitrogen balances are positive under all three scenarios (Table 6-1). At the national level, the nitrogen applied to croplands in synthetic fertilizers is roughly the same as that obtained in harvested crops (not including crop residues). Nitrogen balances range from 6 million to 9 million metric tons (6.7 million to 10 million tons) under the low and high scenarios, respectively. Under the high scenario, the nitrogen balance is nearly equal to the amount of nitrogen purchased in synthetic fertilizer. The results reported under the high scenario in Table 6-1 are similar to those reported by Power (1981) and Follett and colleagues (1987) for nitrogen mass balance in 1977 (Table 6-6).

These aggregate mass balances must be interpreted with caution. As discussed earlier, not all of the estimated nitrogen inputs are available for crop growth. A positive balance should, therefore, be expected; and a positive balance of 7 million metric tons (7.8 million tons) of nitrogen, such as estimated under the medium scenario in Table 6-1, does not mean that fertilizer nitrogen applications can be reduced by this same amount.

TABLE 6-6   Estimated Nitrogen Balance for Crop Production in the United States, 1977

| Nitrogen Output or Input | Metric Tons of Nitrogen | Percentage of Total Nitrogen Input Mass |
| --- | --- | --- |
| Output in 1977 in | | |
| Harvested crops | 7.6 | 36 |
| Crop residues | 4.3 | 20 |
| Total | 11.9 | 56 |
| Inputs to cropland | | |
| As commercial fertilizer | 9.5 | 45 |
| As symbiotically fixed N | 7.2 | 34 |
| In crop residues | 3.0 | 14 |
| In manure and organic wastes | 1.4 | 7 |
| Total | 21.1 | 100 |

SOURCE: Adapted from J. F. Power. 1981. Nitrogen in the cultivated ecosystem. Pp. 529–546 in Terrestrial Nitrogen Cycles—Processes, Ecosystem Strategies and Management Impacts, F. E. Clark and T. Rosswall, eds. Ecological Bulletin No. 33. Stockholm, Sweden: Swedish Natural Science Research Council.

The magnitude of the estimated positive balances, however, does help to explain the prevalence of elevated nitrate concentrations in surface water and groundwater in intensive agricultural watersheds. The magnitude of the positive nitrogen balance and the portion of that balance lost to surface water, groundwater, or the atmosphere, however, vary greatly by region and between farms.

The amount of nitrogen taken up (the output term) varies from crop to crop and with crop yield. This variation is evident in the aggregate mass balances among the states (Tables 6-3 and 6-4). Such aggregate differences, however, do not account for the disparity between nitrogen additions and removals for selected crops. For example, as many large-scale balances would suggest, the harvested crop nitrogen output is slightly greater than the fertilizer nitrogen input. However, more than 35 percent of the nitrogen output in harvested crops is accounted for by various legumes that receive very little nitrogen fertilizer. Major commodities, including corn, cotton, potatoes, rice, and wheat account for more than 80 percent of the fertilizer-N applied. The nitrogen output in harvested grain from these commodities, however, accounts for only about 57 percent of their fertilizer-N input. If all legume inputs and outputs are taken out of the national balance, the remaining harvested crops output is only equivalent to about 35 to 40 percent of the fertilizer- and manure-N inputs. In 1987, approximately 41 percent of total fertilizer-N used was applied to corn, whereas approximately 26 percent of the nitrogen in all harvested crops was in corn.

These data illustrate one part of the nitrogen balance problem; at current rates of nitrogen application, some crop management systems are not as efficient as was once presumed. Nitrogen recovery, even apparent nitrogen recovery, by agronomic crops is seldom more than 70 percent, and average values are closer to 50 percent (Keeney, 1982). Furthermore, some of the nitrogen recovered by crop plants is returned to the soil nitrogen pool as crop residues and roots and becomes part of the nitrogen pool in the soil. The amount of nitrogen actually removed from the system in harvested portions of the crop can be more in the range of 35 percent or less, particularly for continuous cropping of corn or other grains that receive large additions of nitrogen (Meisinger et al., 1985; Sanchez and Blackmer, 1988; Timmons and Baker, 1991; Varvel and Peterson, 1990).

Peterson and Frye (1989) obtained a similar result; that is, for U.S. corn production, the amount of nitrogen fertilizer used has exceeded the amount of nitrogen harvested in grain by 50 percent every year since 1968. This situation is even more striking because the data do not account for any other nitrogen additions—from manure, legumes in rotation, or soil nitrogen mineralization—that are common in corn production.

## Losses to the Environment

As discussed, the residual or balance term in the nitrogen balance is an estimate of the amount of nitrogen that may go into storage or be lost into the environment. Various cropland studies show that the post-harvest residual of available nitrogen in the soil, both in the fall and following crop season, is proportionately related to the amount of nitrogen applied (e.g., Bundy and Malone, 1988; Jokela, 1992; Jokela and Randall, 1989). In the context of climatic variability and related crop yield variability, some residual nitrogen and some losses into the environment are inevitable. The magnitude of this residual is related to the potential for excessive losses into the environment.

### Losses to the Atmosphere

Nitrogen can volatilize directly from the fertilizers and manure applied to the surface of croplands and can be lost from the soil as nitrogen gases are produced through denitrification. Losses from direct volatilization can be quite large, especially from surface applications of manure. These contributions of nitrous oxides, ammonia, and methane to greenhouse gases is of concern. Recent studies suggest that, except under special conditions, loss of nitrogen through denitrification may be lower than previously thought (Schepers and Fox, 1989). For most rainfed systems of fertilized crops, estimates of nitrous oxide emissions from denitrification and nitrification range from 1 to 20 kg/ha/year (1 to 18 lb/acre/year) (Duxbury et al., 1982; Thomas and Gilliam, 1978) and the proportion of fertilizer-N lost is estimated at 2 to 3 percent per year (Eichner, 1990; Goodroad et al., 1984). Cultivated legumes also contribute to nitrous oxide emissions (Eichner, 1990) and losses from flooded rice production can be quite high for many gaseous forms of nitrogen (Lindau et al., 1990). Part of the unaccounted for nitrogen in Table 6-1 is undoubtedly delivered to the atmosphere, but the probable amount of nitrogen lost to the atmosphere is difficult to estimate.

### Losses to Surface Water and Groundwater

A portion of the unaccounted for nitrogen is delivered to surface water and groundwater through runoff, erosion, and leaching. Larson and colleagues (1983) estimated that 9.5 metric tons (10.5 million tons) of nitrogen was lost with eroded soil in 1982, an amount roughly equivalent to the amount of nitrogen applied in synthetic fertilizers in 1987. In addition, some nitrogen in the form of ammonium (ammonium-N) is lost along with the organic-N attached to soil particles. This ammonium-N contributes to the available nitrogen in surface water. Little soluble nitrogen is lost in true runoff. The majority of soluble nitrogen, nitrate, is lost in leaching through the soil and may move as shallow, subsurface flow or as deeper groundwater into surface waters (Lowrance, 1992a). Most of the nitrate found in surface waters comes from this groundwater component (Hallberg, 1987).

Proportional relationships between nitrogen applications and the nitrogen found in water have been shown in several studies (Hallberg, 1989b; Keeney, 1986a). The amount of nitrate-nitrogen lost in leaching to drainage tiles installed beneath topsoil was related in a nearly linear fashion to the amount of nitrogen applied for lands with application rates that exceeded 50 kg/ha (45 lb/acre) (Baker and Laflen, 1983). Nitrate accumulated in the water of subsoils of three experimental sites in Virginia only after the amount of nitrogen applied exceeded the optimum rate (Hahne et al., 1977). Investigators found a similar pattern in central Nebraska. The concentration of nitrate-nitrogen in groundwater under croplands was found to increase as the rate of nitrogen application increased (Schepers et al., 1991). The groundwater under croplands

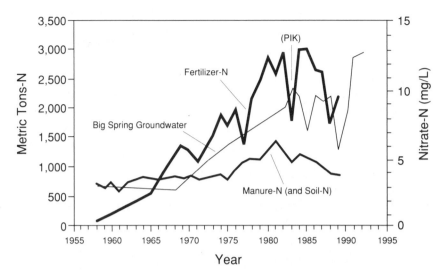

FIGURE 6-2   Amount of fertilizer-N and manure-N applied in relation to annual average nitrate concentration in groundwater in Big Spring Basin, Iowa. PIK, the Payment-In-Kind program initiated by the U.S. Department of Agriculture in 1983 that resulted in taking large acreages of cropland out of production in 1983. Source: G. R. Hallberg. 1989. Nitrate in ground water in the United States. Pp. 35–74 in Nitrogen Management and Ground Water Protection, R. F. Follet, ed. Amsterdam: Elsevier. Reprinted with permission from © Elsevier Science Publishers, B.V.

that received nitrogen at 45 kg/ha (50 lb/acre) less than the recommended application rate had nitrate concentrations of about 10 mg/liter (10 ppm). Concentrations of 18 to 25 mg/liter (18 to 25 ppm) were found under croplands where producers applied nitrogen at up to 168 kg/ha (150 lb/acre) in excess of the recommended amount (Schepers et al., 1991). In this area of central Nebraska, there was no significant difference in yields between fields where 23 kg/ha (50 lb/acre) less than the recommended amount was applied and fields where 168 kg/ha (150 lb/acre) more than the recommended amount of nitrogen was applied.

Data from the Big Spring Basin in Iowa trace the relationship of increasing residual nitrogen and groundwater nitrate concentrations over time (Figure 6-2). The amount of nitrogen applied as commercial fertilizer, manure, and legume-N has increasingly exceeded that harvested in the crop since 1970 (Hallberg, 1987). The concentration of nitrate in groundwater has increased as the difference between the amount of nitrogen applied and the amount of nitrogen harvested has increased and as the number of years of applied nitrogen amounts in excess of harvested nitrogen amounts has increased.

Many studies have shown that the amount of residual nitrogen in cropland soil is closely related to the amount of nitrogen applied as fertilizer manure or provided as legumes. The application of nitrogen in excess of that needed for crop requirements leaves a pool of residual nitrogen in the soil at the end of each growing season. Much of this residual nitrogen is in the form of nitrates. Nitrates are soluble in water and move quickly and easily through the soil profile. It is the residual nitrogen that is most likely to pollute groundwater or surface water. Some of this residual may remain in the root zone and contribute to subsequent crops (Jokella and Randall, 1989), but this residual can readily be lost to pollute groundwater and surface water (Hallberg, 1987; Sanchez and Blackmer, 1988).

## OPPORTUNITIES TO REDUCE NITROGEN LOSSES

The nitrogen mass balances in Tables 6-1, 6-3, and 6-4 illustrate the fact that, under some situations, the mass of unharvested nitrogen can be quite large. The balance between the nitrogen entering and leaving the soil-crop system is the critical factor that must be managed on croplands to prevent unacceptable losses of nitrogen to the environment. The goal is to strike a balance between the amount of nitrogen entering the system and the amount taken up and removed by crops while minimizing the amount of nitrogen left in the system so that the mass of residual nitrogen that may end up in water or the atmosphere, over time, is as small as possible. Reducing the mass of residual nitrogen added to the soil-crop system can improve both economic and environmental performances.

Reducing the mass of residual nitrogen in the soil-crop system can be accomplished by accounting for all sources of nitrogen added to the system, refining estimates of crop nitrogen requirements, refining yield goals, synchronizing the application of nitrogen with crop needs, and increasing seasonal nitrogen uptake in the cropping system.

### Accounting for Nitrogen from All Sources

The nitrogen balances in Tables 6-1, 6-3, and 6-4 suggest the importance of accounting for all nitrogen sources in the farming system when attempting to improve nitrogen management. Regional or farm-level nitrogen balances reveal similar imbalances between nitrogen inputs and outputs.

## Regional Nutrient Balances

Peterson and Russelle (1991) estimated that alfalfa, which occupies only 8 percent of the cropland in the Corn Belt, fixes more than 1 billion kg of nitrogen (2.2 billion lb) annually, whereas 4 billion kg (8.8 billion lb) of nitrogen is applied in the form of commercial fertilizer to croplands in the eight-state Corn Belt region. Alfalfa accumulates nitrogen through symbiotic fixation and the concentration of nitrogen from the soil profile. It contributes some of this nitrogen directly to the soil-crop system when it is plowed under for a succeeding crop through mineralization of plant residues. It contributes nitrogen indirectly through manures from livestock fed alfalfa. Peterson and Russelle (1991) estimate that if the nitrogen contributed both directly and indirectly by alfalfa was accounted for, fertilizer-N applications in the Corn Belt as a whole could be reduced between 8 and 14 percent with no yield loss (Table 6-7). For states with larger areas of alfalfa crops, the potential fertilizer reductions are much greater. In Wisconsin, for example, the range of possible nitrogen application reductions was 37 to 66 percent, in Michigan it was 20 to 36 percent, and in Minnesota it was 13 to 23 percent.

Lowrance and colleagues (1985) estimated nutrient budgets for agricultural watersheds in the southeastern coastal plain. They accounted for nitrogen inputs from precipitation, commercial fertilizer, and legumes and estimated the outputs in stream flows and harvests. The proportion of nitrogen unaccounted for in harvested crops ranged from 47 to 75 percent of total inputs, depending on the watershed and the year studied.

## Farm Nitrogen Balances

Legg and colleagues (1989) estimated nitrogen balances for southeastern Minnesota and found that nitrogen from alfalfa, soybeans, and manure provided, on average, 95 kg/ha (85 lb/acre) or 64 percent of the nitrogen applied in commercial fertilizers. The total nitrogen per hectare applied from all sources was, on average, 72 kg/ha (64 lb/acre) in excess of the nitrogen needed to achieve yield goals. The importance of accounting for all sources of nitrogen applied to the crop-soil system is even more apparent if the data provided by Legg and colleagues (1989) for four farms in their study area are examined (see Chapter 2, Table 2-2). For farms A, B, and C, respectively, 42, 102, and 29 percent of the nitrogen needed to achieve yield goals was provided by legumes and manure alone. All three farms, however, applied commercial fertilizer in

TABLE 6-7   Potential Reductions in Nitrogen Fertilizer Applied to Corn

| | Area of Corn Following Alfalfa[a] ($10^3$ ha) | Potential Fertilizer Reductions ($10^6$ kg) | | | | | |
| | | After Alfalfa | | With Manure | | Total (percent) | |
| State | | High[b] | Low[c] | High[d] | Low[e] | High[f] | Low[f] |
|---|---|---|---|---|---|---|---|
| Illinois | 86 | 22 | 7 | 16 | 13 | 38 (5) | 20 (3) |
| Indiana | 47 | 12 | 4 | 8 | 7 | 21 (5) | 11 (2) |
| Iowa | 185 | 43 | 14 | 34 | 27 | 76 (9) | 41 (5) |
| Michigan | 162 | 34 | 11 | 27 | 21 | 61 (36) | 33 (20) |
| Minnesota | 219 | 39 | 13 | 36 | 29 | 75 (23) | 42 (13) |
| Missouri | 51 | 11 | 4 | 7 | 5 | 17 (13) | 9 (7) |
| Ohio | 78 | 20 | 7 | 14 | 11 | 33 (12) | 17 (6) |
| Wisconsin | 363 | 57 | 19 | 56 | 45 | 133 (66) | 64 (37) |
| Total | 1,191 | 238 | 79 | 198 | 157 | 435 (14) | 237 (8) |

NOTE: Potential reductions are estimated by adjusting fertilizer application rates to account for the nitrogen supplied by alfalfa by fixation or indirectly from manure produced by livestock fed alfalfa.

[a]Assuming 28.6 percent of the alfalfa area is rotated to corn each year.

[b]Assuming corn does not require nitrogen fertilizer the first year following alfalfa and requires half the average rate the second year after alfalfa.

[c]Assuming corn requires half the average rate the first year following alfalfa and the full average rate the second year after alfalfa.

[d]Assuming 40 percent of the nitrogen in manure is available to corn the first year after application and 40 percent of the remaining nitrogen is available the second year after application.

[e]Assuming 30 percent of the nitrogen in manure is available to corn the first year after application and 30 percent of the remaining nitrogen is available the second year after application.

[f]Fertilizer-N reduction expressed as a percentage of total nitrogen fertilizer applied to corn.

SOURCE: T. A. Peterson and M. P. Russelle. 1991. Alfalfa and the nitrogen cycle in the Corn Belt. Journal of Soil and Water Conservation 46:229–235. Reprinted with permission from © Journal of Soil and Water Conservation.

amounts nearly adequate to achieve yield goals in the absence of any other nitrogen inputs. For these three farms, commercial fertilizer applications could have been reduced by 39, 100, and 19 percent, respectively, without any change in yield goals or loss in yields.

Similar results have been reported elsewhere (Hallberg, 1987; Lanyon and Beegle, 1989; Magette et al., 1989; Olson, 1985). A budget for the state of Nebraska suggests that since the mid-1960s, the amount of nitrogen applied to croplands has exceeded crop requirements by 20 to 60 percent (Olson, 1985). The regional and farm level nitrogen balances reinforce the results of the state and national balances in Tables 6-3 and 6-4.

*Improving Nitrogen Management*

These results clearly suggest that producers have a great opportunity to improve nitrogen management and reduce the mass of residual nitrogen in the soil-crop system by properly accounting for all sources of nitrogen. The importance of accounting for all sources of nitrogen varies greatly from farm to farm and region to region, depending on the relative contributions of various sources of nitrogen to the soil-crop system. Regional variation is apparent in Tables 6-3 and 6-4 (see also Figure 3-1, Chapter 3).

When multiple sources of nitrogen are involved, a proper accounting of all sources may be the single most important step in improving nitrogen management. The amount of nitrogen that needs to be applied to cropland depends on how much nitrogen is already available from all sources. Nitrogen available from manure applications, legumes, soil organic matter, and other sources should be accounted for before recommendations for supplemental applications of nitrogen are made. The importance of carefully accounting for all sources of nitrogen has been repeatedly stressed as a way to improve nitrogen management (see, for example, Bock and Hergert [1991], Peterson and Frye [1989], Schepers and Mosier [1991], and University of Wisconsin-Extension and Wisconsin Department of Agriculture, Trade and Consumer Protection [1989]; U.S. Congress, Office of Technology Assessment [1990]).

Even though crop producers can nearly always reduce their costs by adequately accounting for all sources of nitrogen, the available survey data suggest that such accounting is the exception rather than the rule in current practice. In 1987, El-Hout and Blackmer (1990) evaluated the nitrogen status of first-year corn fields, following alfalfa rotations, in northeastern Iowa using soil and tissue tests. The evaluations showed that most producers were not taking adequate credits for their alfalfa. Fertilizer-N application rates ranged from 6 to 227 kg/ha (5 to 203 lb/acre) and averaged 136 kg/ha (121 lb/acre), yet yields ranged from 9 to 13 metric tons/ha (4 to 6 tons/acre), averaging about 12 metric tons/ha (5 lb/acre). Fifty-nine percent of the fields also received some manure applications. Of the 29 fields, 86 percent had greater concentrations of soil nitrate than were needed for optimal yields; 56 percent had at least twice the critical amount needed, and 21 percent had at least three times this amount. Crop response studies in this region have consistently shown that no fertilizer-N or only a small starter nitrogen application is needed to produce optimal or maximum yields after a multiyear alfalfa stand. Had such recommendations for rotation benefits been followed, the average optimal fertilizer-N rate would have been 13 kg/ha (12 lb/acre), 123 kg/ha (110 lb/acre) less than the rate that was used (El-Hout and Blackmer, 1990).

### Soil Testing

Although soil testing in the fall can be an effective management tool for phosphorus and potassium, this is not the case for nitrogen. Measuring the nitrogen available as nitrate or ammonium in fall soil samples is ineffective for estimating the amount of residual nitrogen available from the soil for the next growing season. Because such nitrogen is readily transformed or leached over the fall, winter, and early spring, the available forms of nitrogen present in the fall often have little bearing on the available nitrogen for the next season in the humid and subhumid Grain Belt states (Jokela and Randall, 1989; Magdoff, 1991a,b). Organic carbon content is sometimes measured by using fall soil samples, and this measure is used to provide an estimate of the amount that may be mineralized in the next growing season. The long-term average amount of mineralized nitrogen contributed is one of many factors implicitly incorporated into long-term nitrogen application rate experiments and, hence, is also implicitly included in nitrogen recommendations based on such studies. New soil testing approaches are showing promise to provide enhanced management, particularly for crop production in the grain belt (Binford et al., 1992; Magdoff, 1991a,b).

## Improving Estimates of Crop Nitrogen Needs

The first stage in nitrogen management is the establishment of the nitrogen requirements and the yield response of the crop to nitrogen. This work is done through field trials by growing the crop using various nitrogen application rates, usually on research plots, and measuring the changes in crop yields. The variability in crop response to nitrogen is accounted for by multiple plot replications of the same nitrogen application rates to integrate the local variability imposed by soil (and imposed by the research methods used on small plots), replication of experiments in different areas to assess the variability caused by different soil and climatic conditions, and replication of experiments over time at the same location to integrate the variability imposed by annual climatic differences. Variability in results is confounded, for example, by genetic improvements in corn hybrids, crop rotations, tillage, and pest and weed management.

Such experiments have been used to establish realistic crop production potentials for various regional (substate) combinations of soil, climate, and management. With all the sources of variance in such data (for example, year-to-year and plot-to-plot variations), determination of optimal fertilization rates involves the fitting of some form of statistical model to the observed crop yield responses to the various rates of fertilizer application over time.

### Economically Optimum Rate of Nitrogen Application

The concept of the economically optimum rate of nitrogen application was developed early in the assessment of the use of fertilizers to enhance crop production (Heady et al., 1955; National Research Council, 1961). Because there is a declining rate of yield increase at increasing rates of application, the economically optimum rate is functionally the point at which the price of the last small increment of fertilizer equals the value of the additional crop produced by this fertilizer. At higher rates the additional crop is worth less than the additional fertilizer. This relationship is affected by changes in fertilizer and corn prices. Many different statistical response models have been used to identify economic optimum rates. Various reports have noted that these models can disagree significantly in identifying optimal rates (Anderson and Nelson, 1975; Cerrato and Blackmer, 1990; Nelson et al., 1985), but these disagreements have received little attention. There is no standard approach for selecting one model over another. Typically, investigators use the best-fitting model, determined by a correlation coefficient, to the given set of data. Corn yield responses to nitrogen most typically have been described by a quadratic equation model and field studies with two to four replications of two to four rates of fertilizer application, particularly for long-term studies.

### More Refined Models Needed

Recent work provides a more rigorous statistical comparison and assessment of such models. Using data from long-term crop rotation studies (with up to 28 years of continuous treatment), Blackmer (1986) illustrated that testing two to four different rates of fertilizer application does not provide enough data to define the economically optimal nitrogen application rate. Cerrato and Blackmer (1990) evaluated the five most widely used response models, and their resultant predictions, from 12 site-years of corn yield data. They used 10 nitrogen application rates for each site and three replications of each treatment. The various models provided similar, significant correlations and predicted similar maximum crop yields. However, the models predicted widely different economically optimum rates of nitrogen fertilization ranging from 128 to 379 kg/ha (114 to 338 lb/acre). Using the standard model the predicted rate was 22 percent greater than the best model indicated by more thorough statistical evaluation of the results.

This illustrates a source of potential error that contributes to excess nitrogen use. Although refinement of such crop response models is hardly as simple as it seems, refinement of such models is important for refining nitrogen input recommendations (Bock and Sikora, 1990).

### Determining Realistic Yield Goals

Ideally, the nitrogen fertilizer application recommendation should be based on the amount of nitrogen that must be made available during the growing season to produce the crop. However, estimates of the amount of nitrogen that the crop needs must be made before the crop is grown and before the weather and other factors that will affect the year's yield are known. Hence, the producer establishes a yield goal: a preseason estimate of the crop yield the producer hopes to realize. The yield goal is then used to project the amount of nitrogen that should be applied on the basis of the projected amount needed to achieve the yield goal.

The importance of setting realistic yield goals as the basis for making both economically and environmentally sound recommendations has been highlighted many times (see, for example, Bock and Hergert [1991]; Peterson and Frye [1989]; University of Wisconsin-Extension and Wisconsin Department of Agriculture, Trade and Consumer Protection [1989]; U.S. Congress, Office of Technology Assessment [1990]). Setting realistic yield goals is particularly important for reducing residual nitrogen. An unrealistically high yield goal will result in nitrogen applications in excess of that needed for the yield actually achieved and will contribute to the mass of residual nitrogen in the soil-crop system. (See Chapter 2 for more discussion of yield goals.)

The most reliable way to set yield goals is to base goals on historical yields, for example, during the past 5 years, actually achieved on a field-by-field basis. Use of a yield achieved under optimal weather conditions that lead to a bumper crop as the goal will lead to the overapplication of nitrogen during most years. This practice increases production costs and residual nitrogen; in addition, many soils, except those low in organic matter, may supply the added nitrogen needed during a bumper crop year because the warm and moist conditions that lead to a bumper crop also increase the amount of nitrogen mineralized from soil organic matter (Schepers and Mosier, 1991).

Another part of the problem is that some producers set yield goals for their whole farm rather than for each field and often fertilize each field similarly. Not only must yield goals be set realistically but, to optimize management, they should also be set on a field-by-field and preferably a soil-by-soil basis (Carr et al., 1991; Larson and Robert, 1991).

## Synchronizing Applications with Crop Needs

The need to improve nitrogen management by synchronizing applications with periods of crop growth has been often highlighted (see, for example, Ferguson et al. [1991]; Jokela and Randall [1989]; Peterson and Frye [1989]; Randall [1984]; Russelle and Hargrove [1989]; University of Wisconsin-Extension and Wisconsin Department of Agriculture, Trade and Consumer Protection [1989]; U.S. Congress, Office of Technology Assessment [1990]).

Nitrogen is needed most during the period when the crop is actively growing. Nitrogen applied before that time is vulnerable to loss through leaching or subsurface flow because of the mobility of nitrates in the soil system. Larger applications of nitrogen are generally used if nitrogen is applied in the fall, in particular, to make up for the nitrogen that is lost or that becomes unavailable in the soil during the period between application and crop growth. However, timed or multiple applications must be carefully evaluated for their economic and environmental efficacy. Simply increasing the number of applications presuming that this will improve crop uptake efficiency may ignore many other factors that affect crop growth (Killorn and Zourarakis, 1992; Timmons and Baker, 1991).

Production and environmental advantages to simple changes in timing of application may be climate and site specific. When timing is coupled with new tools, such as the presidedress soil nitrate test, to gauge the amount of nitrogen available, and hence the additional amount actually needed, significant economic and environmental benefits may be possible.

## New Tools for Nitrogen Management

New tools and management methods are needed to accurately assess available residual nitrogen and to reduce the producer's uncertainty in estimating a crop's nitrogen needs. As discussed, typical soil test methods are inadequate. In practice, nitrogen recommendations rely on evaluating general soil types and using the state's (extension-experiment station) recommended rate for a given yield goal for the soil types in that region. This approach, in part, has led to the blanket nitrogen applications that are part of the current problems and inefficiencies. Practical and accurate soil and plant testing methods that allow refined assessment of crop nitrogen needs, in relation to the nitrogen available (through soil mineralization, available residual, rotation, and manure additions) in a particular growing season, are needed to reduce the uncertainty and risk involved with nitrogen fertilizer applications.

Various plant and tissue tests, such as petiole testing for potato (Wescott et al., 1991), have proved valuable to more efficient nitrogen management for high value vegetable and citrus crops. Such methods must be refined and implemented for the major row crops, such as corn, that consume the majority of nitrogen applied to croplands. Many methods are being tested across the Corn Belt (see, for example, Binford et al., 1992; Binford and Blackmer, 1993; Blackmer et al., 1989; Cerrato and Blackmer, 1991; Fox et al., 1989; Magdoff, 1991a; Meisinger et al., 1992; Motavelli et al., 1992; Piekielek and Fox, 1992; Roth et al., 1992; Tennessee Valley Authority, National Fertilizer Development Center, 1989). One of the methods showing promise is the presidedress soil nitrate test (PSNT). The soil testing is done at a specified time after crop emergence and measures the amount of nitrate-N available in the upper 0.3 to 0.6 m (1 to 2 feet) of the soil profile. The PSNT provides a measure of whether or not supplemental nitrogen is actually needed given the estimate of nitrogen that is already available to the crop. In a project to implement and evaluate the PSNT with fertilizer dealers in Iowa, replicated on-farm trials produced equivalent crop yields but reduced nitrogen applications an average of 42 percent using the PSNT to refine nitrogen applications. The test saved money for producers and significantly reduced environmental loading of nitrogen (Blackmer and Morris, 1992; Hallberg et al., 1991).

Implementation of soil or tissue tests requires that producers sidedress a significant portion of their nitrogen. Few producers, however, currently sidedress their nitrogen applications. Further work is needed on an early spring test that might be useable for preplant applications. In this regard, development of monitoring and modeling systems to help estimate nitrogen availability from the soil and annual carryover, related to climatic, soil, and crop conditions are also needed. Such systems could help to provide forecasts to producers about carryover and availability for them to consider in their annual nutrient and fertilizer application plans.

## Obstacles to Better Nitrogen Management

The measures described above, if implemented, would greatly improve the efficiency with which nitrogen is now used in current cropping systems. Many of these measures could be implemented immediately; others require the development of refined tools such as better soil tests or improved crop response models. In the short term, efforts to improve nitrogen management in current cropping systems should be the priority and should have the potential to result in immediate gains in both economic and environmental performance.

In the long term, however, it is unclear whether such improvements in nitrogen management alone will be sufficient to reduce nitrogen loadings to levels where damages are acceptable. There are elements of the nitrogen problem that suggest that, in the long term, changes in cropping systems that will allow producers to capture more of the available nitrogen may be necessary to adequately reduce nitrogen loadings to surface water and groundwater in some environments.

### Economic Obstacles

Producers face a management dilemma because the effectiveness and the efficiency of nitrogen management cannot be assessed, economically or environmentally, until the growing season is over. A crop that produces poor yields because of inclement weather will result in poor nitrogen use efficiency and uptake, potentially leaving large amounts of nitrogen to be lost to the environment, no matter how carefully a management plan was designed. Since producers must make nitrogen applications without being able to predict weather and crop yields, the potential for being wrong is always present and will always occur in some years. Current recommendations of crop nitrogen needs are based on long-term assessments designed to average the many sources of variance in the nitrogen-yield response. This method also averages the recoveries of residual nitrogen carried over from a previous year or the greater amounts that may be mineralized and available under optimal climatic conditions.

In addition, the nature of the crop response to nitrogen and its resulting effect on the economically optimal rate of nitrogen application also constrain the extent to which improvements in nitrogen management alone may reduce nitrogen losses from current cropping systems.

The first stage in current management is to establish the nitrogen requirements of a crop under various soil and climatic conditions. Figure 6-3 shows the yield response of corn to nitrogen for various soils under continuous corn, and Figure 6-4 shows the nitrogen-yield response for corn for three crop rotations on the same soils. The relationships in Figures 6-3 and 6-4 illustrate the benefits of nitrogen fertilization, up to a certain point, in increasing crop yield, particularly in continuous corn.

Figure 6-4 also illustrates the need for less supplemental nitrogen in crop rotations with legumes; there is no yield benefit to nitrogen use following alfalfa. Figure 6-3 illustrates the inherent variability among soils in their capacity to supply nitrogen from mineralization. These factors vary not only among soils and cropping systems but from year to year as well. Figure 6-5 reexamines the relative efficiencies of the typical nitrogen-yield response relationship from the data in Figure 6-4. The nature of the relationship between the nitrogen application rate and

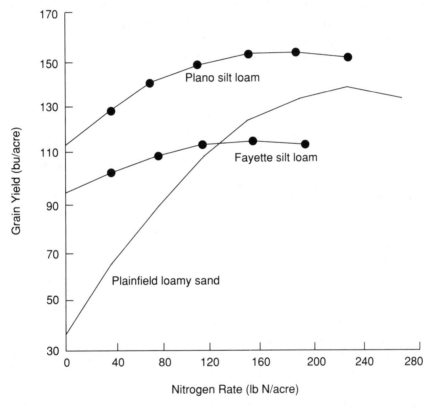

FIGURE 6-3  Yield response of corn to nitrogen applied to three soils. Fayette silt, Fayette silt loam (fine-silty, mixed, mesic Typic Hapludalfs); Plano silt, Plano silt loam (fine-silty, mixed, mesic Typic Argiudolls); and Plainfield ls, Plainfield loamy sand (mixed, mesic Typic Udipsamments). Source: S. L. Oberle and D. R. Keeney. 1990. A case for agricultural systems research. Journal of Environmental Quality 20:4–7. Reprinted with permission from © American Society for Agronomy, Crop Science Society of America, and Soil Science Society of America.

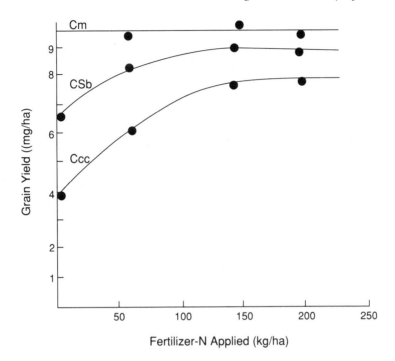

FIGURE 6-4   Yield response of corn to fertilizer for three crop rotations. Ccc, continuous corn; CSb, corn-soybeans-(corn oats); Cm, corn-oats-meadow-meadow (meadow-alfalfa brome mix). Source: Adapted from A. M. Blackmer. 1984. Losses of fertilizer N from soil. Report No. CE-2081, Ames, Iowa: Iowa State University, Cooperative Extension Service.; A. M. Blackmer. 1986. Potential yield response of corn to treatments that conserve fertilizer-N in soil. Agronomy Journal 78:571–575; and J. R. Webb. 1982. Rotation-fertility experiment. Pp. 16–18 in Annual Progress Report Northwest Research Center. Ames, Iowa: Iowa State University.

yield for continuous corn illustrates that, at some point, additional increments of nitrogen application become less efficient. For every additional kilogram of nitrogen applied, less grain is produced, and hence, less of that increment of nitrogen is taken up by the plant. This result is illustrated by the shaded area and dashed lines in Figure 6-5. As the rate of nitrogen application increases, less is recovered in the harvested grain (or in plant residues) and more nitrogen remains as residual nitrogen, potentially to be lost into the environment.

The shaded areas in Figure 6-5 represent a range of values, for perspective. The apparent nitrogen recovery is calculated from the grain yield of a particular increment on the continuous corn yield curve, using

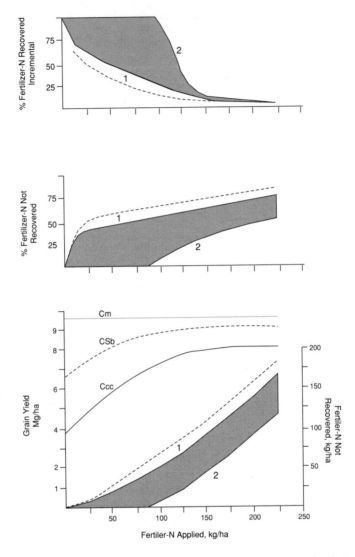

FIGURE 6-5 Nitrogen recovery related to fertilization rate. FN, fertilizer-N; Ccc, continuous corn; CSb, corn-soybeans-(corn-oats); Cm, corn-oats-(meadow-alfalfa brome mix). Source: Adapted from A. M. Blackmer. 1984. Losses of fertilizer-N from soil. Report No. CE-2081, Ames, Iowa: Iowa State University, Cooperative Extension Service.; A. M. Blackmer. 1986. Potential yield response of corn to treatments that conserve fertilizer-N in soils. Agronomy Journal 78:571–575; and J. R. Webb. 1982. Rotation-fertility experiment. Pp. 16–18 in Annual Progress Report Northwest Research Center. Ames, Iowa: Iowa State University.

various standard assumptions. The first assumption is that the grain contains 1.5 percent nitrogen (and weighs 25 kg/ha [56 lb/bu]); the second assumption is that the corn plant requires 0.54 kg of nitrogen per kg of grain (1.2 lb/bu) produced, and the nitrogen is proportioned at 60 percent nitrogen into the grain and 40 percent into the stover. These and the other assumptions given below provide a set of curves enclosed by the shaded envelopes in Figure 6-5. The upper boundary (line 1 of the shaded areas), indicating the lower recovery of fertilizer-N for a given fertilizer application rate, was estimated by subtracting the nitrogen recovered by the unfertilized corn (yield, about 4 metric tons/ha [64 bu/acre]) from the total nitrogen recovered for a given fertilizer application rate. The nitrogen recovered by the unfertilized corn provides a measure of the average amount of nitrogen provided from the soil system (mineralization, including crop residues, and precipitation). The lower boundary provides a conservative estimate that is based on the total amount of nitrogen recovered in the grain but uncorrected for yields from unfertilized areas. The dashed lines (near the line 1 boundary in Figure 6-5) show the upper bound estimated from the corn yields in the corn-soybean rotation.

The values for the incremental fertilizer recovery illustrate how fertilizer-N recovery declines rapidly as the crop approaches optimum and maximum yields. At the maximum yield, recovery effectively reaches zero; at the economically optimum yield, recovery of the last increment of fertilizer-N is less than 10 percent. Even under the more conservative second assumption, less than 50 percent of fertilizer-N is recovered at the economically optimum yield for continuous corn.

Hence, even with economically optimum yields, there is considerable potential for nitrogen losses into the environment. Because of the form of the nitrogen-yield response, the potential for nitrogen losses is very sensitive at high nitrogen application rates when plant uptake of nitrogen is limited. Decreasing the economically optimum yield goal by 5 percent reduces the unrecovered fertilizer-N by about 20 to 30 percent for the continuous corn and reduces the unrecovered amount even more for the corn-soybean rotation. Attempts to push for a last small yield increment can greatly contribute to nitrogen losses. The fate of this nitrogen can follow many paths in the nitrogen cycle; some is immobilized, but other portions may be leached into groundwater or otherwise lost.

### Seasonal Obstacles

In addition to the economic incentives, elements of nitrogen dynamics in the soil-crop system may constrain the gains from improved management of nitrogen inputs alone.

The application of nitrogen in the spring is followed by immobilization of nitrogen by plants and microbes in the spring and summer. This immobilization period is followed by mineralization of the nitrogen from plant and microbial tissues in the fall. The seasonal dynamics are such that nitrate levels in the soil are very low during the late summer and early fall (Boone, 1990; Magdoff, 1991b). Following harvest, crop residues, root tissues, and microbial cells begin to mineralize and nitrify, often leading to high soil nitrate concentrations that are susceptible to loss through leaching or runoff during the fall, winter, and spring (Gold et al., 1990). Thus, the nitrate that is lost from cropping systems is not simply nitrogen that has not been used by the crop but includes nitrogen that has been cycled through plant and microbial tissues during the growing season.

Fine-tuning nitrogen input management will reduce losses of nitrogen but may not provide sufficient reductions in nitrate losses from mineralization of crop residues, root material, and microbial cells following harvest. In some settings, the only way to manage this residual nitrogen may be to keep it tied up in plant or microbial tissues by preventing mineralization or to provide a sink for this nitrogen in plants or microbes once it is mineralized. Mineralization can be inhibited by controlling the substrate quality of the residue (for example, residues with a high carbon-to-nitrogen ratio do not release much nitrogen). Use of cover crops or relay crops to take up the nitrogen mineralized following harvest is a mechanism for storing nitrogen in plants. In many environments, it is likely that techniques for managing residual nitrogen will need to be used along with refined input management, or nitrate losses may remain unacceptably high.

### Cropping Systems as a Nitrogen Management Tool

The development of cropping systems that prevent the buildup of residual nitrogen during the dormant season has been a focus of research in the past 10 years. The major emphasis has been on the use of cover crops planted after crop harvest (for reviews, see Hargrove, 1988, 1991). Although cover crop techniques have demonstrated abilities to reduce erosion, surface runoff, and leaching into groundwater, several problems limit their widespread use and effectiveness. Langdale and colleagues (1991) report that the cover cropping systems are better developed in the southeastern United States than in other parts of the country and that because of the fragmentation of research efforts and the short-term economic policy structure of the U.S. agricultural system, cover crop use in other regions is prohibitive. The drawbacks and concerns associated with cover crop use include depletion of soil water by cover crops, slow release of the nutrients contained in cover crop biomass, and difficulties in establishing and killing cover crops, especially in northern areas of the United States (Frye et al., 1988; Lal et al., 1991; Wagger and Mengel, 1988).

Several aspects of the effects of cover crops on total crop system function are poorly understood. Cover cropping changes organic matter pools and microbial nutrient cycling patterns, affecting crop nutrient uptake and fertilizer use efficiency. It likely takes several years for these changes to stabilize and create a new equilibrium of organic matter and nutrient dynamics in soil (Doran and Smith, 1991). More important, the fate of the nutrients absorbed by subsequent cover crops is not clear. Studies with isotopically labeled nitrogen, as well as more conventional nitrogen budget studies, have shown that less than 50 percent of the nitrogen contained in cover crop tissues is absorbed by subsequent crops (Ladd et al., 1983; M. S. Smith et al., 1987; Varco et al., 1989). In many cases, recovery of cover crop nitrogen has been found to be lower than recovery of fertilizer-N (Doran and Smith, 1991). It is critical to determine whether cover crops continually recycle the nitrogen that they absorb or whether they merely act as a temporary sink for the residual nitrogen that ultimately ends up in groundwater or surface water.

# 7

# Phosphorus in the Soil-Crop System

Phosphorus is an essential plant nutrient and a necessary input for acceptable crop yields. The beneficial effect of phosphorus on crop yields has been known for well over a century (Kamprath and Watson, 1980). Viets (1975) estimated that one-third to one-half of modern yields are attributable to fertilizer additions and that maintenance of present production levels without fertilizer would require a 20 to 29 percent increase in the area of cultivated land.

However, when phosphorus enters surface waters in substantial amounts it becomes a pollutant, contributing to the excessive growth of algae and other aquatic vegetation and, thus, to the accelerated eutrophication of lakes and reservoirs. (Eutrophication is the process by which a body of water becomes, either naturally or by pollution, rich in dissolved nutrients and, often, seasonally deficient in dissolved oxygen.) Development of strategies to reduce phosphorus loadings to surface water requires an understanding of phosphorus inputs and outputs and the transport mechanisms that deliver phosphorus to surface water. Simplistic solutions may exacerbate trade-offs. Simply eliminating phosphorus additions might bring marginal lands into production, increasing the amount of erosion on such lands (Sharpley and Menzel, 1987). Practices to reduce phosphorus loadings must be based on an understanding of phosphorus sources, a balance between inputs and outputs, and transport processes.

## THE PROBLEM OF PHOSPHORUS DELIVERY TO SURFACE WATERS

Excessive nutrient loads in surface water bodies lead to accelerated eutrophication. Algal blooms are one result of accelerated eutrophication and can result in oxygen depletion, fish kills, and other water quality problems.

Phosphorus is most often the limiting nutrient in freshwater aquatic systems and was thought to be the major contributor to nuisance algal blooms in Wisconsin lakes in the late 1940s (Sawyer, 1947). In a variety of Japanese and U.S. lakes, Dillon and Rigler (1974) found a consistent relationship between the phosphorus concentration in the water and the size of the algal standing crop. It is not clear whether this phosphorus limitation is universal. Schindler (1977) maintains, however, that all freshwater lakes will eventually be phosphorus limited because other nutrients have an atmospheric pathway in their biogeochemical cycles and are thus more subject to internal regulation, whereas phosphorus cycling is strictly geologic and thus more sensitive to external factors.

Relatively low concentrations of phosphorus in surface waters may create eutrophication problems. Sawyer (1947) estimated a critical level of 0.01 mg of soluble inorganic phosphorus per liter (0.01 parts per million [ppm]); other investigators have not been as ready to assign specific critical levels (Viets, 1975), although a range of 0.01 to 0.03 mg/liter (0.01 to 0.03 ppm) seems to be accepted (Baker et al., 1978).

## SOURCES OF PHOSPHORUS

Phosphorus can enter surface water from a variety of sources including municipal wastes, industrial wastes, animal feedlots, and runoff from croplands.

### Point Sources

Point sources of pollutants, such as municipal wastewater treatment facilities or industrial wastewater outlets, were formerly the major sources of phosphorus input to surface waters, with agricultural and other diffuse or nonpoint sources playing a relatively minor role (Bjork, 1972; Sawyer, 1947). In nonindustrialized countries where sewage treatment is limited, this dominance of point sources is still the case (Gilliam et al., 1985); but in the United States and Canada, nonpoint sources are increasingly important because of more effective point source control.

Overall trends for U.S. rivers indicate that there are about equal numbers with increasing and decreasing phosphorus loads. In general, the decreases are linked to point source reductions, whereas the increases appear to be due to nonpoint source increases (R. A. Smith et al., 1987). The increases in total phosphorus loads were associated with increased suspended sediment loads and with some measures of agricultural land use, such as the proportion of fertilized land and cattle population density.

By 1978, about 45 to 50 percent of the total phosphorus load to the Great Lakes was attributed to diffuse sources, primarily agricultural activities (Groszyk, 1978; Johnson, 1978). In 1979, an estimated 28 to 40 percent reduction in the diffuse phosphorus load to Lake Erie was required to meet water quality goals (Logan et al., 1979).

### Agricultural Sources of Phosphorus

The potential for phosphorus delivery to surface waters varies widely among different agricultural practices, and cost-effective solutions should target the systems with the greatest potential phosphorus delivery reductions per dollar spent on control measures. The general categories of agricultural phosphorus sources are croplands, lands in pasture or forage crops, and livestock wastes.

Most of the phosphorus load to surface waters is due to row crops, particularly on fine-textured soils near watercourses (Groszyk, 1978). In one intensive study in Canada (Coote et al., 1978), soil clay content and the area of a watershed that was in row crops were two of the most important variables explaining the total phosphorus load in the watershed.

Most of the phosphorus lost from croplands is not in solution but is bound to eroded soil particles. Sediment-bound phosphorus is not 100 percent available for plant uptake, but sediment control in itself is desirable from a number of standpoints, including the fact that many of the pesticides lost from fields are sediment-bound (Johnson, 1978).

Sediment and total phosphorus loads from pasturelands are generally lower than those from croplands, but more of the phosphorus lost is in the more available dissolved form (Baker et al., 1978). This result has been ascribed to lack of fertilizer incorporation and leaching of phosphorus from foliage and animal wastes on pastures (Baker et al., 1978; Viets, 1975).

Manure from livestock waste disposal may be a significant source of phosphorus loads in water; one estimate (Moore et al., 1978) is that about 5 percent of the phosphorus excreted by livestock annually ends up in surface waters. If manure is spread on frozen ground, losses of phosphorus through runoff from manure may be severe. In the Great Lakes region, 30 to 38 percent of the total livestock waste phosphorus load is lost through runoff from manure on frozen ground (Moore et al., 1978). Moore and colleagues (1978) found that most of the rest of the phosphorus load from wastes (44 to 50 percent) in water was due to runoff from dairy cattle operations. In other regions, such as the southeastern United States, swine waste is a potentially large source of phosphorus (K. R. Reddy et al., 1978).

In some cases, reduction of sediment phosphorus losses can result in increases of soluble phosphorus loss (Sharpley and Menzel, 1987), so the answer to phosphorus loading problems is not as simple as sediment control and is likely to involve trade-offs.

## Forms and Bioavailability of Phosphorus

Phosphorus occurs in many forms in both the solution phase and, in particular, the solid phase. These forms are little understood, even though there are many data in the literature concerning the chemistry of phosphorus in water, soils, and sediments. The relative bioavailability of various forms of phosphorus varies, but there is no standard method of determining this important quantity.

### Soluble Phosphorus

Soluble phosphorus is arbitrarily defined as phosphorus that will pass through a 0.45-μm-pore filter. Soluble reactive phosphorus is that fraction of phosphorus that is reactive with molybdate, according to the Murphy-Riley procedure or its variants. This fraction has been assumed to consist of orthophosphate, but there is evidence that some organic phosphorus is included (Rigler, 1968); for this reason, molybdate-reactive phosphorus is usually referred to as soluble reactive phosphorus or dissolved reactive phosphorus rather than orthophosphate.

Not all of the dissolved reactive phosphorus in lake water is completely available for algal growth (Sharpley and Menzel, 1987). The relative difference in dissolved reactive phosphorus and bioavailable phosphorus in water is greater in waters with low levels of phosphorus and is less in solutions with higher dissolved reactive phosphorus concentrations (Sharpley and Menzel, 1987). Even though not all of the phosphorus in water is available to algae, there is often a close relationship between the total amount of phosphorus in water and the standing algal crop (Dillon and Rigler, 1974).

### Particulate Phosphorus

Phosphorus is strongly bound to sediments by anion adsorption reactions. These reactions probably account for the rapid removal from water of phosphorus that is in contact with lake sediments (Syers et al., 1973). Much of this adsorbed phosphorus is not easily desorbed, and the amount that is desorbable decreases with the age of the sediment-adsorbed phosphorus complex (Syers et al., 1973).

Particulate phosphorus is associated with iron, aluminum, and manganese in sediments (Bortleson and Lee, 1974; McCallister and Logan, 1978; Syers et al., 1973), although the association with manganese may be artifactual because of the coprecipitation of manganese and iron in nodules (Syers et al., 1973). In this regard, iron seems to be most commonly associated with phosphorus, aluminum and manganese are less so, and calcium carbonate is not commonly associated with phosphorus (Syers et al., 1973). The fraction of iron phosphorus seems to be associated with is the oxalate-extractable fraction, referred to as short-range-order or amorphous oxides. Oxalate extraction reduces or eliminates the phosphorus sorption capacity (Syers et al., 1973).

For the most part, discrete phosphorus compounds have not been found in lake sediments, although there are exceptions (Syers et al., 1973). Most emphasis has been on the phosphorus fractions removed by a number of extractants that remove phosphorus that is more or less tightly bound. The nonspecificities of extractants for phosphorus removal and potential reprecipitation of phosphorus make this work difficult to interpret (Syers et al., 1973).

Estimates of the fraction of sediment-bound phosphorus that is available for biological uptake vary according to the methods used to obtain the estimate, and the estimates obtained by different methods are difficult to interpret, making some standard means of obtaining bioavailability estimates desirable (Sharpley and Menzel, 1987). The bioavailability of phosphorus in sediments as measured by a variety of methods usually does not exceed 60 percent of the total phosphorus in the sediment (Sonzogni et al., 1982) and varies with the source of the sediment (Logan et al., 1979).

A bioassay that measures phosphorus uptake by algae is the standard by which most chemical extractants are measured. The various assay methods include exchange with a hydroxy-aluminum-coated resin or phosphorus-32 or extraction with ammonium fluoride, sodium hydroxide, or nitriloacetic acid (Sharpley and Menzel, 1987). Opinions vary, but the best chemical extractant for measuring bioavailable phosphorus seems to be 0.01 M sodium hydroxide (Dorich et al., 1985; Sharpley et al., 1991; Williams et al., 1980), even though none of the chemical extractants appears to remove the specific fraction of phosphorus removed in an algal assay (Dorich et al., 1980).

The relevance of the algal assay to total potentially bioavailable phosphorus may be questioned for several reasons. The phosphorus uptake mechanisms of algae may be different from those of rooted aquatic plants, and in many cases, rooted aquatic plants are a more serious consequence of eutrophication than algal blooms (Sharpley and Menzel, 1987). In terrestrial systems, some plants can solubilize phosphorus from sources usually considered to be unavailable to plants (Jayman and Sivasubramaniam, 1975). Similar mechanisms may exist in aquatic systems. Use of phosphorus sediments by rooted plants also releases the phosphorus bound in sediments to the water column. In one study (Carignan and Kalff, 1980), rooted aquatic plants derived 72 to 100 percent of their phosphorus nutrition from sediments, making them potential nutrient pumps to the open water.

Algal assays also do not account for possible phosphorus release when sediments are subjected to anoxic conditions. Anoxic conditions cause the release of phosphorus, which is thought to be due to the reduction of iron ions ($Fe^{3+}$ to $Fe^{2+}$) (Sharpley and Menzel, 1987; Syers et al., 1973). The amount of phosphorus in solution may increase manyfold when the sediments are subjected to reducing conditions (Mortimer, 1940, 1941).

Regardless of the validity of assay techniques, the availability of sediments for nutrient exchange with the overlying water is important. Physical mechanisms such as the rate of settling of phosphorus-containing particles affect the availability of sediments, as does the thickness of the sediment layer that interacts with the overlying water. This layer may be only a few millimeters or up to a few centimeters thick (Sharpley and Menzel, 1987) and can be affected by physical mixing or aquatic organisms burrowing in the sediment (McCall et al., 1979). One study in a shallow, well-mixed area of a lake noted significant reductions in the sediment phosphorus concentrations in association with spring algal blooms (Wildung et al., 1974).

Phosphorus associated with sediments may remain a problem years after excess phosphorus inputs cease. Lake Trummen in Sweden experienced nuisance algal blooms 10 years after nutrient inputs were reduced. The algal blooms were eliminated only after removal of the enriched sediments (Bjork, 1972). Desorption of phosphorus from sediments is estimated to contribute about 10 percent of the total phosphorus load to Lake Erie (Sharpley and Menzel, 1987).

### Total Phosphorus

Although soluble and particulate phosphorus are discussed separately, they are closely related. The equilibrium concentration of soluble phosphorus is controlled by the concentration of sediment-bound phosphorus. Furthermore, both soluble and particulate phosphorus contribute to water quality problems. A focus solely on reductions in soluble or particulate phosphorus can lead to trade-offs because, in some cases, reductions in sediment-bound phosphorus losses can result in increased soluble phosphorus losses (Sharpley and Menzel, 1987). Total bioavailable phosphorus is a more useful measure of phosphorus loadings, but it can be difficult to estimate because of the problems with estimating bioavailability discussed earlier. Total phosphorus can serve as a useful proxy for total bioavailable phosphorus, and reductions in total phosphorus loadings should be the goal of phosphorus control programs.

## PHOSPHORUS IN THE SOIL-CROP SYSTEM

Like nitrogen and other plant nutrients, the phosphorus added to the soil-crop system goes through a series of transformations as it cycles through plants, animals, microbes, soil organic matter, and the soil mineral fraction. Unlike nitrogen, however, most phosphorus is tightly bound in the soil, and only a small fraction of the total phosphorus found in the soil is available to crop plants.

### The Phosphorus Cycle

Figure 7-1 is a simplified illustration of the phosphorus cycle in the soil-crop system. Most of the phosphorus in soil is found as a complex mixture of mineral and organic materials. Organic phosphorus compounds in plant residues, manures, and other organic materials are broken down through the action of soil microbes. Some of the organic phosphorus can be released into the soil solution as phosphate ions that are immediately available to plants. Much of the organic phosphorus is taken up by the microbes themselves. As microbes die, the phosphorus held in their cells is released into the soil. A considerable amount of organic phosphorus is held in the humic materials that make up soil organic matter. A portion of this organic phosphorus is released each year as these humic materials decay.

The phosphate ions released from the decomposition of organic phosphorus compounds or added directly in inorganic phosphorus-containing fertilizers readily react with soil minerals and are immobilized in forms that are unavailable to plants for growth. Phosphorus retention in soils is generally considered to be due to adsorption, although some evidence of direct phosphorus precipitation from solution exists (Martin et al., 1988).

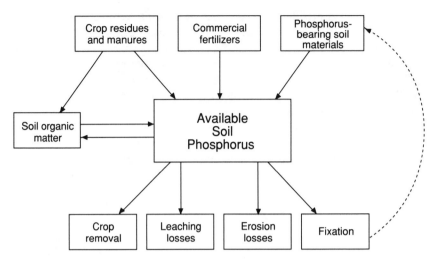

FIGURE 7-1 The phosphorus cycle. Source: H. O. Buckman and N. C. Brady. 1969. The Nature and Properties of Soils, 7th Ed. London: Macmillan. Reprinted with permission from © Macmillan Inc.

Aluminum appears to be more widely involved than sediments in phosphorus sorption in soils. Amorphous aluminum compounds (Kawai, 1980), peat-aluminum complexes (Bloom, 1981), and aluminum-substituted goethite (an iron hydrogen oxide) (Karim and Adams, 1984) have been implicated in phosphorus adsorption in soils.

Phosphorus adsorption is related to iron oxides as well, but the more crystalline forms of the oxides (citrate-dithionite-bicarbonate [CDB] or CDB-extractable iron) (Karim and Adams, 1984; Solis and Torrent, 1989a) rather than the amorphous (oxalate-extractable) forms that are important for sediments appear to be involved. CDB-extractable iron and phosphorus are correlated (Solis and Torrent, 1989b), and by direct observation phosphorus has been seen to be enriched in some iron-enriched nodules (McKeague, 1981).

Phosphorus is relatively enriched in finer soil fractions, so it is perhaps expected that phosphorus adsorption is correlated with the clay content of soils (Solis and Torrent, 1989b) and with the soil surface area (Olsen and Watanabe, 1957).

Some organic acids reduce the adsorption capacity of phosphorus, perhaps by competition for anion adsorption sites (Kafkafi et al., 1988; Lopez-Hernandez et al., 1986; Violante et al., 1991). Solubilization of phosphorus and aluminum by organic acids has been noted (Fox et al., 1990a,b; Jayman and Sivasubramaniam, 1975) and appears to be related

TABLE 7-1  Phosphorus Inputs and Outputs in the
United States, 1987

| Input or Output | Metric Tons of Phosphorus |
|---|---|
| Inputs | |
|    Fertilizer-P | 3,570,000 (79) |
|    Manure-P | 655,000 (15) |
|    Crop residues | 272,000 (6) |
|      Total inputs | 4,500,000 (100) |
| Outputs | |
|    Harvested crops | 1,320,000 (29) |
|    Crop residues | 272,000 (6) |
|      Total outputs | 1,600,000 (36) |
| Balance | 2,900,000 (63) |

NOTE: Values in parentheses are percentage of total phosphorus
input mass. See the Appendix for a full discussion of the methods
used to estimate phosphorus inputs and outputs.

to the stability of the aluminum-organic acid complex (Fox et al.,
1990a,b).

Under anaerobic conditions, phosphorus release to solutions low in
phosphorus concentration is increased, the adsorption capacity of the
anaerobic soil increases when the solution has a high phosphorus
concentration. This increase is thought to be due to the increased surface
areas of the reduced iron oxides (Khalid et al., 1977; Patrick and Khalid,
1974).

Phosphate ions added to soils from either organic or inorganic sources
enter into this complex series of precipitation or sorption reactions.
These reactions greatly reduce the amount of phosphate ions that are in
the soil solution and available to plants. The equilibrium level of
dissolved phosphorus in the soil solution is controlled by the chemical
environment of the soil (Nelson and Logan, 1983).

### Mass Balance

Phosphorus is added to croplands in crop residues and manures, in
synthetic fertilizers, and from phosphorus-bearing minerals in the soil
(Figure 7-1). Part of the phosphorus entering the soil-crop system is
removed with the harvested crop; the balance is immobilized in the soil,
incorporated into soil organic matter, or lost in surface or subsurface
flows to surface water or groundwater.

Table 7-1 provides estimates of national phosphorus mass balances as
the mass of phosphorus applied to croplands as synthetic fertilizers

(fertilizer-P) and crop residues and voided in manures (manure-P). (See the Appendix for a full explanation of mass balance estimates.) Much of the total mass of phosphorus voided in manures is not economically recoverable for use as an input in annual crop production systems because it is deposited on pasturelands or rangelands, from which collection is impossible. Furthermore, a substantial portion of the phosphorus voided in manures is lost in surface runoffs from pasturelands, rangelands, and handling and storage facilities. Only that portion of total phosphorus voided in manures that can be economically recovered for use on croplands was used in Table 7-1 as an estimate of phosphorus inputs from manure. The difference between phosphorus inputs and phosphorus outputs in crops and crop residues is reported as phosphorus balances. A more detailed analysis of phosphorus inputs and outputs to croplands helps to identify opportunities for reducing phosphorus loadings to surface water from farming systems.

### Phosphorus Inputs

The phosphorus in fertilizer-P is the single most important source of phosphorus added to croplands in the United States (Table 7-1). The majority of this fertilizer-P was added to annual crops, and the amount of phosphorus added in synthetic fertilizers varies from crop to crop and region to region. Corn consumes more phosphorus than any other single crop, followed by wheat, soybeans, and cotton, in that order. Phosphorus application rates also vary between crops, with corn again receiving the greatest rates of phosphorus application per unit area; this is followed by cotton, soybeans, and wheat, in that order. Of the phosphorus applied to croplands in the United States, 42 percent is applied to land planted in corn, and 67 percent of the total phosphorus applied to U.S. croplands is planted in four crops: corn, cotton, soybeans, and wheat. These differences in phosphorus application rates combine with regional differences in crop mixes to produce the state-to-state variability in the total amount of phosphorus applied in synthetic fertilizers (Tables 7-2 and 7-3).

The amount of recoverable manure-P is small compared with that supplied in synthetic fertilizers at the national level (Table 7-1). The phosphorus in manure represents only 15 percent of phosphorus inputs. The total mass of phosphorus voided in manures is much larger than that which is economically recoverable. The recoverable phosphorus represents less than half of the total mass of phosphorus in manure. Locally, the proportion of phosphorus supplied by manures can be large. Recoverable phosphorus in manure, for example, supplies 65 percent of total phosphorus inputs in Vermont (Table 7-3).

## Phosphorus Outputs

The fraction of total phosphorus inputs lost to erosion and runoff can be substantial, but it is difficult to estimate the amount. Larson and colleagues (1983) estimated that 1.74 million metric tons (1.92 million tons) of phosphorus, or about 50 percent of the estimated total phosphorus balance in Table 7-1, was lost in eroded sediments in 1982. Additional phosphorus can be lost in solution (see below).

The importance of animal manures as a potential source of phosphorus loadings can be seen in the difference between total phosphorus and that which is recoverable in manure. The 662,000 metric tons (730,000 tons) of recoverable phosphorus accounted for in Table 7-2 represents only 49 percent of the total estimated 1,349,000 metric tons (1,487,000 tons) of phosphorus excreted in animal manures. A substantial fraction of this difference between the total amount of phosphorus excreted in manures and the amount that can be recovered for use in crop production may represent direct losses of phosphorus in runoffs from pastures, feedlots, and manure storage facilities.

The majority of the total and recoverable phosphorus balance on agricultural lands is immobilized in either the mineral or the organic fractions of the soil. The potential for buildup of phosphorus levels in soil over time has important implications for efforts to reduce phosphorus loadings to water (see below).

## Phosphorus Buildup in Soils

Relatively small annual additions of phosphorus may cause a soil buildup of phosphorus as illustrated in (Figure 7-2) (McCollum, 1991). Some of the phosphorus added in excess of crop needs remains as residual plant-available phosphorus, but not all of the added phosphorus will be available to crops; the amount of extractable phosphorus declines with time because of the slow conversion of phosphorus to unavailable forms (McCollum, 1991; Mendoza and Barrow, 1987; Sharpley et al., 1989; Yost et al., 1981). The rate of decline in extractable phosphorus (discounting plant uptake) varies with the phosphorus adsorption properties of the soil and the initial level of phosphorus in the soil (that is, the relative saturation of adsorption capacity) and with the amount of applied phosphorus.

The phosphorus level in the soil is the critical factor in determining

TABLE 7-2 State and National Phosphorus Inputs and Outputs (metric tons)

| State | Inputs | | | | Outputs | | | Balance |
|---|---|---|---|---|---|---|---|---|
| | Fertilizer-P | Recoverable Manure-P | Crop Residues | Total | Harvested Crops | Crop Residues | Total | |
| Alabama | 49,700 | 9,280 | 851 | 59,800 | 7,130 | 851 | 7,980 | 51,800 |
| Alaska | 1,060 | 0 | 6 | 1,060 | 93 | 6 | 100 | 964 |
| Arizona | 20,700 | 6,520 | 432 | 27,700 | 3,040 | 432 | 3,470 | 24,200 |
| Arkansas | 46,300 | 15,200 | 5,710 | 67,200 | 28,200 | 5,710 | 33,900 | 33,300 |
| California | 142,000 | 40,800 | 3,680 | 187,000 | 24,700 | 3,680 | 28,500 | 158,000 |
| Colorado | 14,500 | 19,600 | 4,450 | 38,500 | 23,200 | 4,450 | 27,600 | 10,900 |
| Connecticut | 3,540 | 2,250 | 7 | 5,790 | 430 | 7 | 438 | 5,350 |
| Delaware | 5,640 | 3,180 | 392 | 9,210 | 1,780 | 392 | 2,170 | 7,040 |
| Florida | 111,000 | 6,510 | 290 | 118,000 | 3,160 | 290 | 3,450 | 114,000 |
| Georgia | 76,400 | 13,700 | 1,840 | 91,900 | 11,900 | 1,840 | 13,700 | 78,200 |
| Hawaii | 7,770 | 786 | 0 | 8,560 | 14 | 0 | 14 | 8,580 |
| Idaho | 69,400 | 9,330 | 3,380 | 82,100 | 22,600 | 3,380 | 26,000 | 56,100 |
| Illinois | 335,000 | 18,700 | 34,800 | 389,000 | 142,000 | 34,800 | 177,000 | 212,000 |
| Indiana | 238,000 | 17,000 | 18,300 | 273,000 | 75,500 | 18,300 | 93,800 | 179,000 |
| Iowa | 259,000 | 43,700 | 36,400 | 339,000 | 149,000 | 36,400 | 185,000 | 154,000 |
| Kansas | 138,000 | 45,000 | 17,000 | 200,000 | 75,700 | 17,000 | 92,700 | 107,000 |
| Kentucky | 97,300 | 8,600 | 3,220 | 109,000 | 20,500 | 3,220 | 23,700 | 85,400 |
| Louisiana | 37,300 | 1,610 | 2,620 | 41,500 | 13,100 | 2,620 | 15,800 | 25,800 |
| Maine | 8,950 | 3,000 | 108 | 12,100 | 1,700 | 108 | 1,810 | 10,200 |
| Maryland | 24,900 | 6,950 | 1,150 | 33,000 | 5,880 | 1,150 | 7,030 | 26,000 |
| Massachusetts | 4,780 | 1,600 | 15 | 6,400 | 602 | 15 | 619 | 5,800 |
| Michigan | 119,000 | 13,500 | 5,880 | 139,000 | 28,300 | 5,880 | 34,200 | 104,000 |
| Minnesota | 204,000 | 31,300 | 20,000 | 255,000 | 90,900 | 20,000 | 111,000 | 144,000 |
| Mississippi | 35,000 | 5,410 | 2,280 | 42,700 | 13,300 | 2,280 | 15,500 | 27,200 |
| Missouri | 135,000 | 18,700 | 10,400 | 164,000 | 58,400 | 10,400 | 68,900 | 94,900 |
| Montana | 53,500 | 4,290 | 4,630 | 62,500 | 28,000 | 4,630 | 32,600 | 29,900 |

| | | | | | | |
|---|---|---|---|---|---|---|
| Nebraska | 116,000 | 37,900 | 23,600 | 177,000 | 90,200 | 114,000 | 63,600 |
| Nevada | 5,840 | 1,040 | 46 | 6,920 | 2,430 | 2,500 | 4,440 |
| New Hampshire | 1,190 | 926 | 2 | 2,120 | 375 | 377 | 1,740 |
| New Jersey | 12,700 | 655 | 270 | 13,600 | 1,710 | 1,980 | 11,600 |
| New Mexico | 7,380 | 5,200 | 602 | 13,200 | 3,740 | 4,340 | 8,840 |
| New York | 51,600 | 24,400 | 1,820 | 77,900 | 15,600 | 17,400 | 60,400 |
| North Carolina | 81,600 | 15,600 | 2,740 | 99,900 | 14,800 | 17,600 | 82,300 |
| North Dakota | 131,000 | 4,090 | 9,500 | 144,000 | 50,500 | 60,000 | 84,500 |
| Ohio | 160,000 | 17,100 | 11,900 | 189,000 | 56,300 | 68,200 | 121,000 |
| Oklahoma | 85,700 | 9,190 | 2,970 | 97,900 | 22,000 | 25,000 | 72,900 |
| Oregon | 30,800 | 4,120 | 1,400 | 36,300 | 11,700 | 13,100 | 23,200 |
| Pennsylvania | 38,100 | 27,600 | 2,870 | 68,600 | 18,200 | 21,000 | 47,600 |
| Rhode Island | 668 | 0 | 0 | 668 | 37 | 38 | 629 |
| South Carolina | 26,900 | 2,920 | 979 | 30,800 | 5,680 | 6,660 | 24,200 |
| South Dakota | 60,600 | 13,200 | 8,740 | 82,500 | 44,200 | 52,900 | 29,600 |
| Tennessee | 96,500 | 6,970 | 2,110 | 106,000 | 14,500 | 16,600 | 89,000 |
| Texas | 188,000 | 50,700 | 11,300 | 250,000 | 50,400 | 61,700 | 189,000 |
| Utah | 10,800 | 3,910 | 427 | 15,200 | 5,010 | 5,440 | 9,730 |
| Vermont | 3,240 | 6,060 | 25 | 9,320 | 1,990 | 2,020 | 7,300 |
| Virginia | 47,800 | 8,850 | 970 | 57,700 | 8,870 | 9,840 | 47,800 |
| Washington | 42,400 | 8,600 | 3,360 | 54,400 | 21,300 | 24,600 | 29,800 |
| West Virginia | 10,300 | 2,270 | 92 | 12,700 | 1,810 | 1,900 | 10,800 |
| Wisconsin | 119,000 | 53,700 | 8,230 | 181,000 | 48,500 | 56,800 | 124,000 |
| Wyoming | 3,830 | 3,410 | 484 | 7,730 | 5,510 | 5,990 | 1,730 |
| United States | 3,570,000 | 655,000 | 272,000 | 4,500,000 | 1,320,000 | 1,600,000 | 2,900,000 |

NOTE: See the Appendix for a full discussion of the methods used to estimate phosphorus inputs and outputs.

TABLE 7-3    State and National Phosphorus Inputs and Outputs as
Percentage of Total Mass of Phosphorus Inputs

| | Percentage of Total Input Mass | | | | | |
| | Inputs | | | Outputs | | |
| State | Fertilizer-P | Recoverable Manure-P | Crop Residues | Harvested Crop | Crop Residues | Balance |
|---|---|---|---|---|---|---|
| Alabama | 83 | 16 | 1 | 12 | 1 | 86 |
| Alaska | 99 | 0 | 1 | 9 | 1 | 90 |
| Arizona | 75 | 24 | 2 | 11 | 2 | 87 |
| Arkansas | 69 | 23 | 9 | 42 | 9 | 49 |
| California | 76 | 22 | 2 | 13 | 2 | 84 |
| Colorado | 38 | 51 | 12 | 60 | 12 | 28 |
| Connecticut | 61 | 39 | <1 | 7 | <1 | 92 |
| Delaware | 61 | 35 | 4 | 19 | 4 | 76 |
| Florida | 94 | 6 | <1 | 3 | <1 | 97 |
| Georgia | 83 | 15 | 2 | 13 | 2 | 85 |
| Hawaii | 91 | 9 | 0 | <1 | 0 | 99 |
| Idaho | 85 | 11 | 4 | 28 | 4 | 68 |
| Illinois | 86 | 5 | 9 | 37 | 9 | 54 |
| Indiana | 87 | 6 | 7 | 28 | 7 | 65 |
| Iowa | 76 | 13 | 11 | 44 | 11 | 45 |
| Kansas | 69 | 23 | 9 | 38 | 9 | 54 |
| Kentucky | 89 | 8 | 3 | 19 | 3 | 78 |
| Louisiana | 90 | 4 | 6 | 32 | 6 | 62 |
| Maine | 74 | 25 | 1 | 14 | 1 | 85 |
| Maryland | 76 | 21 | 4 | 18 | 4 | 78 |
| Massachusetts | 75 | 25 | <1 | 9 | <1 | 90 |
| Michigan | 86 | 10 | 4 | 20 | 4 | 75 |
| Minnesota | 80 | 12 | 8 | 36 | 8 | 56 |
| Mississippi | 82 | 13 | 5 | 31 | 5 | 63 |
| Missouri | 82 | 11 | 6 | 36 | 6 | 57 |
| Montana | 86 | 7 | 7 | 45 | 7 | 47 |
| Nebraska | 65 | 21 | 13 | 51 | 13 | 35 |
| Nevada | 84 | 15 | 1 | 35 | 1 | 64 |
| New Hampshire | 56 | 44 | <1 | 18 | <1 | 82 |
| New Jersey | 93 | 5 | 2 | 13 | 2 | 85 |
| New Mexico | 56 | 39 | 5 | 28 | 5 | 67 |
| New York | 66 | 31 | 2 | 20 | 2 | 77 |
| North Carolina | 82 | 16 | 3 | 15 | 3 | 82 |
| North Dakota | 91 | 3 | 7 | 35 | 7 | 58 |
| Ohio | 85 | 9 | 6 | 30 | 6 | 64 |
| Oklahoma | 88 | 9 | 3 | 23 | 3 | 74 |
| Oregon | 85 | 11 | 4 | 32 | 4 | 63 |
| Pennsylvania | 56 | 40 | 4 | 27 | 4 | 69 |
| Rhode Island | 100 | 0 | <1 | 6 | <1 | 94 |
| South Carolina | 87 | 10 | 3 | 18 | 3 | 78 |
| South Dakota | 74 | 16 | 11 | 54 | 11 | 35 |

TABLE 7-3 *(Continued)*

| State | Percentage of Total Input Mass | | | | | |
|---|---|---|---|---|---|---|
| | Inputs | | | Outputs | | |
| | Fertilizer-P | Recoverable Manure-P | Crop Residues | Harvested Crop | Crop Residues | Balance |
| Tennessee | 91 | 7 | 2 | 14 | 2 | 84 |
| Texas | 75 | 20 | 5 | 20 | 5 | 75 |
| Utah | 71 | 26 | 3 | 33 | 3 | 64 |
| Vermont | 35 | 65 | <1 | 21 | <1 | 78 |
| Virginia | 83 | 15 | 2 | 15 | 2 | 82 |
| Washington | 78 | 16 | 6 | 39 | 6 | 54 |
| West Virginia | 81 | 18 | 1 | 14 | 1 | 85 |
| Wisconsin | 66 | 30 | 5 | 27 | 5 | 68 |
| Wyoming | 50 | 44 | 6 | 71 | 6 | 22 |
| United States | 79 | 15 | 6 | 29 | 6 | 63 |

NOTE: See the Appendix for a full discussion of the methods used to estimate phosphorus inputs and outputs.

actual loads of phosphorus to surface water and the relative proportions of phosphorus lost in solution and attached to soil particles.

Solution and sediment-bound phosphorus losses are closely interrelated. The equilibration of a sediment-solution mixture produces a certain solution phosphorus concentration, which depends on the nature of the material and the percentage of the phosphorus sorption capacity of the sediment. The amount of phosphorus on the soil particles controls the solution phosphorus concentration and is related to the phosphorus application history of the site (Sharpley and Menzel, 1987). This solution concentration is referred to as the equilibrium phosphorus concentration (EPC) (Gilliam et al., 1985; Sharpley and Menzel, 1987). Since the solution phosphorus concentration is particularly important regarding potential water quality effects, an increase in EPC because of the increasing phosphorus content of the soil is an undesirable situation with regard to water quality.

EPC increases with increasing phosphorus additions, regardless of the source of the added phosphorus. Increasing synthetic fertilizer applications increase the EPC (Gilliam et al., 1985; Logan and MacLean, 1973), as do all other sources of added phosphorus (G. Y. Reddy et al., 1978; K. R. Reddy et al., 1978). Manure additions, in some cases, raise EPC more than equivalent additions of chemical fertilizer do (G. Y. Reddy et al., 1978). In addition, eroded sediment generally supports a higher EPC than the source soil does (Gilliam et al., 1985), and the EPC

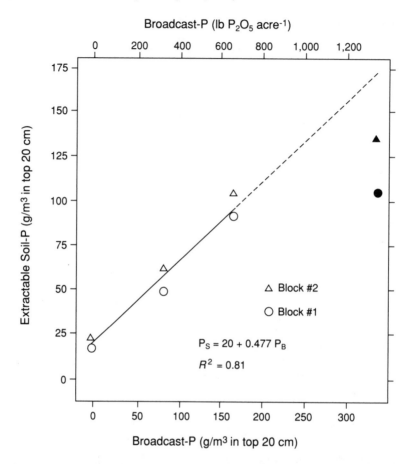

FIGURE 7-2  Relationship between broadcast phosphorus ($P_B$) and extractable soil phosphorus ($P_S$). Phosphorus levels in soil were measured, using the Mehlich 1-extractable soil phosphorus method, after 1 year of equilibration; each symbol is the average of 15 observations; solid symbols were not used in the prediction equation. Source: Derived from R. E. McCollum. 1991. Buildup and decline in soil phosphorus: 30-year trends on a Typic Umprabuult. Agronomy Journal 83:77–85. Reprinted with permission from © American Society for Agronomy, Crop Science Society of America, and Soil Science Society of America.

of runoff sediment varies inversely with soil loss (Sharpley and Menzel, 1987).

Increased residual phosphorus levels in the soil lead to increased phosphorus loadings to surface water, both in solution and attached to soil particles.

The estimated magnitude of phosphorus unaccounted for (see Table 7-1) suggests that the potential for buildup of phosphorus in soil is great under current rates of application. Regional and farm-level phosphorus mass balances suggest similar conclusions. Lowrance and colleagues (1985) estimated phosphorus inputs from precipitation and fertilizer and phosphorus outputs in harvested crops for four watersheds in Georgia. Harvested phosphorus accounted for 20.1 to 40.8 percent of phosphorus inputs, depending on the watershed and the year studied. Stinner and colleagues (1984) developed phosphorus budgets for conventional and no-till sorghum. The proportion of harvested phosphorus ranged from 34.4 to 39.8 percent of phosphorus inputs from fertilizer, seed, and precipitation. Management of phosphorus inputs for the prevention of unnecessary buildup of soil phosphorus levels should be part of programs to reduce phosphorus loadings to surface water.

## TRANSPORT PROCESSES

Phosphorus can be lost from the soil-crop system in soluble form through leaching, subsurface flow, and surface runoff. Particulate phosphorus is lost when soil erodes. Understanding the relative importance of transport pathways and the processes regulating these transport pathways helps to design measures to reduce phosphorus losses.

### Leaching and Subsurface Flow

In general, phosphorus loss by leaching to groundwater is not a problem (Gilliam et al., 1985). Phosphorus is bound to soil particles by adsorption and, perhaps, precipitation reactions, and most added phosphorus remains near the surface.

Exceptions to this generality are organic soils and sandy soils, both of which lack the iron and aluminum oxide fractions important for phosphorus retention. Organic soils with low mineral content allow phosphorus to leach readily under laboratory conditions (Fox and Kamprath, 1971; Larsen et al., 1958), and field losses from intensively cropped organic soils may be large (Duxbury and Peverly, 1978). Similarly, substantial downward movement of phosphorus has been found in soils of sand to sandy loam texture both in the laboratory (Mansell et al., 1985) and in the field (G. Y. Reddy et al., 1978; Spencer, 1957).

Significant losses of phosphorus to groundwater generally do not occur, but losses in drainage water from drain tiles installed in the soil can be substantial (Duxbury and Perverly, 1978). The more usual situation, however, is for losses from drain tiles to be small. The tile drainage phosphorus concentration is related to the available phosphorus content and the phosphorus sorption capacity of the soil layer where the tiles are installed (Duxbury and Peverly, 1978; Hanway and Laflen, 1974); in some cases, drainage waters may be depleted of phosphorus relative to the levels in input water (Carter et al., 1971).

Organic phosphorus may be more subject to leaching than are other forms (Gilliam et al., 1985). One waste management study noted a greater tendency for the downward movement of phosphorus as applied in fresh liquid manure compared with that of the phosphorus in weathered barnyard manure (Pratt and Laag, 1981).

## Surface Flow

The majority of phosphorus lost from agricultural lands is through surface flow, both in solution (soluble phosphorus) and bound to eroded sediment particles (particulate phosphorus). Particulate phosphorus is not as readily available to organisms as soluble phosphorus (Gilliam et al., 1985; Sharpley and Menzel, 1987), but particulate phosphorus can be a long-term source of phosphorus once the particulate is delivered to surface water.

### Soluble Phosphorus Losses

Soluble phosphorus losses are greater from pasturelands than from croplands (Baker et al., 1978). Losses from pasture or forage crops increase after freezing of the foliage in the fall (Wendt and Corey, 1980).

If manure is applied to the forage crop, soluble phosphorus losses increase even more. In one study (Young and Mutchler, 1976), alfalfa with unincorporated manure lost four times as much of the added soluble phosphorus as did corn with incorporated manure. Corn with unincorporated manure lost an intermediate amount of soluble phosphorus, but all alfalfa with manure treatments lost more soluble phosphorus than did corn. In general, manure appears to provide a more soluble form of phosphorus than do chemical fertilizers (K. R. Reddy et al., 1978), so soluble phosphorus losses from lands to which manure is applied may be generally higher than those from lands treated with chemical fertilizers.

Soluble phosphorus losses from croplands are often less than those from pasturelands, but they may be substantial. Leaching losses from foliage may contribute from 20 to 60 percent of total phosphorus in runoff, varying with plant age, whereas leaching losses can be up to 90 percent of the total in phosphorus-deficient plants (Sharpley, 1981).

Fertilizer additions also increase soluble phosphorus losses, even though total losses of fertilizer-P are thought to be less than 1 percent (Nelson et al., 1978; Viets, 1975). The soluble phosphorus concentration in runoff water increases with increasing fertilizer addition rates (Ryden et al., 1974), and the relationship is approximately linear (Romkens and Nelson, 1974). This relationship is important because, as noted below, much phosphorus fertilizer is added to soils for which no crop yield increase would be expected.

Solution phosphorus concentrations also increase as the sediment load of the runoff decreases, varying inversely with the logarithm of the sediment concentration (Sharpley et al., 1981). This relationship holds because sediment in the runoff buffers the solution phosphorus concentration, and decreased sediment in the runoff decreases the buffering power of the system. This phenomenon is part of the reason that runoff control as a measure to decrease phosphorus loss involves trade-offs.

## Sediment and Sediment-Bound Phosphorus Losses

Most of the total phosphorus loss from croplands is in sediment-bound form (Gilliam et al., 1985; Sharpley and Menzel, 1987; Viets, 1975). As with soluble phosphorus, particulate phosphorus losses also increase with increasing fertilizer additions, with sediment-extractable phosphorus increasing approximately linearly with the fertilization rate (Sharpley, 1981). Stream-suspended sediments in agricultural watersheds derive mainly from surface soils rather than from stream bank erosion on the basis of mineralogical and other characteristics (Wall and Wilding, 1976). However, stream sediments are relatively enriched in clay, particularly fine clay, compared with the source soils (Rhoton et al., 1979). This enrichment is due to the preferential erosion of fine and lighter particles. The finer soil particles also adsorb phosphorus to a greater extent than do coarse particles, so that sediments are enriched in phosphorus in comparison with source soils (Massey and Jackson, 1952; Rogers, 1941; Sharpley, 1980, 1985; Stoltenberg and White, 1953). The ratio of phosphorus content in sediment to that in soil is referred to as the phosphorus enrichment ratio (ER). As the sediment load increases, thus including relatively more of the coarse soil fractions, the ER decreases. There is a well-documented negative log-log relationship between ER and soil loss (Massey and Jackson, 1952; Sharpley, 1985).

Several other findings regarding ERs are notable. The ER is relatively greater for bioavailable forms of phosphorus than it is for less available forms (Sharpley, 1985), and ERs increase with increasing additions of fertilizer (Sharpley, 1980).

### Changes During Transport

During transport from agricultural fields to streams and lakes, various changes in the forms of phosphorus are likely to occur (Sharpley and Menzel, 1987). The changes depend on, for example, the relative phosphorus concentrations in water and sediment and EPC. There is an inverse linear correlation between the soluble phosphorus in stream water and the logarithm of the sediment concentration in runoff (Sharpley and Menzel, 1987). The general trend is for phosphorus to be converted to less available forms during transport from a field's edge to a lake (Sharpley and Menzel, 1987).

## POSSIBLE MANAGEMENT METHODS FOR PHOSPHORUS LOSS REDUCTION

There are two primary ways to reduce the amount of phosphorus lost from agricultural production: reduce phosphorus levels in the soil and reduce erosion and runoff from croplands.

### Procedures to Establish Threshold Levels

The level of phosphorus in soil (soil-P) is an important determinant of the amount of phosphorus lost to surface water and groundwater from cropping systems. Higher soil-P levels lead to increased losses of both soluble and particulate phosphorus. Phosphorus management to reduce unnecessarily high soil-P levels should be part of efforts to reduce phosphorus losses from cropping systems.

Phosphorus applications in excess of that harvested in the crop leads to the buildup of soil-P. In addition, some soils have naturally high levels of phosphorus, and the addition of phosphorus to these soils can exacerbate already high levels of soil-P. Phosphorus applications at levels that lead to a buildup in soil have been encouraged, in part, because of difficulty in predicting crop responses to phosphorus applications.

Although the response of crops to phosphorus additions has been known for well over a century and attempts to define the crop-available soil-P date at least from 1894, there is still no universally useful extractant for soil-P, and crop responses to recommended phosphorus applications are erratic. Crop yields and/or cumulative phosphorus uptake are often not predicted well by soil tests (Prabhakaran Nair and Mengel, 1984; Yang and Jacobsen, 1990; Yerokun and Christenson, 1990).

Kamprath and Watson (1980) note that there is a problem in the interpretation of soil test results as a result of various soil-P buffer capacities, usually referred to as quantity-intensity relationships; that is, soils vary in their ability to replenish soil solution phosphorus when it is depleted by plant uptake. For example, in North Carolina, a threefold greater phosphorus level, as determined by soil tests, is needed to supply a crop's needs in a sandy soil than is needed in a finer-textured soil (Cox et al., 1981; Kamprath and Watson, 1980). This difference in a soil's capacity to supply phosphorus to a growing crop is related to the soil's phosphorus adsorption capacity. Some findings that may be of use in improving recommendations for phosphorus applications are summarized below.

The texture effect noted above occurred in a greenhouse study of phosphorus test predictions (Lins and Cox, 1989). In that study, the clay content and surface area of soil were the variables that best improved phosphorus soil test yield predictions. As noted above, the clay content and surface area of soil are correlated with phosphorus adsorption capacity.

Another study (Kuo, 1990) found that the best variable that can be used to predict plant phosphorus uptake by several soils that vary in their amorphous aluminum contents was the fraction of phosphorus adsorption sites in soil that were occupied. As in the previous study (Lins and Cox, 1989), the variables related to the soil's phosphorus adsorption capacity improved the accuracy of the predictions. It may be useful to include some measure of a soil's phosphorus buffer capacity in routine soil tests.

Recommendations for fertilizer use include a safety factor to compensate for the uncertainty of predictions of crop responses to phosphorus. Because of either a history of phosphorus applications in excess of that harvested or naturally high soil-P levels, or both, soil-P levels have increased in many U.S. soils (Thomas, 1989), and many now have high levels of phosphorus. Table 7-4 lists the percentage of soil tests in each state reading high to very high or medium or less for phosphorus.

Soil-P is often at levels above which a crop yield increase from additional phosphorus would be predicted (McCollum, 1991; Novais and Kamprath, 1978; Yerokun and Christenson, 1990). Mallarino and colleagues (1991) cited several studies reporting that increases in soybean or corn yields are small or nonexistent when soil test levels for phosphorus are within the medium category (Grove et al., 1987; Hanway et al., 1962; Million et al., 1989; Obreza and Rhoads, 1988; Olson et al., 1962; Rehm, 1986). Phosphorus additions to soils with high soil-P test results should not produce increased corn and soybean yields in the Corn Belt (Bharati et al., 1986; Hanway et al., 1962; Olson et al., 1962; Rehm, 1986). This phenomenon suggests that applications of additional phosphorus to 56, 63, 78, 68, and 35 percent of the soils tested in Iowa, Illinois, Indiana, Ohio, and Missouri, respectively, would be expected to produce no increase in yields.

Similar situations exist in the southeastern United States. Kamprath (1967, 1989) and McCollum (1991) have shown that corn and soybeans grown on Piedmont and Coastal Plain soils testing high in available phosphorus do not respond to phosphorus fertilizer additions. On the basis of the soil test data presented in Table 7-4, no response to phosphorus would be expected on approximately half of the soils in the southeastern United States. In North Carolina, phosphorus recommendations for soybeans grown on soils testing medium for phosphorus are greater than the amount of phosphorus removed in the grain (Kamprath, 1989). Thus, current recommendations will lead to soil-P levels greater than those needed for corn or soybean production.

The rate of increase in soil-P with fertilizer additions is either linear or quadratic (Figure 7-2) (Cox et al., 1981; McCollum, 1991), and the rate of decrease is exponential. The rate constant is soil dependent (Cox et al., 1981) and increases with higher initial soil-P levels (McCollum, 1991). The decline approximates the kinetics of a first-order chemical reaction (McCollum, 1991). The increasing loss rate with increasing initial level essentially means that overfertilization is a waste of money because the phosphorus is converted to unavailable forms. That is, doubling of the initial phosphorus application rate does not double the residual phosphorus effect (McCollum, 1991).

Mallarino and colleagues (1991) studied the effect on yields of phosphorus additions to a soil testing high for phosphorus. They reported occasional positive yield responses to fertilization, but these positive responses were not, in most cases, sufficient to pay for the cost of the added phosphorus. In the 11 years of the study, phosphorus applications to this soil that tested high for phosphorus showed appreciable positive economic returns in only 1 year for corn and for no year for soybeans. The addition of phosphorus resulted in negative returns in most years for both corn and soybeans, with losses in 1 year being greater than $49/ha ($20/acre) for corn (Figure 7-3).

Some of the phosphorus added in excess of crop needs remains as residual plant-available phosphorus, but not all of the added phospho-

TABLE 7-4  Soil Tests Reporting Very Low to Medium or High to Very High for Soil-P (percent)

| State | Very Low to Medium | High to Very High |
|---|---|---|
| Alabama | 65 | 35 |
| Arizona | 49 | 51 |
| Arkansas | 86 | 14 |
| California | 59 | 41 |
| Colorado | 57 | 43 |
| Connecticut | 49 | 51 |
| Delaware | 35 | 65 |
| Florida | 55 | 45 |
| Georgia | 62 | 38 |
| Idaho | 40 | 60 |
| Illinois | 37 | 63 |
| Indiana | 22 | 78 |
| Iowa | 44 | 56 |
| Kansas | 61 | 39 |
| Kentucky | 58 | 42 |
| Louisiana | 63 | 37 |
| Maine | 49 | 51 |
| Maryland | 36 | 74 |
| Massachusetts | — | — |
| Michigan | 27 | 73 |
| Minnesota | 24 | 76 |
| Mississippi | 66 | 34 |
| Missouri | 65 | 35 |
| Montana | 59 | 41 |
| Nebraska | 69 | 31 |
| Nevada | 52 | 48 |
| New Hampshire | — | — |
| New Jersey | — | — |
| New Mexico | — | — |
| New York | 62 | 38 |
| North Carolina | 33 | 67 |
| North Dakota | 70 | 30 |
| Ohio | 32 | 68 |
| Oklahoma | 52 | 48 |
| Oregon | 51 | 49 |
| Pennsylvania | 56 | 44 |
| Rhode Island | — | — |
| South Carolina | 40 | 60 |
| South Dakota | 56 | 44 |
| Tennessee | 51 | 49 |
| Texas | 63 | 37 |
| Utah | 40 | 60 |
| Vermont | 75 | 25 |
| Virginia | 42 | 58 |
| Washington | 46 | 54 |
| West Virginia | — | — |
| Wisconsin | 34 | 66 |
| Wyoming | 62 | 38 |

NOTE: Dashes indicate no data reported.

SOURCE: Adapted from Potash and Phosphate Institute. 1990. Soil test summaries: Phosphorus, potassium, and pH. Better Crops with Plant Food 74(2):16–18.

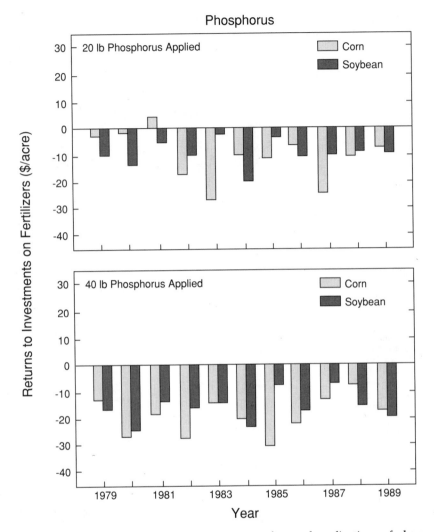

FIGURE 7-3 Economic returns on investments of annual applications of phosphorus (P) fertilizers. Source: A. P. Mallarino, J. R. Webb, and A. M. Blackmer. 1991. Corn and soybean yields during 11 years of phosphorus and potassium fertilization on a high-testing soil. Journal of Production Agriculture 4:312–17. Reprinted with permission from © American Society for Agronomy, Crop Science Society of America, and Soil Science Society of America.

rus is available to crops because the amount of extractable phosphorus declines with time because of the slow conversion of phosphorus to unavailable forms (McCollum, 1991; Mendoza and Barrow, 1987; Sharpley et al., 1989; Yost et al., 1981). The rate of decline in extractable phosphorus (discounting plant uptake) varies with the soil-P adsorption properties and the initial soil-P level, that is, relative saturation of adsorption capacity, and with the amount of applied phosphorus.

Several studies have investigated the buildup of soil-P under continuous phosphorus fertilization conditions (McCallister et al., 1987; Schwab and Kulyingyong, 1989), and others have documented the loss of soil-P under continuous cropping with only residual phosphorus available for crop uptake (Novais and Kamprath, 1978). Both buildup and decline phases have been studied as well (Cope, 1981; McCollum, 1991; Meek et al., 1982), but relatively few studies (Cope, 1981; McCollum, 1991) have been conducted over long time spans (several decades). Results of these few studies may provide some of the best information that can be used to aid in predicting residual phosphorus effects and actual phosphorus fertilization needs. Many soils can be cropped for a decade or more without the soil-P reaching a level at which fertilizer additions would result in a crop yield increase (Figure 7-4). A crop yield increase would not be expected until soil-P levels fall below 22 $g/m^3$, which would occur only after several years of cropping, depending on the initial soil-P level.

The data and studies available suggest that the amount of phosphorus added to cropping systems could be reduced without decreasing crop yields on a significant portion of the nation's croplands. In those soils testing high for soil-P, phosphorus applications other than for starter fertilizer could be suspended without yield losses, depending on the soil, crop, and climate.

Despite weaknesses in the ability to predict crop responses to phosphorus applications, most states have soil test procedures that, although not perfect, can be used to establish the threshold levels of soil-P beyond which no crop response is predicted. Given the importance of reducing soil-P levels for reducing phosphorus losses to surface waters, such thresholds should be established. Applications of additional phosphorus, except for small starter applications, should be discouraged once that threshold level of soil-P is reached. In extreme cases, when damages to surface water are great, a second threshold level of soil-P— beyond which no additional phosphorus should be applied—should also be established. Once established, such threshold values should become a routine part of phosphorus application recommendations supplied by public and private organizations.

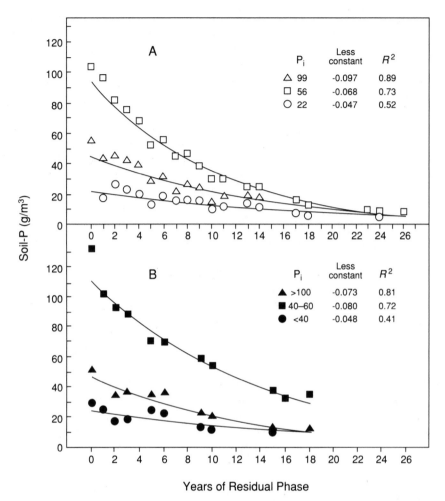

FIGURE 7-4   Decrease of soil-P over time, measured as Mehlich 1-extractable phosphorus, on Portsmouth soil during the residual phase. A, initial soil-P ($P_i$) at year 0 established by one broadcast application; B, $P_i$ at year 0 is the result of 8 previous years of active buildup via annual, banded applications. Source: R. E. McCollum. 1991. Buildup and decline in soil phosphorus: 30-year trends on a Typic Umprabuult. Agronomy Journal 83:77–85. Reprinted with permission from © American Society for Agronomy, Crop Science Society of America, and Soil Science Society of America.

## Methods for Reducing Erosion and Runoff

The most studied methods for reducing phosphorus losses from fields involve a variety of reduced-tillage methods. There is usually a trade-off between sediment and solution phosphorus losses, and the chosen management method should be one that reduces the total bioavailable phosphorus load in runoffs (Sonzogni et al., 1982).

A number of reduced-tillage, no-tillage, and other alternative tillage systems have been tried; the general result is that total phosphorus loss is less with any system that reduces soil exposure (Andraski et al., 1985; McDowell and McGregor, 1984; Romkens et al., 1973; Sharpley and Menzel, 1987). Concentrations of soluble phosphorus in runoffs are often higher, and total losses of soluble phosphorus are sometimes higher. The amount of bioavailable phosphorus, which should be the true measure of a system's effectiveness (Sonzogni et al., 1982), is not always measured, but it generally tends to be lower in reduced-tillage systems (Andraski et al., 1985). If manure rather than chemical fertilizer is applied, the advantage of reduced tillage may be negated because of the leaching of phosphorus from the unincorporated manure, and bioavailable phosphorus losses may be greater than those from conventional tillage systems (Mueller et al., 1984).

Other potential adverse effects of reduced-tillage practices include the stratification of phosphorus near the surface because of a lack of incorporation and the cycling of phosphorus to the surface by plants (Mackay et al., 1987). This phenomenon may increase surface losses of phosphorus because of the increased amount of available phosphorus at the surface and may also restrict nutrient uptake and rooting to surface layers, particularly early in the season.

Reduced-tillage systems or other systems that increase soil cover and that effectively reduce both runoff and soil erosion generally reduce total losses of bioavailable phosphorus to surface waters. Reduced-tillage systems do, however, appear to increase the proportion of bioavailable phosphorus lost in soluble form. The effectiveness of reduced-tillage systems in reducing bioavailable phosphorus losses would be increased if parallel efforts were undertaken to reduce phosphorus concentrations in surface soils. Effective efforts to reduce phosphorus loadings to surface water should simultaneously reduce soil-P levels, erosion, and runoff. Efforts to reduce any of those three factors without reducing the others may exacerbate trade-offs between soluble and particulate phosphorus loadings.

A number of the findings summarized above, as well as a number of others not mentioned, have been incorporated into models that predict soil-P adsorption properties, crop responses to fertilizer-P, and surface losses of phosphorus. These models and their supporting data have been developed by Sharpley and coworkers over a number of years. The models are empirical rather than mechanistic, but they appear to provide useful predictions of phosphorus losses (Sharpley et al., 1991; Smith et al., 1991). These models may be helpful in choosing which of several phosphorus loss management methods will be most effective.

## Buffer Strips

Buffer strips may be helpful in reducing both the particulate and soluble phosphorus fractions, especially when vegetation or crop residues are present. Nutrient load reductions of more than 70 percent have been achieved with several types of buffer strip-surface cover combinations (Alberts et al., 1981; Thompson et al., 1978). Under high flow conditions, the efficiency of the buffer strips diminishes and phosphorus losses may be greater than those under control conditions.

Even after the runoff leaves the field, nutrient loads, particularly sediment-bound fractions, may be dropped by sedimentation near the field's edge above the watercourse. In one study (Cooper et al., 1987), 50 percent of the total sediment load was deposited within 100 m (109 yards) of the field's edge, and 80 percent was removed from above the creek floodplain. Another study involved a managed distribution system that applied beef feedlot runoff to a small wooded watershed (Pinkowski et al., 1985). More than 99 percent of the input phosphorus was retained in the watershed, although phosphorus concentrations and total losses increased relative to those under the baseline conditions. Also, there was a large net loss of nitrogen from the watershed, and the investigators encountered problems with tree mortality because of excessive soil moisture.

Once in the watercourse, nutrient loads may still be reduced. Wetlands have been proposed to act as nutrient filters, but they may be only small sinks for nitrogen and phosphorus and may, in fact, be net exporters of some nutrients (Peverly, 1982).

Buffer strips and protection of riparian areas should help to reduce phosphorus loadings to surface waters. These measures, however, cannot substitute for efforts to reduce soil-P levels, runoff, and erosion. Use of buffer strips and protection of riparian areas should increase the effectiveness of programs to reduce phosphorus loadings if they are part of a larger effort.

## Inclusion of Extreme Weather as Loss Factor

All potential management options must be considered in the context of natural events (Sharpley and Menzel, 1987), that can cause large losses of soil and nutrients from even the best managed systems. Intensive individual events may cause greater losses of phosphorus and sediments than those that occur during years of normal runoff. For example, Cahill and colleagues (1978) attributed 86 percent of total phosphorus loss to a 1-month period in late winter. Nelson and colleagues (1978) found that large runoff events contributed 58 to 78 percent of the total sediment phosphorus load over 2 years. Burwell and colleagues (1975) found that the period that included the planting date plus the 2 months that followed caused the greatest sediment and total phosphorus loss for conventionally tilled systems. Schuman and colleagues (1973) found that a few major rainfall events caused 80 percent of the total phosphorus loss. Hubbard and colleagues (1982) found that 64 to 86 percent of total sediment loss was due to one storm. These examples emphasize the fact that no management technique, no matter how well designed and implemented, will be 100 percent effective. Augmenting phosphorus management with buffer zones or using different cropping systems that provide greater soil protection can reduce the damage caused by extreme events.

### New Cropping Systems

Immediate gains in reducing phosphorus loads to surface waters can be accomplished by simultaneous efforts to reduce soil-P levels, erosion, and runoff from current cropping systems. These improvements can be increased by efforts to incorporate buffer strips and protect riparian vegetation to trap phosphorus in the sediments and runoff from current cropping systems. In the long term, it is unclear whether such improvements will be sufficient to meet water quality goals. Changes in cropping systems, such as the use of cover crops, multiple crops, or other approaches to increasing soil cover and reducing soil-P levels, may be needed to further reduce phosphorus losses from cropping systems.

Cover crops, where applicable, appear to hold the promise of substantially reducing phosphorus losses from cropping systems. Sharpley and Smith (1991) reported that the addition of a winter rye cover crop to conventionally tilled corn reduced total phosphorus losses by 70 percent in Georgia. The combination of an alfalfa-timothy hay cover crop with no-tilled corn in Quebec reduced total phosphorus losses by 94 percent from that with conventionally tilled corn with no cover crop. In Alabama, phosphorus losses were 30 percent less in no-tilled than in conventionally tilled corn. A winter wheat cover crop in combination with no tillage reduced phosphorus losses by 68 percent.

# 8

---

# Fate and Transport of Pesticides

The agricultural production systems of the United States are capable of producing a bountiful supply of food and fiber, but at some cost to the nation's water, soil, and air resources. As agricultural production intensified, the natural pest-predator relationship that keeps many crop pests in check was disturbed. This contributed to the increasing use of pesticides. Chemical control of pests and diseases escalated in the mid-1950s with the discovery of new synthetic organic compounds (for example, dichlorodiphenyltrichloroethane [DDT] and 2,4-dichlorophenoxyacetic acid [2,4-D]). The consumer began to expect attractive-looking food products without blemishes or insects. However, Rachel Carson's book *Silent Spring* (Carson, 1962) gave rise to public concern about the threat of pesticide contamination of the environment. By 1980, agriculture used 72 percent of all pesticides applied in the United States, and herbicides and insecticides made up 89 percent of the pesticides used by agriculture.

About 50,000 pesticide products are now registered for use with the U.S. Environmental Protection Agency (EPA), but the number of those used extensively is smaller. These pesticides are commonly classified according to their intended target organism (for example, insecticides, herbicides, fungicides, nematicides, rodenticides, and miticides) and according to their intended use (for example, defoliants, desiccants, fumigants, and plant growth regulators).

Before World War II, pesticides consisted of products from natural sources such as nicotine, pyrethrum, petroleum and oils, and rotenone, as well as inorganic chemicals such as sulfur, arsenic, lead, copper, and lime. During and then after World War II, phenoxy herbicides and organochlorine insecticides were widely used. In the mid-1960s, their use declined because they were replaced by triazine and amide herbicides and carbamate and organophosphate insecticides. Some pesticides (for example, DDT and dibromochloropropane [DBCP]) have been banned from use mainly because of their toxicities.

In the past, agriculture was mostly concerned with on-site measures that could be used to enhance crop and livestock production. In the 1960s, investigators became more aware of the off-site effects of farming operations, such as the degradation of surface water quality. In the 1980s, investigators became acutely aware of groundwater contamination. Water pollution, for instance, was initially a local problem created mainly by identifiable and easily regulated point sources of contamination. However, with widespread pesticide applications the problem has spread regionally, nationally, and globally. Recent assessments (Garner et al., 1986; Holden, 1986; National Research Council, 1989a; U.S. Congress, Office of Technology Assessment, 1990; U.S. Environmental Protection Agency, 1990b) of pesticide contamination of waters indicate that contamination is widespread, although at low concentrations.

This chapter evaluates the fate and transport of pesticides in agroecosystems, opportunities for the prevention of water pollution from such systems, and assessment of the knowledge base relative to policy implications.

## FATE AND TRANSPORT PROCESSES

Figure 8-1 (Sawhney and Brown, 1989) shows the interactions and loss pathways of organic chemicals in soils. Figure 8-2 (Cheng, 1990) shows similar and additional features of the environmental fates of pesticides applied to croplands. Pesticides are formulated in a variety of ways (as liquids, gases, and solids) and are applied by a number of methods (aerial or canopy spraying, incorporation or injection into the soil, and with water). Pesticides applied to cropping systems can be degraded by microbial action and chemical reactions in the soil. Pesticides can also be immobilized through sorption onto soil organic matter and clay minerals. Pesticides can also be lost to the atmosphere through volatilization. Pesticides that are taken up by pests or crop plants either can be transformed to degradation products (which are often less toxic than the original compound) or, in some cases, can accumulate in plant or animal tissues. A certain portion of the pesticides applied are also removed when the crop is harvested.

Pesticides that are not degraded, immobilized, detoxified, or removed with the harvested crop are subject to movement away from the point of

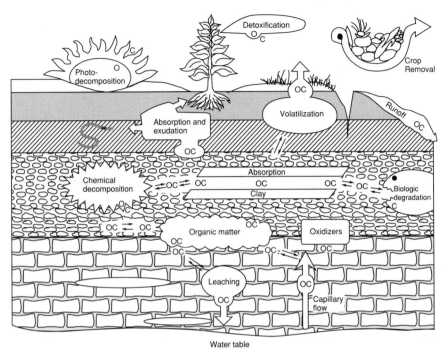

FIGURE 8-1  Interactions and loss pathways of organic chemicals (OCs) in soils. Source: B. L. Sawhney and K. Brown. 1989. Reactions and Movement of Organic Chemicals in Soils. Special Publication No. 22. Madison, Wis.: Soil Science Society of America. Reprinted with permission from © American Society for Agronomy, Crop Science Society of America, and Soil Science Society of America.

application. The major loss pathways of pesticides to the environment are volatilization into the atmosphere and aerial drift, runoff to surface water bodies in dissolved and particulate forms, and leaching into groundwater basins.

The fate and transfer pathways of pesticides applied to croplands are complex, requiring some knowledge of their chemical properties, their transformations (breakdown), and the physical transport process. Transformations and transport are strongly influenced by site-specific conditions and management practices.

### Pesticide Properties

Chemical-specific properties influence the reactivities of pesticides (Porter and Stimman, 1988). Pesticides that dissolve readily in water are

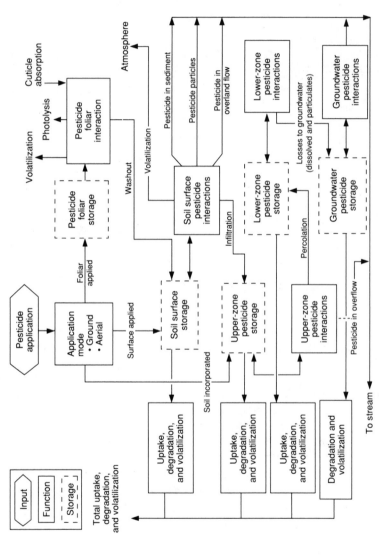

FIGURE 8-2 Pesticide transport and transformation in the soil-plant environment and the vadose zone. Source: H. H. Cheng, ed. 1990. Pesticides in the Soil Environment: Processes, Impacts, and Modeling. Soil Science Society of America Book Series No. 2. Madison, Wis.: Soil Science Society of America. Reprinted with permission from © American Society for Agronomy, Crop Science Society of America, and Soil Science Society of America.

considered to be highly soluble. These chemicals have a tendency to be leached through the soil to groundwater and to be lost as surface water runoff from rainfall events or irrigation practices.

Pesticides with high vapor pressures are easily lost to the atmosphere during application, and shortly thereafter they are lost from the soil through gaseous diffusion. Some highly volatile pesticides, however, may also move downward into aquifers.

Pesticides may be sorbed to soil particles, particularly the clays and soil organic matter. Strongly sorbed pesticides do not readily leach through the soil profile but may be bound to the sediments discharged from croplands.

Pesticides may be degraded (transformed) by chemical and biological processes. Chemical degradation occurs through such reactions as photolysis (photochemical degradation), hydrolysis (reaction with water), oxidation, and reduction. Biological degradation may also occur as soil microbes consume or break down pesticides. These microbes are most prevalent in the top several centimeters of soil. The extent of degradation may range from the formation of metabolites (daughter products) to the formation of inorganic decomposition products.

Once a pesticide enters the soil, its fate is largely dependent on sorption and persistence (Rao and Hornsby, 1989). Sorption is commonly evaluated by use of a sorption (partition) coefficient ($K_{oc}$) based on the organic carbon content of soils. Persistence is commonly evaluated in terms of half-life, which is the time that it takes for 50 percent of a chemical to be degraded or transformed. Pesticides with low sorption coefficients are likely to leach. Pesticides with long half-lives could be persistent.

In Table 8-1 Rao and Hornsby (1989) provide a list of the pesticides as well as their sorption coefficients and half-lives. Pesticides are classified as nonpersistent if they have half-lives of 30 days or less, moderately persistent if they have half-lives longer than 30 days but less than 100 days, and persistent if they have half-lives longer than 100 days. Within these persistence classes, the pesticides are listed in ascending order of their sorption coefficients. Threshold values indicating the potential of a chemical for groundwater contamination have been proposed by the EPA (1986a). A pesticide is likely to contaminate groundwater (leach) if its sorption coefficient is low, its half-life is long, and its water solubility is high. It should be noted that it is difficult to predict the half-life of a chemical in the field because of dependent variables such as soil temperature and moisture, microbial populations, and soil types.

The pesticide residues most commonly found in U.S. groundwaters include alachlor, aldicarb, atrazine, bromacil, carbofuran, cyanazine,

TABLE 8-1   Partition Coefficients and Half-Lives of Pesticides Used in Florida

| Pesticide (common name) | Sorption Coefficient (ml/g of organic chemical) | Half-Life (days) |
|---|---|---|
| *Nonpersistent* | | |
| Dalapon | 1 | 30 |
| Dicamba | 2 | 14 |
| Chloramben | 15 | 15 |
| Metalaxyl | 16 | 21 |
| Aldicarb | 20 | 30 |
| Oxamyl | 25 | 4 |
| Propham | 60 | 10 |
| 2,4,5-T | 80 | 24 |
| Captan | 100 | 3 |
| Fluometuron | 100 | 11 |
| Alachlor | 170 | 15 |
| Cyanazine | 190 | 14 |
| Carbaryl | 200 | 10 |
| Iprodione | 1,000 | 14 |
| Malathion | 1,800 | 1 |
| Methyl parathion | 5,100 | 5 |
| Chlorpyrifos | 6,070 | 30 |
| Parathion | 7,161 | 14 |
| Fluvalinate | 100,000 | 30 |
| *Moderately Persistent* | | |
| Picloram | 16 | 90 |
| Chlormuron-ethyl | 20 | 40 |
| Carbofuran | 22 | 50 |
| Bromacil | 32 | 60 |
| Diphenamid | 67 | 32 |
| Ethoprop | 70 | 50 |
| Fensulfothion | 89 | 33 |
| Atrazine | 100 | 60 |
| Simazine | 138 | 75 |
| Dichlorbenil | 224 | 60 |
| Linuron | 370 | 60 |
| Ametryne | 388 | 60 |
| Diuron | 480 | 90 |
| Diazinon | 500 | 40 |
| Prometryn | 500 | 60 |
| Fonofos | 532 | 45 |
| Chlorbromuron | 996 | 45 |
| Azinphos-methyl | 1,000 | 40 |
| Cacodylic acid | 1,000 | 50 |
| Chlorpropham | 1,150 | 35 |
| Phorate | 2,000 | 90 |
| Ethalfluralin | 4,000 | 60 |

TABLE 8-1   (*Continued*)

| Pesticide (common name) | Sorption Coefficient (ml/g of organic chemical) | Half-Life (days) |
|---|---|---|
| Chloroxuron | 4,343 | 60 |
| Fenvalerate | 5,300 | 35 |
| Esfenvalerate | 5,300 | 35 |
| Trifluralin | 7,000 | 60 |
| Glyphosphate | 24,000 | 47 |
| | *Persistent* | |
| Fomesafen | 50 | 180 |
| Terbacil | 55 | 120 |
| Metsulfuron-methyl | 61 | 120 |
| Propazine | 154 | 135 |
| Benomyl | 190 | 240 |
| Monolinuron | 284 | 321 |
| Prometon | 300 | 120 |
| Isofenphos | 408 | 150 |
| Fluridone | 450 | 350 |
| Lindane | 1,100 | 400 |
| Cyhexatin | 1,380 | 180 |
| Procymidone | 1,650 | 120 |
| Chloroneb | 1,653 | 180 |
| Endosulfan | 2,040 | 120 |
| Ethion | 8,890 | 350 |
| Metolachlor | 85,000 | 120 |

SOURCE: P. S. C. Rao and A. G. Hornsby. 1989. Behavior of Pesticides in Soils and Waters. Soil Science Fact Sheet SL 40 (revised). Gainesville: University of Florida.

DBCP, dimethyltetrachloroterephthalate (DCPA), 1,2-dichloropropane, dinoseb, dyfonate, ethylenedibromide (EDB), metolachlor, metribuzon, oxamyl, simazine, and 1,2,3-trichloropropane (U.S. Environmental Protection Agency, 1986a).

Those pesticides that are strongly sorbed to soil clays and organic matter may be subject to removal by surface runoff. Pesticides that exhibit such behavior and that are present in surface waters include the organochlorine DDT and its metabolites dichlorodiphenyldichloroethane) (DDD) and dichlorodiphenyldichloroethylene (DDE), dieldrin, endosulfan, toxaphene, lindane, heptachlor, chlordane, and difocol. Other pesticides that are weakly sorbed and have high water solubilities may be lost in the dissolved state. Erosion control practices will have little effect on such losses (Wauchope, 1978). Examples of pesticides found in the water phase of agricultural runoffs include the herbicides 2,4-D, dicamba, dinoseb, (4-chloro-2-methylphenoxy)acetic acid (MCPA), and molinate (Wauchope, 1978).

## Soil Properties

Soil properties have significant influences on the fate and transport of pesticides in croplands (Porter and Stimman, 1988). In general, the infiltration rate and hydraulic conductivity (soil permeability) of coarser-textured soils are greater than those of finer-textured soils. A chemical that readily infiltrates into the soil is less likely to be lost in surface runoff but is more likely to be leached into groundwater. The travel time of soil water and its associated dissolved pesticide is shorter in coarser- than in finer-textured soils. Soil permeability may have some influence on the rate at which volatile gases are lost. Moreover, the sorptive capacity of fine-textured soils is greater than that of coarse-textured soils because of the higher clay and organic matter contents of fine-textured soils; hence, pesticides are less vulnerable to leaching.

pH is an important soil property for those pesticides participating in hydrolysis reactions. For instance, DBCP is chemically degraded into its metabolites by the substitution of chloride and bromide at the halogenated sites by hydrogen ions. The hydrolysis or dehalogenation of DBCP occurs in the soil at a faster rate in the alkaline pH range.

Soil structure is another property that reflects the manner in which soil particles are aggregated and cemented. A soil with a weak structure is more likely to be eroded and have lower infiltration rates, and hence, sorbed pesticides are more likely to be discharged through runoff. Recent evidence indicates that at times soil macropores and cracks have a major effect on the movement of pesticides in soils. Macropores are formed by earthworms and decayed root systems, while cracks are formed by soil shrinkage. Under particular water application rate conditions, both water and chemicals in the dissolved and particulate forms tend to preferentially move through the macropores and cracks and reach the water table in a shorter period of time.

## Site Conditions

Other site conditions affect runoff and leaching of pesticides (Porter and Stimman, 1988). In general, the groundwater table is shallower in humid regions than in more arid regions. A shallow depth to the groundwater offers less opportunities for pesticide sorption and degradation. The travel time of the pesticide to the water table may range from days to a week if the depth to the water table is shallow, the soil is permeable, and the amount of rainfall exceeds the water-holding capacity of the soil. In contrast, the travel time may be on the order of decades in arid regions where the water table is tens of meters below the land surface.

Hydrogeologic conditions (underground plumbing) beneath the soil profile may dictate the direction and rate of chemical movement. The presence of impermeable lenses or layers in the soil profile and underlying strata may limit the vertical movement of pesticides. Such impermeable layers may, however, contribute to the lateral flow of shallow groundwaters and to the eventual discharge of groundwaters and its contaminants into surface waters. On the other hand, the presence of high-permeability earth materials such as sands and gravel may greatly accelerate the vertical and horizontal flows of contaminants. Of particular concern is the presence of karsts (limestone) and fractured geologic materials that generally transmit water and chemicals rapidly to the groundwater body.

Climatic and weather conditions other than rainfall may also influence the fate of pesticides. Warmer temperatures tend to accelerate physical, chemical, and biological processes such as volatility, water solubility, and microbial degradation, respectively. High winds and high evaporation rates may accelerate volatilization and other processes that contribute to gaseous losses of pesticides.

## Management Practices

Management practices such as the rate and timing of pesticide applications and the mode of pesticide application also affect pesticide transport processes. The recommended practices (Porter and Stimman, 1988; U.S. Congress, Office of Technology Assessment, 1990) include pesticide use only when and where it is necessary and in amounts adequate to control pests. Those who use pesticides should carefully follow the directions on the label to minimize harmful effects to the applicator as well as potential losses to the environment. Pesticide users should select pesticides that are less likely to leach. Irrigation should be avoided shortly after pesticide application, to reduce losses through runoff and leaching. The best management practices for pesticide use are highly specific to crops and locations.

Following is a list of variables that affect and conditions that increase the likelihood of pesticides leaching into groundwater:

- pesticide properties
  —high solubility
  —low adsorption
  —persistence

- soil characteristics
  —coarse texture, high permeability
  —low organic matter content
  —presence of macropores
- site conditions
  —high permeability of vadose region
  —shallow depth to groundwater
  —wet climate or heavy irrigation
  —low soil temperatures
- management practices
  —pesticide injection or incorporation into soil
  —poor timing of chemical application with rainfall or irrigation

A number of universities and agencies in various locations (for example, Hawaii, New York, Pennsylvania, Minnesota, and Florida) are using geographic information systems to identify aquifers vulnerable to contamination by pesticides. The management practices that can be used to reduce pesticide pollution of surface water and groundwater are discussed below in greater detail.

### Mass Balance

Figure 8-2 presents a comprehensive scheme of the fates of pesticides applied in agroecosystems. Despite the vast knowledge base for the reactivity and transport of pesticides, a complete mass balance of the fate of any field-applied pesticide does not exist in the literature.

Investigators have difficulty obtaining mass balances of the fates of pesticides for a number of reasons. Pesticides include a broad class of agrichemicals with widely ranging properties and behaviors that defy generalizations. There are technical difficulties and high costs associated with measuring over time the fraction of pesticides present in the various multimedia compartments and subcompartments in Figure 8-2. Some processes, such as volatilization, sorption, photolysis, foliar washouts, and surface runoffs, occur over short time intervals (for example, hours and days), whereas others occur over long time intervals (for example, months and decades), such as hydrolysis, microbial degradation, and transport through the vadose region for cases in which the water table is many tens of meters below the land surface (the vadose region is that part of the soil above the permanent groundwater level). Recognizing this dilemma of acquiring an adequate mass balance, a concerted effort to obtain mass balances is being made through modeling (see below).

Some researchers have estimated that only 1 to 2 percent of insecticides applied to foliage is absorbed by the target pest. They base this estimate on a synthesis of conceptual mass balance. Figure 8-3, for example, is a mass balance for a typical aerial spray-foliar application of an insecticide. The hypothetical mass balance in Figure 8-3 indicates problem areas where the efficacies of pesticide applications can be improved.

Even though a complete mass balance may not be available for a specific field case study, a few examples of the measured fates of pesticides from numerous literature sources would give some perspective. During the application stage of pesticide use, considerable losses may occur through spray drift and volatilization. Spray drift constitutes about 3 to 5 percent of the loss under quiescent wind conditions, but it is typically much greater (40 to 60 percent for many insecticides). Loss from volatilization ranges from 3 to 25 percent, but it may be as great as 20 to 90 percent for methylparathion, for example, depending on weather conditions. With regard to the efficacy of pesticide applications, losses to soil and peripheral nontarget foliage may be as high as 60 to 80 percent for most sprays (Cheng, 1990).

In contrast, pesticide losses from soil-incorporated application methods are much lower. Field measurements of pesticides applied by such practices reveal that the portion of pesticide volatilized is 2 to 12 percent for most pesticides but could be as high as 50 to 90 percent for volatile chemicals such as trifluralin. Seasonal losses of pesticides in surface runoffs are typically in the range of less than 1 to 5 percent (Wauchope, 1978); the lower losses are for foliar-applied organochlorines like toxaphene, and the higher losses are for wettable powders such as triazine.

Pesticide loss through leaching into groundwater is another major component in the mass balance. Factors contributing to the vulnerability of groundwater contamination were discussed above. The mass flux for leaching is sometimes taken at some prescribed soil depth, like the bottom of the root zone or the surface of the groundwater table. It should be noted that within the crop root zone and in the vadose region the pesticide is subject to numerous degradation and immobilization mechanisms.

A study of the fate of DBCP applied to cropland in California's San Joaquin Valley provides an example of the impact of these multiple processes. The peak concentration of DBCP at the time of application was about 1,500 mg/L (1,500 ppm), but at the 30-cm depth in the surface soil its peak concentration was only about 1.5 mg/L (1.5 ppm). Based on a transport model that considered diffusive and mass flows as well as sorption and decay (Tanji, 1991a), the peak concentration is expected to be about 0.03 mg/L (0.03 ppm) as the DBCP reached the water table after passing through 30 m of the vadose zone. That peak concentration of

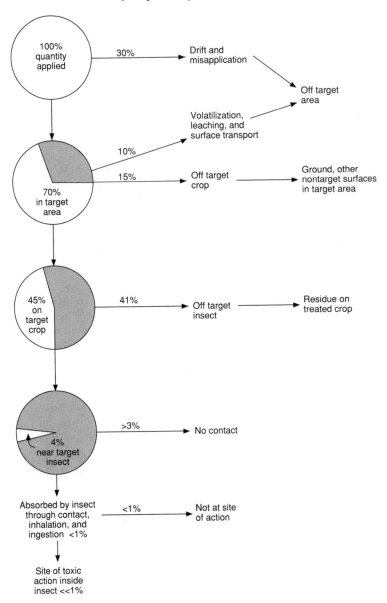

FIGURE 8-3   Mass balance of a hypothetical aerial foliar-spray application of an insecticide. Source: U.S. Congress, Office of Technology Assessment. 1990. Beneath the Bottom Line: Agricultural Approaches to Reduce Agrichemical Contamination of Groundwater. Report No. OTA-4-418. Washington, D.C.: U.S. Government Printing Office.

DBCP at the bottom of the vadose region was expected to be diluted to about 0.009 mg/L (0.009 ppm) in the surface of the horizontally flowing groundwater. The cumulative loss of DBCP from the soil surface to the groundwater, therefore, was about 99.4 percent of the 1.5 mg/L (1.5 ppm) at the 30-cm soil depth and nearly 100 percent of the 1,500 mg/L (1,500 ppm) at the soil's surface. These calculations, which involved 31 years of travel time, indicate that even a persistent chemical like DBCP is subject to huge sink losses, but because the maximum contaminant level set by EPA is so low 0.02 $\mu$g/L (0.02 ppb) DBCP in drinking water in California, the minuscule leaching loss to the groundwater system is of growing concern.

Given the difficulty of obtaining mass balances on the fates of pesticides applied to croplands and the fact that minuscule losses to particular portions of the environment may be hazardous, a concerted effort must be made to increase the efficiency of pesticide use. Since everything must go somewhere, source control is the best approach.

## Modeling Efforts

Computer simulation models are used as tools to evaluate more fully the fate and transport of pesticides in agricultural systems.

### Spray Pesticide Models

Spray pesticide models (Cheng, 1990) based on diffusion or ballistics are available for the design and evaluation of spray application systems to minimize aerial drift and volatile losses as well as to elevate accurate targeting of the spray.

### Pesticide Runoff Models

Pesticide runoff models (Cheng, 1990) from the small plot to the watershed scales are being used to develop best-management practices. Considerable efforts were initially made toward determining the hydrologic processes that contribute to runoff and erosion. The reactivities of pesticides were then coupled to runoff models. For the most part, these runoff models are more useful for evaluation of alternative management strategies and are less useful for predictive purposes.

### Pesticide Leaching Models

More recently, a concerted effort has been made to model pesticide leaching. Modeling of pesticide behavior and the leachabilities of pesti-

326 / Soil and Water Quality: An Agenda for Agriculture

cides has taken two general approaches. Conceptual screening models for pesticides typically consider solubility, sorption, persistence, volatility, and mobility. Such models rank the behavior of the pesticide and its potential movement in soil. One example is that prepared by Rao and Hornsby (1989) (see Table 8-1). A more comprehensive approach is that of Jury and colleagues (1984), in which chemicals are screened under idealized, standardized scenarios. However, screening models are not environmental fate prediction models and are inappropriate outside the idealized conditions that lead to their derivation.

*Process-Based Simulation Modeling*

Process-based simulation modeling for pesticide reactivity and transport has received a more intensive effort. These research-oriented models require extensive input data and have mainly been tested in laboratory soil columns and small-scale research plots. Recently, they have been extended and/or applied for management of larger-scale field environments, for example, the erosion/productivity impact calculator (EPIC) (Sharpley and Williams, 1990), groundwater loading effects of agricultural management systems (GLEAMS) (Leonard et al., 1987), pesticide root zone model (PRZM) (Carsel et al., 1984), and leaching estimation and chemistry model—pesticides (LEACHM-P) (Wagenet and Hutson, 1989).

*Model Performance*

Jury and colleagues (1988) and Green and colleagues (1986) have pointed out some difficulties in predicting groundwater contamination by chemicals even with state-of-the-art simulation models. They point out that the convection-dispersion models appear to be unable to predict pesticide transport in the vadose zone. The reasons contributing to this dilemma include the spatial variability in the hydraulic properties usually encountered in field soils, the potential nonequilibrium sorption in the field, the depth dependency of biodegradation, and preferential flow through macropores. Such considerations need to be incorporated into pesticide leaching models for improved model performances.

Most recently, Pennell and colleagues (1990) compared the performances of five simulation models for simulating the behaviors of aldicarb and bromide from a given field study. The models tested were the chemical movement in layered soils model (CMLS) (Nofziger and Hornsby, 1986), the method of saturated zone solute estimation (MOUSE) (Steehuis et al., 1987), the pesticide root zone model (PRZM)

Groundwater in artesian aquifers have enough pressure to flow all the way to the surface. The detection of pesticides in aquifers that are hundreds of feet deep has increased concern about the eventual fate of pesticides applied to croplands. Credit: Agricultural Research Service, USDA.

(Carsel et al., 1984), GLEAMS (Leonard et al., 1987), and LEACHM-P (Wagenet and Hutson, 1989). GLEAMS and MOUSE underestimated bromide and aldicarb dissipations, whereas the other models proved satisfactory in predicting both the depth of the solute's center of mass and the amount of pesticide degradation. None of the models, however, accurately predicted the pesticide concentrations measured throughout the soil profile. In addition to possible deficiencies in the model, the investigators pointed out the potentially large sampling error in the field because of spatial variability.

### Appraisals of Models

Mathematical models of surface runoff and leaching of pesticides have been constructed, tested, and used with varying degrees of success. The formulation of each model varies according to the objectives of the modeling exercises and the professional training and biases of the model developer. The result has been a collection of approaches applicable to descriptions of surface runoff processes and a second body of efforts that have focused on leaching processes. Investigators have not often attempted to make comprehensive simultaneous descriptions, and when they have, the results have been complex, data-intensive models that cannot easily be used by anyone other than the developer (Wagenet and Rao, 1990).

A number of models simulate surface runoff and the resultant pesticide loading of surface waters (Adams and Kurisu, 1976; Bruce et al., 1975; Donigian and Crawford, 1976; Donigian et al., 1977; Frere, 1978; Haith, 1980, 1986; Leonard and Wauchope, 1980; Wauchope and Leonard, 1980). In almost all cases, the models represent a compromise between the available data, which are often quite sparse and variable, and the need for a predictive tool that can be used across different soils, climates, and pesticides. Investigators have obtained mixed results with these models. To date there is apparently no increased predictive capability obtained by using models that are more mechanistic and data intensive than using models that provide less of an understanding of the field-scale processes related to pesticide loss including surface hydrological processes.

Soil leaching models of pesticide fates contain similar problems, although the basic physical, chemical, and biological processes in the soil are perhaps better defined than surface hydrology processes. Useful field-scale models exist in both mechanistic (Carsel et al., 1984; Wagenet and Hutson, 1989) and nonmechanistic (Nofziger and Hornsby, 1986; Rao et al., 1976) forms, although care must be used in choosing the situations to which these models are applied. Neither mechanistic nor empirical models have been widely tested under field conditions. The empirical versions are generally intended for qualitative educational purposes rather than quantitative regulatory purposes. A number of solute transport models that are intermediate between the mechanistic and nonmechanistic extremes have been proposed (reviewed by Addiscott and Wagenet, 1985), but they have yet to be applied to pesticide leaching by water.

The spatial variability of soil processes also has generated interest in stochastic or probabilistic approaches to describing chemical leaching in soil (Jury et al., 1988) or surface loss of pesticides (Mills and Leonard, 1984). These approaches may prove to be the most useful because they show promise as descriptors of spatially variable processes, yet they are neither as mathematically cumbersome nor as computationally demanding as current mechanistic models (Wagenet and Rao, 1990). Stochastic or probabilistic approaches can also account for the stochastic nature of precipitation and its effect on leaching or runoff (Hornsby, 1988).

One excellent source of advancements in groundwater modeling software is the International Ground Water Modeling Center, Colorado School of Mines. Although not all of these groundwater models would be suitable for simulating pesticide transport, some should be directly applicable. These new modeling efforts, however, are not typically incorporated into pesticide leaching models; hence, incorporation will require considerable effort on the part of modelers researching the transformations and transport of pesticides in agricultural systems.

## REDUCTION OF PESTICIDE POLLUTION

The management practices that can be used to reduce environmental pollution from pesticide use in agroecosystems can be broadly categorized into

- selection of proper pesticides and formulations;
- timing of and improvement in pesticide application methods to minimize drift and volatile losses;
- use of erosion and runoff control measures to reduce losses through runoff and leaching;
- use of nonchemical pest control measures such as crop rotations and management; and
- integrated pest management, which embodies most of the recommended practices cited earlier.

Source control or reducing the amounts of pesticides used should be the first line of action.

## Selection and Formulation of Pesticides

Pesticides should be used only when and where they are necessary and only in amounts adequate to control the target pest. If a potential pesticide user can choose among a number of available pesticides, the user should select those that will be least harmful to the environment. For many conditions, the characteristics of a selected pesticide should include low water solubility, high sorptive capacity, low vapor pressure, higher potential for chemical and microbial degradation, and shorter overall half-life in the field.

Although pesticides are formulated mainly for ease of application, the natures of the formulations do have some impacts on potential losses to the environment. For instance, use of pesticides in the granular, pelleted, or emulsified form results in less drift and volatile losses during application. Pesticides in the form of dusts, wettable powders, or fine liquid sprays are more subject to drift losses. Pesticides applied as liquid mixtures or concentrated solutions have greater potential for loss through volatilization. Those pesticides in wettable powders are more susceptible to runoff losses.

## Timing and Pesticide Application Methods

For maximum efficacy, pesticides should be applied at the right time. Irrigation shortly after application may result in excessive runoff losses. On the other hand, some pesticides, especially those for soilborne pests, are irrigated into the soil for more uniform application or deeper placement. At times, repeated applications of a given pesticide may become ineffective, perhaps because of an increase in the transformation rate.

Pesticide users should follow the directions on the pesticide label. Pesticides should be carefully measured, and the application equipment should be properly calibrated and maintained. Pesticides are applied by aerial and ground methods and through irrigation systems. During application, the pesticide should be directed only to the target site or pest. Aerial application methods generally result in higher drift losses than those from ground application. If conversion from aerial to ground spraying is not possible, aerial applications should be accomplished when the potential for drift, volatile losses, and runoffs are the least, that is, under calm conditions and cooler temperatures and not when rain is likely to occur.

Ground application of pesticides is favored, but substantial losses can also occur by use of this application method. Losses may be reduced with improved application technologies such as controlled droplet applicators, drift-shielded applicators, ultra-low-volume equipment, electrostatic sprayers, and computer-controlled equipment. Other means of controlling losses include the use of formulations that thicken the spray, such as oil emulsions and foliage-wetting agents.

Pesticides may be also introduced into surface irrigation streams or pressurized systems such as sprinklers and drip/trickle irrigation devices. Surface water applications, such as furrow and basin methods, tend to be less uniform in distributing water to the field and, therefore, often affect the uniformity of distribution of water and chemicals. Use of overhead sprinklers can result in the same losses that occur with aerial spray applications.

### Erosion and Runoff Control Practices

Environmental losses of pesticides through surface runoff, leaching, and volatilization may be reduced by erosion and runoff control practices (Wagenet and Rao, 1990; Wauchope, 1978), essentially the same practices recommended for the control of other agricultural nonpoint source pollutants. In some instances, however, practices that result in lower losses by one pathway (for example, runoff) may result in greater losses through another pathway (for example, leaching).

Conservation practices can reduce surface runoff and soil losses through tillage practices, including conservation tillage and no-till practices, contouring and strip-cropping, and use of cover crops, grassed waterways, and filter strips. Structural practices include the use of land leveling, terraces, subsurface drainage, improved application systems, and sediment retention ponds.

There are some differences of opinion about the potential benefits of conservation and no-tillage practices with regard to pesticides. For instance, conservation tillage practices have great potential for reducing erosion and sediment production. Such reductions, in turn, would reduce the discharge of sediment-bound pesticides. In reducing surface water runoff, however, some pesticides may be subjected to greater losses through leaching. Increasing the soil organic matter content may reduce the erosion hazard, but it would increase sorption, making the chemical less bioavailable, or it would increase the rate of microbial degradation of pesticides. Herbicide application rates are generally greater in no-tillage systems. In some conservation tillage practices, incorporation of pesticides into soil may be more difficult.

There is great potential for reducing runoff and leaching losses in irrigated agricultural systems. Improved furrow irrigation systems that recycle drainage water may eliminate runoff losses. Subsurface drainage, and hence, excessive leaching, may be reduced with improved water distribution uniformities, irrigation scheduling by using agroclimatic data, and use or management of the shallow groundwater.

## Nonchemical Control Measures

Nonchemical pest control methods may involve such crop management practices as crop rotation, intercropping, and manipulation of planting and harvesting dates to aid in controlling pest populations. For the foreseeable future, a mixture of chemical and nonchemical practices cannot be avoided. Future developments in breeding of pest-resistant crop species or genetic engineering of organisms that prey on current pests will play a large role in determining the extent to which nonchemical control measures can be adopted. For the foreseeable future, nonchemical control measures that complement crop management practices intended for pest control will need to be developed. This approach is the general strategy of integrated pest management programs.

### Integrated Pest Management

Integrated pest management (IPM) has the potential to reduce the need for pesticides and reduce the use of pesticides that might become pollutants. IPM involves understanding the pest in question, its host crop, and its natural predators so that ecologically and economically sound pest control techniques can be realized (Flint, 1989; Holden, 1986; U.S. Congress, Office of Technology Assessment, 1990).

An IPM strategy involves a number of guidelines.

- Determine the economic threshold of damaging pests. This economic threshold is defined as the point at which the cost of pest control equals the value of the crop lost because of pest damage.
- Lower the equilibrium position of the pest below the economic threshold. The equilibrium position is the average pest density in a field as determined over many years. Lowering the equilibrium position may be achieved by encouraging or establishing the pest's natural enemies such as parasites and predators, using pest-resistant or pest-free plant varieties, and modifying the pest's environment by using crop rotations and eliminating overwintering sites.
- Use the least environmentally damaging pesticide.

- Monitor the pest populations to decide when to apply pesticides or when to adjust integrated pest management strategies.

There is a need for integrated management of not only pesticides but also fertilizers. There must also be proper soil, water, and crop management.

## ASSESSMENTS OF THE KNOWLEDGE BASE

There is considerable laboratory-based knowledge about the physical and chemical properties of pesticides, their chemical and biological degradation mechanisms and persistence, their tendency to be sorbed by soil particles, and their movement in the dissolved and gaseous states in soils. In contrast, investigators' abilities to predict the behavior and transport of pesticides under field conditions appear to be weak. Part of this weakness may be attributed to the spatial and temporal variabilities that are part of every field soil, introducing much uncertainty into the interpretation of sampling and monitoring studies. This unpredictability applies to all of the agricultural nonpoint source pollutants, but it is especially so for pesticides because of the diversity of pesticides in use as well as their different behaviors in physical, chemical, and biological processes.

Nevertheless, it seems that the existing knowledge base from research and practical field experiences is not being fully disseminated or used to protect the environment. On the basis of chemical-specific properties and vulnerable site conditions, investigators should be able to assess whether a given pesticide will be a leacher that contaminates the underlying groundwater body. As monitoring of groundwaters for pesticides is aggressively pursued and a larger data base is accumulated, investigators may be able to confirm candidate leachers. The same applies to pesticide losses via surface runoff. Wauchope and colleagues (1992) developed an extensive data base that provides referenced data on sorption, degradation vapor pressure, and aqueous solubilities. This data base can be used to select pesticides that are less vulnerable to leaching or runoff.

It is of interest to examine DDT and DBCP. DDT is an organochlorine insecticide that is strongly sorbed by soils and has a low water solubility and a long half-life. Yet, it is found in some California wells, even though investigators predicted that it would take thousands of years for DDT to reach the water tables of those wells. The mode of entry into these groundwaters was probably not from passage through the soil profile and substrata but, perhaps, through the well casing. In contrast,

DDT and its metabolites are still found in suspended sediments eroded from furrow irrigated lands in some portions of California. With a half-life of about 3,800 years and a sorption coefficient of about 2,400, it is understandable that residues of DDT and its metabolites still exist in surface soils and sediments.

DBCP is a nematicide that was banned from use in 1979 because of its toxicity to humans. This compound has a high water solubility, moderate vapor pressure, low sorption coefficient, and short photolysis half-life, but it has a long hydrolysis half-life in the vadose zone. Although much of this volatile pesticide is dissipated shortly after application, it is highly subject to leaching losses. The presence of DBCP was detected in California and Florida wells in 1979. After 15 years of well sampling in California, 2,500 of 4,500 wells showed detectable concentrations of DBCP. The maximum contaminant level of 1 $\mu$g/liter (1 ppb) for DBCP in California was lowered to 0.02 $\mu$g/liter (0.02 ppb) in 1989, and numerous drinking water wells have been shut down recently. The behavior and accumulation of DBCP in well waters are now more clearly understood because of the gain in the knowledge base of this pesticide since the 1960s.

## PROPER USE OF PESTICIDES

There appears to be little chance of discovering a perfect pesticide—one that is precise enough to attack the target pest and then suddenly dissipate and accurate enough to reach the target pest and not move past the root zone. Given the difficulty of predicting the fate and transport of pesticides with certainty, efforts to reduce pesticide losses by reducing the total mass of pesticides used, reducing pesticide losses through runoff and erosion, improving the efficacies of pesticide applications, and matching the pesticide selection to site conditions must go forward at the same time that investigators improve their understanding of pesticide behavior in the environment. Currently available technologies, farming systems, and farming practices allow significant reductions in pesticide losses while sustaining profitability. Aggressive efforts to adopt and adapt these available technologies, systems, and practices must be pursued. The research required to develop alternative pest control strategies and to develop farming systems based on alternative pest control practices should be accelerated. Long-term efforts to reduce the need for environmentally damaging pesticides is the most promising approach to reducing environmental damages from pesticides.

Pesticides are perhaps the only toxic substances that are purposefully applied to the environment, a rather unique permit given the present-day regulations covering toxic compounds. Although the benefits derived from the use of pesticides are considerable, increasing numbers of them are expected to be regulated for only restrictive use or banned outright as the public becomes increasingly aware of the risks to humans and the ecological environment.

# 9

# Fate and Transport of Sediments

Sediments—eroded soil deposited in streams, rivers, drainage ways, and lakes—can be defined as a soil resource out of place. As such, it is widely recognized as one of the major environmental concerns worldwide.

## EFFECTS OF EROSION AND SEDIMENTATION

Erosion reduces the productivity of the land resource. Sediment degrades water quality and often carries soil-absorbed polluting chemicals. Sediment deposition in stream channels, irrigation canals, reservoirs, estuaries, harbors, and water conveyance structures reduces the capacities of these water bodies to perform their prime functions and often requires costly treatments.

In many instances, the sediments removed from upland areas and channels and subsequently transported downstream carry adsorbed chemicals that exacerbate water quality problems at points downstream. Such chemicals exacerbate the effect of sediments on aquatic habitats and may destroy fish spawning grounds (Alonzo and Theurer, 1988). Sediments may also reduce water conveyance capacity (increasing flooding) and water storage capacity in reservoirs.

Agriculture has a great impact on sediments deposition. Judson (1981) estimated that river-borne sediments carried into the oceans increased from 9 billion metric tons (10 billion tons) per year before the introduction of intensive agriculture, grazing, and other activities to between 23 billion and 45 billion metric tons (25 billion and 50 billion tons) thereafter. Dudal (1981) reported that the current rate of agricultural land degradation, primarily because of soil erosion, is leading to an irreversible loss in productivity on about 6 million ha (15 million acres) of fertile land a year worldwide. Crop productivity on about 20 million ha (49 million acres) each year is reduced to zero or becomes uneconomical because of soil erosion and erosion-induced degradation (Lal, 1988). Since humans first began cultivating crops on a yearly basis, soil erosion has destroyed about 430 million ha (1,063 million acres) of productive land globally (Lal, 1988). Buringh (1981) estimated that the annual global loss of agricultural land is 3 million ha (7 million acres) because of soil erosion and 2 million ha (5 million acres) because of desertification. Of the total 0.9 billion metric tons (1 billion tons) of sediment carried by rivers from the continental United States, about 60 percent is estimated to be from agricultural lands (National Research Council, 1974). The off-site damages ("off-site" refers to locations where damages are due primarily to deposition of eroded material) caused by sediments in the United States are exorbitant. For example, several million cubic meters of sediment are washed into U.S. rivers, harbors, and reservoirs each year, and dredging of these sediments requires significant financial resources.

Wind erosion problems are especially acute in more arid regions. As with water erosion, wind abrasion destroys many young crops; it also causes severe air quality problems. Reduced vision caused by wind erosion has been identified as the cause of numerous multiple-vehicle accidents in the southwestern United States and has resulted in the loss of many lives. Blowing dust from fallow fields has been identified as the cause of many breathing problems for humans. There are, however, even fewer research data on the definition and control of the wind erosion process than there are for water erosion (Lal, 1988). Although the basic principles governing wind erosion process and control are similar to those governing water erosion, the specific cause-effect relationships and the effectiveness of wind erosion control practices have not been as widely investigated as have those for water erosion.

## SEDIMENTATION PROCESSES

Erosion and sedimentation by water and wind embody the processes of detachment, transport, and deposition of soil particles (sediment) by the erosive forces of wind, the impacts of raindrops, water and wind shear, and water runoff over the soil surface. Detachment is the dislodging of soil particles by the erosive agents. Transportation is the entrainment of the sediment in wind or water and movement of the sediment from its original position. Sediment travels in wind or water from its point of origin through air or stream systems until it reaches a point where the wind and water energy is insufficient to continue the movement and deposition occurs. The ultimate sites of deposition are the oceans, but in general, only a small portion of the eroded sediment moves that far without interruption by storage. Deposition occurs at a wide variety of sites, including the bottom of a hill slope; the edge of a field; in a windbreak, lake, or reservoir; and on a floodplain. This process is sedimentation.

### On-Site Processes

Precipitation is a key resource that makes land productive. Each year, roughly 378 million to 8 billion liters (100 million to 2 billion gallons) of rain falls on each square kilometer of U.S. land. This water is essential for crop production, but it may also cause soil erosion and flooding.

Rain falls as drops averaging less than 0.3 cm (one-eighth inch) in diameter, but each drop strikes the land as a tiny bomb. Every year throughout most of the United States, more than a quadrillion rain drops strike each square kilometer of land with an impact energy of thousands of metric tons of TNT. The impact energy of rain falling on the state of Mississippi, for example, annually equals the energy of 1,000 1-megaton bombs, or 0.9 billion metric tons (1 billion tons) of TNT (Meyer and Renard, 1991).

When raindrops fall on unprotected soil, they detach soil particles from the soil layer and the particles are then transported down slope by the runoff. (Runoff is the rainfall excess that is not absorbed by the soil.) Not only does this runoff carry raindrop-detached soil and cause additional erosion itself but the water is also lost for use in crop production. Runoff from fields and forests to streams and rivers in Mississippi, for instance, averages nearly 76 trillion liters (20 trillion gallons) annually.

The factors affecting erosion can be expressed by the following equation Renard and Foster (1983):

$$E_r = f(C_i, S_p, T_o, SS, M), \qquad \text{[Eq. 9-1]}$$

where $E_r$ is erosion, $C_i$ is climate, $S_p$ is soil properties, $T_o$ is topography, $SS$ is soil surface conditions, and $M$ is human activities.

The universal soil loss equation (Wischmeier and Smith, 1978) and the revised universal soil loss equation (Renard et al., 1991) are essentially expressions of the functional relationship shown in Equation 9-1. In the

Up on the hillside the evidence of eroded soil is obvious. The path of the soil's movement can be traced by the rills that cut through the vegetation on the slope and into the mud at the bottom of the hill's slope. Credit: U.S. Department of Agriculture.

revised universal soil loss equation, $R$, the rainfall-runoff factor, expresses the climate; $K$, the soil erodibility factor, and $C$, the cover and management factor, reflect the soil properties; $L$, the slope length factor, and $S$, the slope-steepness factor, reflect the topography; $C$, the cover management factor, reflects the soil surface conditions; and $C$ and $P$, the support practice factor, reflect human activities.

As with water erosion, the factors in the wind erosion equation (Skidmore and Woodruff, 1968) essentially express the same functional relationship shown in Equation 9-1. The wind erosion equation's $C$ factor represents the climate; $I$, the soil erodibility factor, and $K$, the soil ridge roughness factor, represent soil properties; $L$, the unsheltered median travel distance of the wind across a field, represents the topography; $K$ and $V$ represent the soil surface conditions and human activities, respectively.

Although erosion processes can be complex, the principles of controlling erosion are relatively simple. Effective water erosion control requires simultaneous efforts to increase the degree to which and the length of the season during which the soil is covered by plants or plant residues and to decrease the volume and energy of runoff water. Effective wind erosion control also depends on increased amounts of soil cover and reductions in the energy of wind that is in contact with soil particles.

## In-Channel Processes

Agricultural production practices have led to radically altered water flow regimes within agricultural watersheds. Modification of virgin (noncultivated) land often involved deforestation and drainage activities. In combination with cropping and grazing practices, these disruptions of the natural vegetation and soil resulted in loss of the land's water storage capacity. As local water tables lowered, many perennial springs ceased to flow, and streams that were once permanent flowed only intermittently. In addition, the rapid loss of water from the land produced a sharply exaggerated seasonal flow regime, increasing the frequency, severity, and unpredictability of high-volume flows (Menzel, 1983). Commonly, these alterations were well established within the first 40 to 50 years of agricultural development of an area (Trautman, 1957, 1977). These changes in flow regimes have been compounded by intensified land drainage (Menzel, 1983).

The flow regime changes have had serious disruptive effects on streams and rivers by substantially increasing discharge peaks and the erosive powers of streams. The net effect has been to create conditions of serious channel morphology disequilibrium in many drainage systems (Leedy, 1979). As a consequence, headwater channels of the region commonly experience active bed degradation. Whereas in their natural condition the channels were deep and narrow, they have now broadened extensively through active bank and bed erosion. The rapidity of this erosive process can be dramatic, as exemplified by Four Mile Creek in Tama County, Iowa, where average cross-sectional areas increased 54 percent between 1967 and 1979; the channel width increased at an average rate of nearly 0.4 m/year (16 inches/year) and the depth increased at a rate of 0.03 m/year (1.2 inches/year) (Kimes et al., 1979; Menzel, 1983).

The kind of channel erosion described above can become an important contributor to the sediment load in streams and rivers. In Four Mile Creek, 25 percent of the average annual loading of sediment is contributed by channel erosion (Menzel, 1983). Forty percent or more of the sediment load of streams and rivers in Iowa and Illinois has been estimated to arise from channel erosion (Glymph, 1956; Leedy, 1979). Measurements on Goodwin Creek in Mississippi revealed that over 60 percent of the sediment consisting of the silt and clay size fractions came from channel and gully erosion, whereas practically all sand came from gullies and channels (Grissinger et al., 1991).

Channel instabilities cause sedimentation problems in many areas of the United States. They vary from the filling in of channels to entrenchment and bank failures associated with erosion within channels. The problems are especially severe along the bluff of the Mississippi and Yazoo river floodplain, where steep gradients promote channel erosion and the low gradients on the flood plain cause the sediment to be deposited. Flood control reservoirs also induce sediment deposition within upstream channels.

Reductions in sediment movements to streams and rivers from adjacent croplands may not result in comparable improvements in sediment loads unless the distortions of watershed flow regimes are addressed. There are two ways to address this problem: (1) reduce the runoff energy of the water and sediments entering stream channels, and (2) protect or restore bottomlands in agricultural watersheds. These solutions are discussed more fully in the section on treatment technology.

## SEDIMENT ESTIMATION AND PREDICTION TECHNOLOGIES

The ability to design programs and policies to control sedimentation in surface water depends on investigators' abilities to predict erosion, sediment delivery, and in-stream processes. Understanding of these phenomena is often incomplete because of the unknown cause-effect implications of human influences.

## Erosion Estimation and Prediction

Research begun in the 1930s under the leadership of Hugh Hammond Bennett, the father of modern soil conservation, ultimately led to the universal soil loss equation (Wischmeier and Smith, 1978) and the wind erosion equation (Skidmore and Woodruff, 1968). This technology has been widely used in the United States and worldwide for several decades for conservation planning efforts. Although the technology is old, investigators initiated an effort to update the universal soil loss equation; and the new technology, the revised universal soil loss equation, is being installed in the Soil Conservation Service of the U.S. Department of Agriculture for conservation program planning (Renard et al., 1991). Although the revised equation retains the six factors of the universal soil loss equation for estimating soil loss, the algorithms used for factor values are different, with the result that soil loss estimates for specific in situ scenarios obtained by the revised equation differ from those obtained by the universal soil loss equation.

Similarly, the wind erosion equation predicts wind erosion rates as the product of five values representing soil erodibility, soil ridge roughness, climate, field length, and vegetative cover. Since the original and revised universal soil loss equations and the wind erosion equation were first introduced, their systematic approach has had a tremendous effect on erosion technology and conservation planning.

During the 1960s and 1970s, fundamental research designed to create a better understanding of the principles and processes of soil erosion by water and wind received increased emphasis. Researchers analyzed and quantified the companion but very different processes of erosion caused by raindrops and runoff. Investigators defined the aerodynamics of wind in relation to soil detachment and transport.

During the 1980s, the knowledge gained from past experiments and fundamental studies provided the basis for developing mathematical models to describe erosion over a wide range of specific conditions and to improve erosion prediction and control methods. At about the same time, the environmental movement gave impetus to an expanded research effort to understand the off-site effects of soil erosion and the potential for chemical pollution resulting from it. Models such as the chemicals, runoff, and erosion from agricultural management systems (CREAMS) model (Knisel, 1980) were formulated. The CREAMS model included hydrologic, erosion, pesticide, and nutrient components. Such models incorporated major advances in describing the physical processes involved in soil erosion, sediment transport and deposition, and chemical transport.

Most recently, the water erosion prediction project (WEPP) model (Lane and Nearing, 1989) was developed by the Agricultural Research Service, the Soil Conservation Service, and the U.S. Forest Service of the U.S. Department of Agriculture (USDA); the Bureau of Land Management of the U.S. Department of the Interior; and cooperating universities. WEPP was developed as the next-generation water erosion prediction model to replace the original and revised universal soil loss equations. The more versatile WEPP model incorporates many of the scientific advances that have been made since development of the universal soil loss equation and is based on the physical principles and processes of soil erosion by water.

A companion effort by other Agricultural Research Service, Soil Conservation Service, and university investigators is under way to improve the predictive capability of sediment loss from wind erosion, culminating in the wind erosion prediction system (WEPS) model. The WEPS model is based on the principles of wind erosion physics associated with climate, soil, topography, and cropping and management systems that affect sediment detachment, transport, and deposition by wind.

These and similar models should assist conservation planners well into the twenty-first century. However, land management agencies in the United States and worldwide will need to expend numerous resources (time, labor, and money) developing new data bases to fully implement the technology.

Given currently available modeling capacities, it should be possible to improve conservation planning. Computer technology should give conservationists the capability of using computer models to assist them in deciding between alternatives, thereby improving soil resource protection. The new technology is not without some costs, however, especially because computer-based planning requires personnel with expertise and training different from those classically required by conservationists.

With the new models, it should be possible to improve the definition of highly erodible lands and predict the effects of farming system changes on erosion. More importantly, it will permit the design of erosion control systems and plans by using probabilistic relationships that recognize the uncertainties of erosion hazards. For example, much of the erosion occurs during relatively infrequent precipitation events. If such infrequent events occur during a particular time of the year, it is imperative that the land have good ground cover at that time. Thus, with carefully designed erosion control practices (for example, use of cover crops), it is possible to minimize the erosion hazard.

### Transport and Delivery

Most watershed planning and evaluation methods require the use of some type of computer model. These models, designed for specific and general applications, deal with watersheds as landscape pieces (grid pieces and/or subwatersheds) or deal with processes such as runoff and erosion in upland areas, channel processes such as the route that water runoff takes, and erosion and sediment transport processes. Unfortunately, the currently available models emphasize upland processes or channel hydraulic processes; a model that treats each of these equally is not available.

Analytical models that simulate the transport of sediments and adsorbed chemicals to points downstream have not progressed to the extent that those for on-site erosion have. The kinematic runoff and erosion (KINEROS) model (Woolhiser et al., 1990) and a mathematical model for simulating the effects of land use and management on water quality (ANSWERS) (Beasley et al., 1980) are capable of addressing hydrology, erosion, and sediment transport but are intended for the simulation of discrete events. In other instances there is a need for continuous simulation of runoff and sediment yields from watersheds and river basins where the impacts of conservation practices must be assessed. In such instances, the user must decide among such models as the simulator for water in rural basins (SWRRB) model (Williams et al., 1985), the agricultural nonpoint source pollution (AGNPS) model (Young et al., 1987), the CREAMS model (Knisel, 1980), and the watershed and grid version of the WEPP model that is being developed (Foster and Lane, 1987).

These and other models (Fan, 1988; Renard et al., 1982) represent a wealth of technology, but they generally have limitations when applied to small upland watersheds where rills and ephemeral gullies may be filled in by tillage. Current technology for addressing ephemeral gully erosion involves use of the ephemeral gully erosion model (EGEM) (Woodward et al., 1991), which is an offshoot of the CREAMS model (Knisel, 1980). Sediment yields estimated with EGEM are at best subject to large errors, and the model is considered interim technology until more physically based process models such as the WEPP model become available.

Regression models have been used to estimate sediment yields into small reservoirs (farm ponds) (American Society of Agricultural Engineers, 1977; Anderson, 1975; Dendy and Bolton, 1976; Flaxman, 1972; Pacific Southwest Inter-Agency Committee, 1968). Most of this technology is site specific and does not permit the land manager or conservationist to assess the impact of agronomic and mechanical changes or the impact of hydraulic erosion control structures on sediment yields. Thus, there is a pressing need to replace such technology, even when it is used only for planning and assessment purposes.

The currently available models used to simulate sediment transport and delivery are generally too site specific (suffer from limited validation and verification) to be widely used in different climates and for different land uses. Although detailed models capable of simulating sediment transport exist, they are expensive (because they require many data) and are specific in their application (for example, sediment transport in a major river) but are impractical for such tasks as targeting lands that produce excessive sediments so that the lands might be targeted for a change in the ways they are managed to reduce sediment yields.

## Inadequacy of Technology

Nearing and colleagues (1990) discussed water erosion prediction technology and identified research needs (weaknesses in technology) in four areas: (1) fundamental erosion relationships; (2) soil and plant parameters and their effects on erosion; (3) data bases, user interfaces, and conservation system design; and (4) model development and analysis. The review of Nearing and colleagues (1990) is based on experience through the development of the WEPP hill slope profile erosion model, which is a computer-based technology for estimating rill and interrill soil losses on hill slopes. They stated the following (Nearing et al., 1990:1710):

> Development of process-based erosion prediction technology has required the delineation and description of fundamental erosion processes and their interactions. Further improvement in prediction technology will require further delineation and mathematical descriptions. Some key topics for study include (i) describing headcutting and sidewall sloughing in rills, (ii) replacing or better describing the concept of sediment-transport capacity and its relationships to detachment and deposition processes, (iii) developing theory and data sets to better predict deposition and sediment enrichment on complex slope profiles, and (iv) developing criteria for climate

Rainfall on unprotected soil detaches soil particles from the soil layer transports the particles downslope. This runoff carries rain-detached soil, which causes additional erosion; and the water is also lost for use in crop production. Credit: U.S. Department of Agriculture.

selection to obtain long-term average estimates of soil loss. New technology for describing erosion and sediment movement on complex hillslope profiles is also needed.

Research on soil and plant parameters related to erosion can be divided into that focused on baseline conditions and on temporal changes. Statistical relationships for estimating baseline soil erodibility as a function of time-invariant soil properties exist. A fundamental approach to prediction is needed to further improve baseline erodibility estimation. Fundamental approaches are also needed to predict temporal changes in soil erodibility in response to climatic and cropping and management influences. Our understanding of and ability to characterize temporal changes in soil properties needs much improvement.

Two specific areas that deserve attention are surface roughness effects on erosion and the effects of surface sealing on infiltration (Nearing et al., 1990:1710).

New process-based erosion prediction technology will require an extensive data base to be effective. Innovative techniques for developing model parameters will be required, including expert systems. The new technology also opens new opportunities for refining existing and developing new erosion-control practices. Methods for using the technology as an interactive tool for conservation systems design are needed.

To apply the new process-based technology, we need additional research directed toward developing techniques for modeling natural-resource systems. Validation and sensitivity analysis of the new erosion models must be done. We know erosion is highly variable in time and space. With the new simulation models, we can begin to address more fully temporal and spatial distributions of soil loss and sediment yield, confidence limits for our erosion estimates, and probabilities of meeting conservation goals with given management systems.

Larson and colleagues (1990) discussed the tools for erosion control options available for conservation planning. Among other things, they discussed the following (Larson et al., 1990:62-63):

Parallel to the developments in computer hardware have been developments in computer software such as geographic information systems (GIS), digital elevation models (DEM), and expert systems (ES). These tools will allow development and display of alternatives by conservationist/operators. Combining digital elevation models with soil maps should permit 3-dimensional views of soils on landscapes and display wedges of soil that could be lost as predicted by WEPP and WEPS. However, these software tools are stressing the attribute data of present digital databases such as the soil map which is the base from which all models run. More robust methods of representing the variability of soil properties within polygons (delineations) must be developed, perhaps to present a probabilistic representation of the properties. This same approach could then be extended to fields or watersheds. Combined with climatic probabilities, systems could be developed according to erosion risks and systems designed to control the risks similar to flood control systems.

The analytical tools and expert systems must be able to integrate all ramifications of a resource management system such as the effects of erosion control practices and crop management systems on water quality and the soil ecosystem. These ramifications are so extensive that only a computer will be able to sort them out and present tradeoffs for each conservation system and crop management system.

Inadequate knowledge of the transport of sediments of specific particle size ranges is a major limitation in predicting the fate of adsorbed agricultural chemicals. Furthermore, investigators must be aware of the temporal and spatial variabilities of absorbed chemicals in upland environments in contrast to those in gullies and channels. Concentrations of adsorbed chemicals in upland areas are presumably high, whereas the sediments originating in gullies and channel peripheries conceivably have lower amounts of adsorbed chemicals.

A major problem with modeling upland erosion and sediment transport where concentrated flow begins involves the hydraulic transport process. Most sediment transport processes for upland erosion models are taken from those developed for stream flows. The WEPP model uses a mathematical relationship described by Yalin (1963), as modified by Foster and colleagues (1981), for nonuniform sediments. It is doubtful whether investigators can make significant progress in this area simply by using a different sediment transport formula. Theory must be developed and experiments specifically related to the development of new transport equations must be conducted for shallow rill- and interrill-type flows. Significant advances in characterizing turbulent flows have recently been made by using flow visualization and other techniques, but those studies have not been extended to the shallow flow conditions common to areas with rill and interrill erosion.

More important than which sediment transport equation should be used to predict soil erosion is the issue of what transport capacity means and how it is used. In basic terms, transport capacity is a balance between the entrainment and the deposition rates of the already detached sediment in the flow. The description of the entrainment process does not include a factor for cohesive soil forces, but considers only the gravity forces of the sediment that must be overcome for the particle to be lifted into the flow. The implicit assumption, then, for erosion of cohesive soils is that cohesive forces are negligible once the soil has initially been detached from the in situ soil mass.

The technology currently used to predict wind erosion in the United States is based on variations of the wind erosion equation (WEQ). The technology uses erosion loss estimates that are integrated over large areas and long time scales to produce average annual values. In order to increase the range of conditions to which WEQ technology can be applied in the short-term, a revised wind erosion equation (RWEQ) is under development. The RWEQ embodies improved values for the

WEQ factors and is designed to calculate erosion during periods as short as a month.

As with water erosion, the widespread availability of personal computers and new research has led to results that can be used to adopt flexible, process-based technologies to assess and plan conservation practices for wind erosion control. Thus the USDA also has a major program under way to develop new wind erosion prediction technologies. The wind erosion model development program has two stages. The first stage is development of a wind erosion research model (WERM); the second stage is development of a wind erosion prediction system (WEPS). In this second stage, the submodels of WERM will be reorganized to increase computational speed, data bases will be expanded in size, and a user-friendly input-output section will be added to make the technology of greater utility to users.

WERM is modular and consists of a supervisory program and seven submodels (weather, hydrology, decomposition, crop, management, soil, and erosion). Four databases are needed—soils, climate, crop growth and decomposition, and management. The submodels permit easy testing and updating with new data during development of the technology. Finally, as in the WEPP technology, extensive experimental work is being carried out simultaneously with model development and is devoted to delineating parameter values that facilitate application of the algorithm to both measured and unmeasured processes (Hagen, 1988; Hagen et al., 1988).

As the new wind erosion prediction technology becomes operational, considerable work will need to be done to develop the data bases required for its implementation over the wide range of environmental conditions that occur in the United States and worldwide. As with water erosion, wind erosion prediction technology will require development of associated technologies such as expert systems, digital elevation models, and geographic information systems.

## Future Needs

Despite the advances that have been made in estimating and predicting erosion by wind and water, many questions related to data sources, methods of data collection and extrapolation, and data accuracy and reliability remain unanswered. Soil erosion and sedimentation research is a capital-intensive and time-consuming exercise. Furthermore, extrapolation to the global scale on the basis of the limited data collected by diverse and nonstandardized methods leads to gross approximations. There is an urgent need for methods that can be used to increase the reliability and accuracy of soil erosion and sedimentation data. Current data are often collected with equipment developed decades ago, and such equipment is incompatible with modern computer simulation technologies. Finally, the historical erosion data bases are often developed from data for agricultural crops (varieties, row spacing, management practices) that are different from those planted today. Significant investments in personnel and funds that are in excess of those currently available will be required to overcome such problems.

From a policy standpoint, land managers and conservationists need to be able to (1) target those lands that are most vulnerable to erosion, (2) develop and apply treatments to these vulnerable lands, and (3) predict how changing land uses and conservation practices have an impact on erosion from the new land uses and conservation practices. Finally, the financial implications of those relationships need to be estimated.

With the current state of technology, the objectives described above will be expedited with further development of (1) geographic information systems, to permit assembly and input of the data needed by the evolving models; (2) the data bases required by the new erosion and sedimentation models; (3) fundamental sediment transport relationships appropriate for use in upland farming areas where runoff occurs in small channels and where the hydraulic roughness is large relative to the flow depth; and (4) transport relations that address the particle size ranges of sediments so that assessments of adsorbed agricultural chemical transports can be made.

## TREATMENT TECHNOLOGY

Technologies that can be used to reduce the amount of sediment in surface water focus on two objectives: (1) improving farming practices to reduce erosion and runoff, and (2) improving stream channels and riparian vegetation to reduce erosion of stream banks and streambeds.

### Farming Practices

The effects of different types of plant cover, tillage, and cropping systems have been evaluated on erosion plots and watersheds and by using rainfall simulators and wind tunnels. Various types of conservation tillage practices have been developed and evaluated. They have been found to reduce greatly both water and wind erosion from land during intensive cropping. Scientists have also identified and quantified those soil and sediment characteristics that affect erosion rates and sediment pollution potential.

Farming technologies, in an effort to meet producer needs to preserve soil quality, are designing equipment to meet changing farming systems. This no-till drill has an adjustable down-pressure system that applies constant force on the openers for consistent penetration in varying soil conditions. Credit: Deere & Company.

In most farming systems, the critical period for erosion is the time after harvest but before a new crop is established. During this period, soil is most exposed to wind and water, and, therefore, is most vulnerable to erosion. Efforts have been and are being made to develop farming practices that increase soil cover during this noncrop period (Mills et al., 1991).

Much effort has gone into the development of reduced-tillage systems that increase the amounts of crop residues to provide soil cover after the crop is harvested. Many different systems of conservation tillage have been developed for different farming systems in different regions. Mannering and colleagues (1987) described five kinds of conservation tillage systems in use in the United States, including no-till or slot planting, ridge-till, strip-till, mulch-till, and reduced-till systems (Table 9-1). All of these systems are designed to cover at least 30 percent of the soil at the time of planting.

TABLE 9-1   Conservation Tillage Systems in the United States

| Tillage System | Description |
| --- | --- |
| No-till or slot planting | The soil is left undisturbed prior to planting. Planting is completed in a narrow seedbed about 2- to 8-cm wide. Weed control is accomplished primarily with herbicides. |
| Ridge-till | The soil is left undisturbed prior to planting. About one third of the soil surface is tilled with sweeps or row cleaners at planting time. Planting is completed on ridges usually 10 to 15 cm higher than row middles. Weed control is usually accomplished with a combination of herbicides and cultivation. |
| Strip-till | The soil is left undisturbed prior to planting. About one third of the soil surface is tilled at planting time. Tillage in the row may be done by a rototiller, in-row chisel, row cleaners, and so on. Weed control is accomplished with a combination of herbicides and cultivation. |
| Mulch-till | The total surface is disturbed prior to planting. Tillage tools such as chisels, field cultivators, disks, sweeps, or blades are used. A combination of herbicides and cultivation is used to control weeds. |
| Reduced-till | This system consists of any other tillage and planting system not described above that produces 30 percent surface residue cover after planting. |

SOURCE: Adapted from J. V. Mannering, D. L. Schertz, and B. A. Julian. 1987. Overview of conservation tillage. Pp. 3–17 in Effects of Conservation Tillage on Groundwater Quality, T. J. Logan, J. M. Davidson, J. L. Baker, and M. R. Overcash, eds. Chelsea, Mich.: Lewis Publishers.

The effects of conservation tillage systems on runoff and soil loss can be dramatic. Table 9-2 compares the amount of surface soil cover, soil erosion, and runoff from a rainfall simulator for three different wheat tillage systems. Increased soil cover was found to greatly reduce both water runoff and soil erosion. Table 9-3 gives sediment yields (soil loss) and water runoffs from two watersheds, one under a conservation (ridge-till) tillage system and one under a conventional tillage system. Average runoff was nearly 3 times greater and soil loss was 10 times greater in the conventionally tilled watershed than were those in the conservation tilled watershed. Table 9-4 illustrates the percentage of cropland (nonforage crops) planted under various forms of conservation tillage in 1985.

The National Association of Conservation Districts reports the area of land under conservation tillage annually through its Conservation Tillage Information Center. The 1991 report (National Association of Conservation Districts, Conservation Tillage Information Center, 1991)

TABLE 9-2   Surface Soil Cover, Soil Erosion, and Runoff from Different Wheat Tillage Systems

| Period | System | Cover (percent) | Runoff (cm) | Soil Loss (kg/ha) |
|---|---|---|---|---|
| Fallow after harvest | Bare fallow | 62 | 0.9 | 662 |
| | Stubble mulch | 91 | 1.5 | 803 |
| | No-till | 91 | 0.1 | 718 |
| Fallow after tillage | Bare fallow | 4 | 3.6 | 9,401 |
| | Stubble mulch | 92 | 0.9 | 208 |
| | No-till | 96 | 0.1 | 17 |
| Wheat 10-cm tall | Bare fallow | 26 | 3.5 | 7,246 |
| | Stubble mulch | 38 | 2.4 | 2,576 |
| | No-till | 85 | 0.5 | 550 |
| Wheat 45-cm tall | Bare fallow | 78 | 4.3 | 2,094 |
| | Stubble mulch | 83 | 2.9 | 836 |
| | No-till | 88 | 1.6 | 337 |

SOURCE: J. M. Laflen, R. Lal, and S. A. El-Swaify. 1990. Soil erosion and a sustainable agriculture. Pp. 569–581 in Sustainable Agricultural Systems, C. A. Edwards, R. Lal, P. Madden, R. H. Miller, and G. House, eds. Ankeny, Iowa: Soil and Water Conservation Society. Reprinted with permission from © Soil and Water Conservation Society.

TABLE 9-3   Runoff and Soil Loss from Watersheds under Conventionally and Conservation Tilled Systems

| Year | Conservation Tilled Watershed | | Conventionally Tilled Watershed | |
|---|---|---|---|---|
| | Runoff (mm) | Soil loss (metric tons/ha) | Runoff (mm) | Soil loss (metric tons/ha) |
| 1973 | 27 | 0.2 | 75 | 1.1 |
| 1974 | 2 | 0.0 | 14 | 0.7 |
| 1975 | 3 | 0.0 | 21 | 1.8 |
| 1976 | 10 | 2.5 | 4 | 0.0 |
| 1977 | 8 | 0.2 | 104 | 18.4 |
| 1978 | 40 | 1.1 | 81 | 9.3 |
| 1979 | 76 | 0.2 | 102 | 4.3 |
| 1980 | 50 | 4.5 | 116 | 51.8 |
| Average | 27 | 1.1 | 65 | 10.9 |

SOURCE: J. M. Laflen, R. Lal, and S. A. El-Swaify. 1990. Soil erosion and a sustainable agriculture. Pp. 569–581 in Sustainable Agricultural Systems, C. A. Edwards, R. Lal, P. Madden, R. H. Miller, and G. House, eds. Ankeny, Iowa: Soil and Water Conservation Society. Reprinted with permission from © Soil and Water Conservation Society.

TABLE 9-4   Cropland Area under Various Forms of Conservation
Tillage, 1985

| Region | Percent Cropland Planted in Nonforage Crops by | | | | | | |
|---|---|---|---|---|---|---|---|
| | NT | RiT | ST | MT | ReT | All | Percent Idle |
| Pacific Northwest | 3.6 | 0.1 | 0.0 | 20.7 | 7.9 | 32.2 | 32 |
| Northern Great Plains | 3.6 | 0.1 | 0.0 | 16.0 | 9.5 | 29.3 | 32 |
| Central Great Plains | 3.5 | 2.0 | 1.3 | 24.0 | 14.8 | 45.7 | 27 |
| Southern Great Plains | 1.2 | 0.2 | 0.1 | 16.7 | 12.3 | 30.5 | 24 |
| Northern Corn Belt | 3.1 | 1.4 | 0.2 | 31.5 | 0.5 | 36.6 | 5 |
| Southern Corn Belt | 7.3 | 0.7 | 0.2 | 29.4 | 2.7 | 40.3 | 5 |
| Northeast | 6.2 | 0.1 | 0.0 | 7.0 | 8.9 | 22.2 | 3 |
| Eastern Uplands | 15.4 | 0.0 | 0.1 | 14.9 | 8.4 | 38.8 | 6 |
| Piedmont | 22.7 | 0.3 | 0.3 | 16.4 | 8.0 | 47.7 | 3 |
| Coastal Plains | 8.6 | 0.1 | 0.6 | 13.1 | 2.8 | 25.2 | 6 |
| Associated Delta | 2.0 | 0.2 | 0.0 | 4.4 | 5.3 | 11.7 | 12 |

NOTE: NT, no-till; RiT, ridge-till; ST, strip-till; MT, mulch-till; ReT, reduced-till; All, sum of area under any conservation tillage system.

SOURCE: R. R. Allmaras, G. W. Langdale, P. W. Unger, R. H. Dowdy, and D. M. Van Doren. 1991. Adoption of conservation tillage and associated planting systems. Pp. 53–84 in Soil Management for Sustainability, R. Lal and F. J. Pierce, eds. Ankeny, Iowa: Soil and Water Conservation Society. Reprinted with permission from © Soil and Water Conservation Society.

contains much information regarding efforts to control erosion with conservation tillage practices. Figure 9-1 shows, by U.S. counties, the land area on which conservation tillage systems are used. In 1991, conservation tillage was used on 28.14 percent of all planted lands; this estimate was up from 26 percent in 1990 and 25.6 percent in 1989, indicating increasing adoption of this method of erosion control.

The use of cover crops is an important way of controlling erosion. For example, an extended period without soil cover during periods of potentially high erosion leads to excessive erosion in many areas. Hargrove (1991) recently described research highlighting the advantages of using cover crops for maintaining clean water.

### Channel Management

Channel erosion tends to increase when there are low sediment loads from decreased upland erosion and tends to decrease when there are high sediment loads from increased upland erosion. Thus, erosion control on farm fields and upland areas, such as that which might result from the use of conservation tillage or grassed waterways, may result in excessive channel instability if runoff is also not controlled. Channel

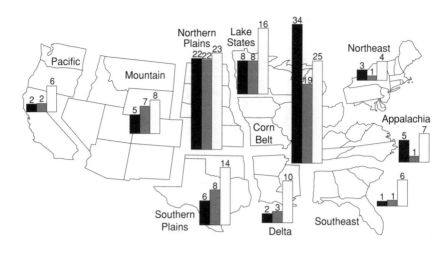

Million Acres Planted by Crop Residue Level

■ >30% residue    ▨ 15–30% residue    ☐ 0–15% residue

FIGURE 9-1 Crop residue levels on planted acreage by region in 1992. U.S. Department of Agriculture, Economic Research Service. 1993. Agricultural Resources: Cropland, Water, and Conservation. Situation and Outlook Report AR-30. Washington, D.C.: U.S. Department of Agriculture, Soil Conservation Service.

erosion damages include instability resulting from steepened banks and storm flows over channel banks in reaches where sand and gravel accumulate, leading to poor floodplain drainage, runoff pollution, and off-site damages. Upland erosion control programs may also reduce runoff rates and amounts, but it may be accomplished at the expense of reduced water supplies downstream.

Channel maintenance programs suffer from the perceived idea that engineering efforts associated with, for example, bank stabilization, grade control, dredging, and energy control are aesthetically unacceptable and destroy wildlife and biological habitats. Therefore, engineering approaches to watershed management (as well as flood control and channel maintenance) are not receiving public support. Yet, there are many scenarios by which a combination of upland and channel treatments are required for ecological, natural resource, and environmental protection or enhancement. Treatment of channel erosion must be based on consideration of the temporal and spatial complexities within the entire watershed, which is a very complex problem at best.

Technologies for channel erosion rectification are available. Such technologies include the use of vegetation, rip-rap (protecting stream banks with rock or other nonerodible materials), and various structural materials for bank stabilization; grade control structures for bed stabilization; and debris removal, channel realignment, and river-training structures (structures that control or direct the flow of water in river channels) for increasing the amount of sediment and water that can be conveyed. Upland treatments may increase the retention times of water runoffs and thus reduce the flood peaks and their associated erosive capacity through channels and limit the delivery of sediments to channels subject to deposition.

Channel maintenance involves control of the energy of the flowing water in such a way that erosion of the stream banks and streambed is minimized in an environmentally and aesthetically acceptable way. Energy control methods involve combinations of agronomic and engineering treatments that enhance channel dynamics and environmental control. Some specific technical problems are described below.

### Sediment Transport

Sediment transport is still poorly defined. Deviations of actual sand transport rate measurements from average trends are large with long time cycles. Consequently, estimates of the sand load for unmeasured streams or for single events in even the most intensively observed channels cannot be made on a reliable basis. Few measurements of the fraction of gravel in the bed material load have been made, but the evidence suggests that gravel accumulation in alternate sand and gravel bars may deflect the water flow and cause erosion of the opposite banks. Consequently, design of control measures remains largely an empirical equilibrium relation with a limited basis to define the probability of failure. Rectification methods vary in cost, so the means to design rectifications to balance construction and maintenance costs with potential losses from failures are needed. Development of new, more economical rectification methods may be feasible. Research is needed to define the relative effectiveness of combinations of various protective measures not only in protecting upstream channels but also in altering the amount of sediment delivered downstream. Development of a systems approach to channel rectification is needed.

### Distribution of Erosive Forces

The distribution of erosive forces between bed and bank materials and between particles of different sizes is poorly understood. More importantly, critical shears are not well understood. Bank material is often affected by electrochemical forces, primarily in the silt-clay fraction, and knowledge of how to treat these forces remains obscure, especially given the heterogeneity of bank materials. Fundamental studies are needed to clarify these uncertainties.

River crossings can be affected by both scour and bed degradation caused by erosion in the river channel. Scouring causes local and often temporary lowering of bed levels over a short distance, whereas degradation causes an extensive and often progressive lowering of the riverbed over a fairly long distance, which also implies a disequilibrium. Whereas scour problems can often be controlled by local protective measures, progressive degradation may be more difficult to control if it is not detected in time. Consequences may include loss of land, exposure of building foundations, stream bank failures, and loss of embankments, dams, or other structures. On the other hand, scour around local structures such as jetties and bridge abutments can proceed independently of the more general degradation process.

### Downstream Impacts

Downstream impacts have been restricted mostly to evaluation of the effects of sediment on channel stability and the filling of reservoirs in relatively small watersheds. Most of this effort has been concentrated in traditionally agricultural areas in the South and Midwest. In addition to climatic variability and hydrologic differences, some areas of the United States have influent rather than effluent streambeds, which change the nature of downstream impacts of channel instability.

### Channel Dynamic Conditions

The channel dynamic conditions must be incorporated into water resource analytic models in sufficient detail to permit meaningful assessment of the role of engineering structures and channel heterogeneity. Whereas past modeling efforts have generally been approached with one-dimensional models, the technology is not appropriate when suspended and bed material move along nonaligned paths.

## Effect of Wetlands on Sedimentation

Wetlands can have important effects on sedimentation. For example, drainage from wetlands can alter the stability and hydrologic balance of downstream channels. Furthermore, the runoff moving from wetlands may be clear of sediment and have sufficient energy to degrade downstream channels. In still another scenario, the runoff storage provided by a wetland may reduce peak discharge rates and reduce downstream flooding and channel deterioration.

What is needed is a wetland classification linked to geographic, aquatic, or ecological features; the relative abundance of wetland types within a region—especially critical watersheds; and the threat of degradation both to the wetland and to off-site resources (Reilly, 1991). Habitat, water quality improvement, flood prevention, and groundwater recharge should be considered.

Bottomlands are a special classification of wetlands that differ considerably between the humid eastern and the semiarid western parts of the United States. The more even distribution of monthly precipitation in humid areas results in infiltration in the upper part of a watershed that often reemerges in bottomlands. The bottomland width is controlled by topography, climate, and geology. Bottomlands tend to discharge groundwater to their associated streams and provide a significant part of the stream flow. Bottomlands in humid areas are often sinks for nutrients and sediments. Thus, the floods that do occur are shallow and of low velocity across heavily vegetated bottomlands that trap most of the upland sediments.

In more arid regions such as those in the western United States, influent streams that result when the water table is below the level of the channel bed result in unique problems in maintaining stable channels. Vegetation in such zones is controlled by the water table depth, which in turn affects above- and belowground biomass and the natural stabilities of the streambeds and stream banks. In other instances, grazing animals have access to the channels (where they go for shade and water) and trample the vegetation, causing bank sloughing and pollution from fecal matter. Thus, downcutting of the channel can lower the water table, have a negative impact on plant physiology and production, and further exacerbate downstream problems of sedimentation, channel stability, water yield, and the ecological balance.

# 10

# Salts and Trace Elements

The salt content (salinity) of soils and the frequently associated drainage problems are pervasive in agriculture. Salinity affects germination as well as seedling and vegetative growth. It also reduces crop yields. Waterlogging causes poor aeration in the root zone, with effects ranging from reduced growth to death of plants. Some constituents of salinity, such as sodium, may have a deleterious impact on the physical condition of the soil, impairing crop production.

About 19 million ha (47 million acres), or 10 percent of U.S. cropland, is irrigated (Postel, 1990). About 5.2 million ha (14 million acres) of this irrigated land is currently affected by salt. Salinity problems are not restricted to irrigated areas and arid climates. Halvorson (1990) estimates that nearly 1 million ha (2.5 million acres) of productive agricultural dryland in the western United States has been salinized through saline seeps. Salinity problems also occur in humid regions through, for example, seawater intrusion into low-lying coastal farmlands.

Soil salinization is a worldwide problem. According to Szabolcs (1989), about 1 billion ha (25 billion acres) of the world's soils are affected by salt. Saline and sodic (sodium-containing) soils cover about 10 percent of the world's arable lands and exist in 100 countries. Szabolcs (1989) reports that, worldwide, some 10 million ha (25 million acres) of irrigated land is abandoned annually because of salinization, sodification, and waterlogging.

Worldwide, in the arid-to-semiarid regions where irrigation is practiced, soil salinization and the frequently accompanying waterlogging problem have plagued agriculture for centuries. Historical records from the past 6,000 years reveal that many ancient civilizations whose existence was based on irrigated agriculture have failed, for example, the Sumerian civilization in the Mesopotamian Plains in Iraq, the Harappa civilization in the Indus Plain region in India and Pakistan, the inhabitants of the lower Viru Valley in Peru, and the Hohokam Indians in the Salt River region in Arizona (Tanji, 1990).

Salinization and waterlogging are not unique to ancient civilizations, however. For example, a critical agricultural and ecological crisis exists in the San Joaquin Valley of California (San Joaquin Valley Drainage Program, 1990). Of the nearly 1 million ha (2.5 million acres) of irrigated cropland on the valley's west side, about 38 percent is waterlogged and 59 percent has increased levels of salt. About 54,000 ha (133,000 acres) of the drainage-impacted lands have tile drainage systems, but only about 21 percent of the entire west side, however, can discharge its saline subsurface drainage into the San Joaquin River for eventual disposal into the Pacific Ocean. Adding to the difficulties in managing irrigation-induced water quality problems was the discovery of selenium and other toxic trace elements in subsurface drainage waters in the San Joaquin Valley's west side (National Research Council, 1989b). Since the 1983 discovery of selenium poisoning of waterfowl at the Kesterson National Wildlife Refuge in the San Joaquin Valley, the U.S. Department of the Interior has implemented the National Irrigation Water Quality Project. Twenty-six sites in 15 western states are being investigated to ascertain whether selenium and other trace elements in irrigation drainage water are causing ecological damage (Engberg et al., 1991). The National Irrigation Water Quality Project has found selenium in detectable quantities in 20 reconnaissance study sites. Of the principal trace elements, selenium appears to pose the greatest potential toxic effects on aquatic biota. The primary source of selenium in the drainage waters of the San Joaquin Valley is the Cretaceous-period marine sedimentary shales of the Coast Range mountains. The ecological hazards of potentially toxic trace elements—such as selenium, boron, arsenic, molybdenum, mercury, vanadium, and uranium—are magnified when agricultural drainage waters are disposed into hydrologically closed basins and sinks and accumulate in the food chain.

This chapter describes the sources of salts and trace elements and their effects on soils and plants. This chapter also explores alternative management options that can be used to minimize the irrigation-induced water quality problems and sustain irrigated agriculture. Because of the complexities of the salinity, drainage, and toxic element problems, the chapter begins with an overview.

Control of salinity is a major problem facing irrigated agriculture in the western United States. A USDA technician in California examines land on which productivity has been seriously damaged by salinity. Credit: U.S. Department of Agriculture.

## OVERVIEW OF SALINITY AND DRAINAGE PROBLEMS

Producers in arid and semiarid regions who irrigate their agricultural lands must always deal with salinity. This well-established fact is often overlooked.

### The Salinization Process

The salinization process begins with snowmelt that runs off mountains in rills, forms streamlets, and then rushes down with ever greater force. It dissolves small amounts of minerals from the riverbed and soils with which it comes into contact, slowly cutting out a channel along its path. Some of the water penetrates the soil or the underlying formations, reemerging as stream flow farther downstream or recharging the groundwater basin; again, some of the minerals that come into contact with the water on the way are dissolved and carried along. Elsewhere, as in the extensive Mancos shale region of Utah and Colorado, percolating waters displace groundwater that, because of its marine origin, contains high levels of salts.

Whatever its origin, water tapped for irrigation contains salt. Sometimes, the amounts are very small, but in other cases they are substantial. When this water is applied to the land to nourish crops, much of it is taken up by plants and is returned to the atmosphere. Since only pure water evaporates from the soil surface or transpires from plant surfaces (evapotranspiration), the evapotranspiration process leaves the salts behind in the soil. Thus, irrigation automatically and relentlessly leads to an accumulation of salts in the soil. Unless provision is made to leach a portion of this accumulating salt out by means of rainfall or the application of irrigation water in excess of crop water needs, the soil will soon become too saline for crop growth. Such excess water will eventually lead to a rising water table and waterlogging unless there is adequate drainage.

Thus, all irrigators must cope with salinity and all irrigated fields need drainage, be it natural or provided by producers. The drainage water tends to be substantially more saline than the original irrigation water. Suppose a producer irrigates his or her land with water containing 250 mg/L of total dissolved solids per liter (250 ppm) and suppose that three-fourths of this water is evapotranspired and one-fourth remains for leaching, then, as a first approximation, the drainage water will contain 1,000 mg/L of total dissolved solids per liter (1,000 ppm).

## Drainage

This simple sketch of the salinization process provides a valid overall picture, but it must be modified in numerous ways to be accurate and useful for devising and understanding practical management schemes. It suffices, however, to make clear that irrigation in arid regions is short-lived unless drainage is provided. As a consequence, irrigation always leads to the need to dispose of drainage water and thus brings about potential off-site water quality degradation. In a typical situation, drainage water is discharged downstream from an irrigation site into a stream that will again be used for irrigation, and this cycle can be repeated a number of times. Each time, more salt is added to less water.

### Sources of Salinity

The occurrence of saline soils is not restricted to irrigated areas. The same process of mineral weathering or dissolution and subsequent concentration because of water evaporation often leads to high salt levels in soils of arid and semiarid regions. The scarcity of rain that makes these areas arid restricts the possibility of leaching and thus leads to salt accumulation. Additional secondary sources of salts include atmospheric deposition and seawater intrusion. An example of increased salinity from atmospheric deposition is found in western Australia, where chloride-dominated salinity has been attributed to substantial deposition by wind over time of salts predominantly made up of sodium chloride derived from the Indian Ocean.

A special case of dryland salinity of particular concern in both North America (Halvorson, 1990) and Australia (Sharma and Williamson, 1984) is that of saline seeps. A saline seep occurs when water in excess of that required by plants percolates below the root zone and, upon encountering some type of barrier or restricting layer, moves laterally downhill and emerges in a seepage area, having picked up dissolved solids in transit. The nature of the restricting layer can vary from a change in soil texture to a coal seam or a change in geologic structure.

Saline seeps are often encountered where farmers practice a wheat-fallow rotation; during dry periods, such a rotation may serve to conserve some water during the noncropped period to aid the following crop, but in somewhat wetter years, the precipitation in excess of that required by plants initiates the process that leads to a seep.

In the U.S. northern Great Plains, the extent of drainage from seeps has been estimated to have affected an area of about 1 million ha (2.5 million acres) (Miller et al., 1981). The problem can sometimes be corrected by installing an interceptor drain (Doering and Sandoval, 1976). The disposal of the drainage water, however, creates a new problem; and the cost of providing an outlet is often excessive. A less costly solution is found by switching to a flexible cropping system in the recharge area (the area of land that is the source of water for the seep) to ensure that the crops grown in sequence use all of the available soil water. The systems must be flexible to adapt to the vagaries of the weather: different cropping sequences are required if precipitation has been above-average than are required if precipitation has been below-average (Black et al., 1981).

## Effects of Salinity

The effects of typical salts on agricultural crops and soils have long been known, and the need for drainage has been advocated for well over a century. Explicit concern about the off-site effects of salinated drainage waters, however, is more recent. Passage of the Colorado River Salinity Control Act (PL 93-320) in 1974 illustrates a formal awakening to the cumulative downstream problems of the stepwise consumption of a river's water. This law prescribed a process for reducing the salinity in the Colorado River to protect downstream users. In studies conducted for that purpose, investigators found that about 37 percent of the salt in the lower part of the river could be attributed to irrigation; they also ascertained that most of the adverse effects from higher salt levels were suffered by those using the water for municipal or industrial rather than for agriculture purposes (Jones, 1984).

Irrigation drainage also may contain nitrates and various pesticides (see Chapters 6 and 8, respectively). The present discussion, however, is restricted to salts and trace elements that are unique to irrigated agriculture.

For well over a century, agricultural producers have been concerned with the adverse effects of salinity on their irrigated agricultural lands, and they have developed detailed management schemes to deal with these problems. For at least 25 years, investigators have given serious attention to the off-site problems (downstream water degradation) associated with irrigation. More recently, however, a new concern has been added: the presence of toxic trace elements in soils and shallow groundwaters.

## Toxic Trace Elements

The term *salinity* has been loosely defined in terms of the major anions and cations found in irrigation water, with careful attention given to some specific ion effects such as those from sodium and chloride. The toxic effects (to plants) of low levels of boron have also been of concern in certain areas; but investigators assumed that, other than boron, trace elements did not occur in sufficient concentrations to be of concern. This changed in 1982 when investigators discovered that the elevated concentration of selenium in the Kesterson Reservoir was causing reproductive failures in aquatic organisms and waterfowl. This reservoir was in fact a set of shallow ponds used to store and evaporate agricultural drainage water (National Research Council, 1989b).

Extensive salinity damage is apparent in this cropland in California's Coachella Valley. Barren areas within the crop rows show where soil is too damaged to sustain plant life. Credit: Agricultural Research Service, USDA.

## Sources of Trace Elements

Since 1982, selenium as well as several other trace metals have been found with some regularity in drainage waters in certain geographic areas (Deason, 1989). Trace elements are natural constituents of the soils or the underlying geologic materials and may be mobilized by irrigation, in contrast to the major salts discussed above. Trace elements also differ from the major salts in terms of their typical concentrations. Whereas an investigator may measure salinity in 100s or 1,000s of milligrams per liter (ppm), a typical selenium concentration may be between 10 and 100 μg/liter (10 and 100 ppb). Furthermore, the effect of excessive salinity is likely to be loss of crop yield or damage to plumbing, whereas high levels of selenium or molybdenum can be toxic to fish and waterfowl, causing substantial harm to the ecosystem and, possibly, humans.

## Reducing the Impacts of Trace Elements

Irrigation practices can be altered in numerous ways to minimize the adverse effects of salinity on agricultural lands, and similarly, management can reduce water quality problems (Suarez and Rhoades, 1977). The same can be said for trace elements, although the specific management or corrective practices may differ. It is not possible, however, to completely eliminate these effects or to make them cost-free.

It is an important principle that irrigation of arid lands always degrades water quality: "irrigation agriculture over time cannot avoid causing an adverse off-site effect. This effect must be acknowledged: it can be minimized, internalized, or rejected, but it cannot be ignored. If irrigation is a desired use of water, then its waste waters must be treated and/or disposal provided for" (National Research Council, 1989b:41).

It seems appropriate to elaborate on this tenet with some examples. The discussion above suggested that irrigation in the upper Colorado River has led to substantial damage to plumbing in Los Angeles. It is possible, for example, to substantially increase the irrigation efficiency in the Colorado River's Grand Valley and to line the delivery canals to reduce the amount of seepage water from ditches, canals, and farm fields that returns to the river. Such actions—now under way—will be effective in reducing the salt contribution to the river from the Grand Valley, estimated at 530 million kg/year (240 million lb/year), because a reduction in seepage flow should lead to a proportional reduction in salt discharge masses (Inman et al., 1984). However, it is not possible to eliminate all salt contributions from the Grand Valley to the Colorado River.

Another situation is encountered in the Imperial Valley of California (Meyer and van Schilfgaarde, 1984). There, the drainage water from the irrigated valley enters the inland Salton Sea. This sea has no outlet; the only water loss is by evaporation. Excessive irrigation (and drainage) raises the level of the sea, damaging the lands adjacent to the sea. Increases in irrigation efficiency lower the sea level but accelerate its rate of salinization. Either way, the salt concentration continues to increase, reducing the Salton Sea's value as a habitat for fish and wildlife. Were it not for irrigation, the sea would be dry. Thus, there is a dilemma. Irrigation originally formed the sea in 1902, but continued irrigation will reduce the sea's biological value. It is unlikely that irrigation is sustainable without environmental insult. Irrigation in the Imperial Valley would be sustainable (that is, it could be continued indefinitely) if society were willing to sacrifice the sea, which would not exist if it were not for irrigation.

Irrigation causes off-site damages in terms of decreased water quality. Society must weigh the benefits accrued from irrigation against the disadvantages associated with and the costs of reducing soil salinization and water pollution (van Schilfgaarde, 1990).

## SOURCES AND EFFECTS OF SALINITY

### Nature of Salinity

The salinity in soils and waters is made up of dissolved mineral salts. The major cations are sodium, calcium, magnesium, and potassium; the major anions are chloride, sulfate, bicarbonate, carbonate, and nitrate. The concentrations of these solutes are reported in milligrams per liter, millimoles per liter, or milliequivalents per liter. Salinity is typically expressed as a lumped salinity parameter: electrical conductivity (EC) or total dissolved solids (TDS). EC is an intensive electroconductivity measure expressed in microSiemens per centimeter ($\mu$S/cm) for soils and waters with lower salinity levels and deciSiemens per meter (dS/m) for soils and waters with higher salinity levels. TDS is an extensive gravimetric measure reported in milligrams per liter, or grams per liter for hypersaline waters. Although no exact relationship exists between EC and TDS, the number of milligrams of TDS per liter can be approximated by multiplying EC (in dS/m) by a factor of 640 for many waters to a factor of 800 for hypersaline waters.

### *Measuring EC and TDS*

Measuring EC and TDS in waters is straightforward, but it is not in soils because salinity is significantly affected by the prevailing soil moisture content. The concentrations of salts in the soil solution do not typically change in direct proportion to changes in soil water content because the major solute species participate in such mechanisms as mineral precipitation and dissolution, cation exchange, and ion association. Moreover, soil salinity is a dynamic property since soluble salts are highly mobile in the soil profile.

Soil salinity is typically measured in the laboratory by obtaining an extract from a soil sample that is moistened with distilled water to a reference saturated soil water content. The EC of the soil saturation extract is reported as $EC_e$ (electrical conductivity of saturation extract). EC in field soils may be obtained by vacuum extraction of the soil solutions from recently wetted soils and is reported as $EC_{ss}$ (electrical conductivity of the soil solution). In situ EC measurements of moist field soils can be also obtained by using the four-electrode salinity probe and the electromagnetic device, which give the bulk EC of soils reported as $EC_a$. The time domain reflectometry is a promising method of measuring salinity and water content independently with the same probe (Dalton et al., 1984).

## Sources of Salinity

A primary source of salts is chemical weathering of the minerals present in soils and rocks. The more important chemical weathering mechanisms include dissolution (contact with water), hydrolysis (reaction with water), carbonation (reaction with dissolved carbon dioxide), acidification (reaction with proton), and oxidation-reduction (transfer of electrons). All of these reactions contribute to an increase in the dissolved mineral load in the soil solution and in waters.

Other important sources of salts include fossil salts (for example, salt domes), secondary deposits from marine or lacustrine (lake) environments (for example, gypsum), wet and dry atmospheric deposition, and seawater intrusion. Anthropogenic (human-derived) sources include salts contained in waters applied to the land, rising groundwater levels, soil and water amendments, animal manures and wastes, chemical fertilizers, sewage effluents and sludges, and oil and gas field brines.

## Effects of Salts on Soils

The physical properties of soils that are conducive to potential high-level crop production include adequate permeability of the soil for water and air and the presence of friable (easily crumbled or pulverized) soil for seed germination and root growth. These two properties are the ones most affected on irrigated lands and are reflected as poor permeability and poor soil tilth (tilth is the state of aggregation of a soil). The major factors affecting these physical conditions of soil are electrolyte content and the sodium content of the applied water or soil. The former is evaluated by EC. The latter is evaluated by the sodium adsorption ratio (SAR), which is defined as

$$SAR = [Na^+]/([Ca^{2+} + Mg^{2+}])^{1/2}$$

(ion concentrations are expressed in millimoles per liter).

### Reduced Water Infiltration

A combination of a low EC and a high SAR results in poor water infiltration rates in many soils (Oster and Rhoades, 1984). The deleterious impact of high-SAR water may be partially overcome by increasing the EC of the applied water. The unfavorable chemistry leading to water infiltration problems may be reflected by slaking (breakdown) of soil aggregates and the dispersion and swelling of clay minerals. Slaking and dispersion lead to reduced permeability and poor soil tilth, which, in turn, result in poor crop establishment, inadequate water intake rates, and increased runoff and erosion.

Swelling reduces the sizes of the interaggregate pore spaces in the soil and, therefore, produces a substantial, although reversible, reduction in the hydraulic conductivities of soils. Swelling is particularly important in soils that contain expandable clay minerals and have SARs of greater than about 15.

Dispersion—the release of individual clay platelets from aggregates—and slaking—the breakdown of aggregates into subaggregate entities—can destroy pore interstices, reducing in an irreversible way the hydraulic conductivity of a soil. Dispersion and slaking can occur at very low SARs if the EC of the soil solution is also very low.

Water penetration problems can develop when a seal with very low hydraulic conductivity forms at the soil surface. Seals are generally considered to be wet, and they reduce infiltration and increase runoff and erosion. Crusts are dry seals and, in addition to slowing water penetration, reduce the ability of seedlings to emerge.

Structural crusts are preferentially formed in soils exposed to the beating action of falling water droplets (for example, from rain or sprinkler irrigation systems). Surface sealing is enhanced by the dispersion and slaking mechanisms that take place when low-EC, high-SAR waters equilibrate with the soil surface.

Depositional crusts form when suspended soil sediments that originated from the detachment of soil materials are deposited at the soil
surface as the water infiltrates the soil. Depositional crusts, like structural crusts, are more likely to form when irrigation waters with low ECs
are applied to the soil because the dispersion-flocculation status of the
suspended sediments determines the hydraulic properties of the depositional crusts that are formed (Shainberg and Singer, 1990).

### Effects of Salinity and Sodicity

Because the effects of salinity and sodicity on soils are interrelated,
both the EC and the SAR of the applied water must be considered
simultaneously when assessing the potential effects of water quality on
soil water penetration. A general and more definite EC-SAR relationship
for all soils cannot be developed because of variations in clay mineralogy; clay, organic matter, and sesquioxide contents; and pH. Figure 10-1
shows that salt accumulation patterns in surface soils vary with the
method of water application (Ayers and Westcot, 1985).

Doneen and colleagues (1960) measured the distribution of salts in the
crop root zone in a lysimeter irrigated with about 2 surface meters of
irrigation water having an EC of 2.9 dS/m. (A lysimeter is a device that
encases a soil profile to provide accurate measurements of water applied
to and draining through the soil, and of chemical changes in the soil.)
The $EC_e$ of the soil profile was initially between 0.4 and 0.6 dS/m. The
distribution of salts in this freely drained, cropped soil profile after
application of 2 m of irrigation water reveals that with each irrigation
there is leaching of salt from the surface soil and subsequent salt
accumulation deeper in the soil. The extent of salt accumulation in the
lower portion of the crop root zone is dependent on the leaching
fraction, which is defined as the ratio of the depth of drainage water past
the root zone to the depth of infiltrated water.

For soils cropped under shallow water table conditions, there is a
tendency for salts to accumulate nearer to the soil surface with a rise in
the water table because of the upward evaporative flux rather than
downward leaching flux in deeply drained soils (Namken et al., 1969).
The extent of soluble salt accumulation and its distribution in the crop
root zone will have an impact on crop growth and yield.

### Effects of Salts on Plants

The adverse effects of salts on plants can be divided into three main
categories (Figure 10-2): effects on water relationships, effects of specific

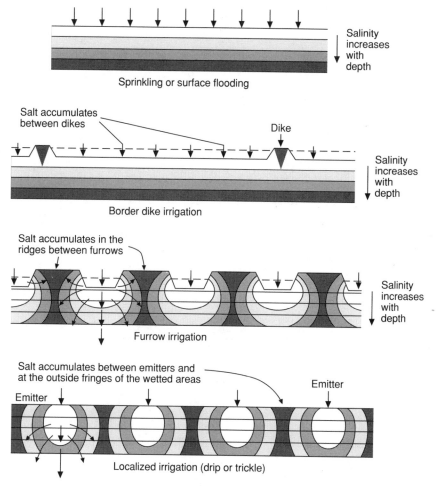

FIGURE 10-1 Typical salt accumulation patterns in surface soils for various methods of water application. Salinity ranges from low (unshaded) to high (darkened). Arrows indicate the direction of soil water flow. Source: R. S. Ayers and D. W. Westcot. 1985. Water Quality for Agriculture. FAO Irrigation and Drainage Paper 29, Rev. 1. Rome: Food and Agriculture Organization of the United Nations.

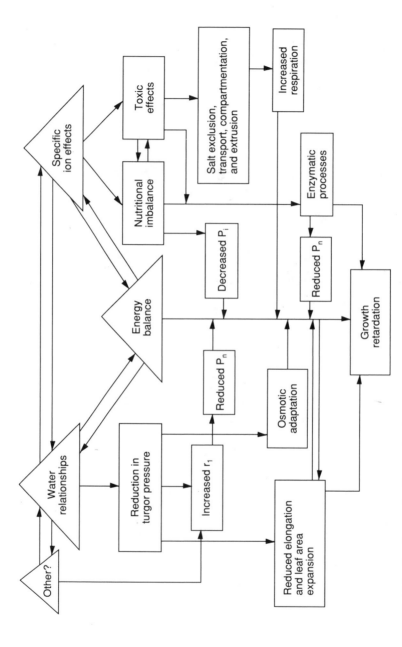

FIGURE 10-2 Detrimental effects of salinity on plant growth. Source: D. Pasternak. 1987. Salt tolerance and crop production—A comprehensive approach. Annual Review of Phytopathology 25:271–291. Reproduced, with permission, from the Annual Review of Phytopathology, Vol. 25, © 1987 by Annual Reviews Inc.

ions, and effects on energy balance (Pasternak, 1987). The relative importance of each of these effects on the overall plant response to salinity may differ sharply from one plant species to another or under different sets of environmental conditions.

### Effects of Water Relationships

Salt in the root zone decreases the osmotic potential of the soil solution and therefore reduces the availability of water to plants. If the osmotic potential of the soil becomes lower than that of the plant's cell, the latter would suffer osmotic desiccation and loss of turgor pressure (turgor pressure is the pressure within a plant cell). To survive, the plant must adjust osmotically to compensate for the lower external water potential. This can be effected by absorption of ions from the medium, synthesis of organic compounds, or both.

The synthesis and transport of organic compounds, the transport of inorganic ions, and the biochemical adjustments necessary for plant survival require the expenditure of metabolic energy. This expenditure results in depletion of the energy pools needed for growth (Figure 10-2).

Specific effects of ions on plants include direct toxicity because of excessive accumulation of ions in the tissues that may affect various physiological processes and a nutritional imbalance caused by an excess of some particular ions.

### Effects of Ions

The potentially toxic effects of certain ions such as boron, chloride, and sodium are associated with their uptake by roots and accumulation in the leaves. Some herbaceous crops and many woody species are susceptible to the toxicities caused by these ions. Some ions, like chloride, can also be absorbed directly into the leaves when moistened during sprinkler irrigation. In addition, many trace elements (see below) are toxic to plants at very low concentrations.

### Effects on Energy Balance

Salinity disrupts the acquisition of mineral nutrients by plants in two ways (Grattan and Grieve, 1992). First, the ionic strength of the substrate can have direct effects on nutrient uptake and translocation, and second, the interactions of major ions in the substrate (that is, sodium and chloride) can have an effect on nutrient ion acquisition and

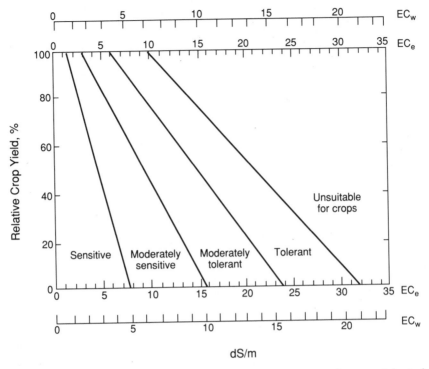

FIGURE 10-3  Relative salt tolerances of agricultural crops. Source: Adapted from R. S. Ayers and D. W. Westcot. 1985. Water Quality for Agriculture. FAO Irrigation and Drainage Paper 29, Rev. 1. Rome: Food and Agriculture Organization of the United Nations.

translocation within the plant. Examples of these effects are sodium ion-induced calcium ion or potassium ion deficiencies and calcium ion-induced magnesium ion deficiencies.

### Effects of Salinity on Crop Yields

Crop plants respond to salinity in widely ranging manners because of differences in their abilities to adjust osmotically, enabling them to extract water from saline soil solutions (Figure 10-3). Typical examples of salt-sensitive crops are bean, onion, almond, peach, orange, and grapefruit; moderately sensitive crops include corn, alfalfa, clover, cabbage, lettuce, potato, and grape; moderately tolerant crops include safflower, soybean, wheat, barley, tall fescue, squash, and olive; and tolerant crops include cotton, sugar beet, Bermuda grass, asparagus, date palm, and guayule.

The yield response function can be defined as $Y = 100 - b(EC_e - a)$, where $Y$ is the relative crop yield, $b$ is the slope of the declining yield with increasing $EC_e$, and $a$ is the salinity threshold value above which crop yield declines. Such yield response functions provide producers with general guidelines on expected yields for specific crops or allow them to pursue leaching of salts to ensure economic crop yields.

Figure 10-3 shows that the salinity threshold value varies widely among crop plants. Moreover, the sensitivity of plants to soil salinity may change from one stage of growth to the next. For instance, most plants are fairly salt tolerant during germination but become quite sensitive to salts during emergence and early seedling growth. Other crops, such as cereals, become quite sensitive to salts in the early reproductive stage, and seed yield may be reduced if soil salinity is at elevated levels. Most crops, however, tend to have increasing salt tolerance with increasing age.

## SOURCES AND EFFECTS OF TRACE ELEMENTS

### Nature of Trace Elements

Trace elements occur in minute concentrations in the environment, for example, less than 100 mg/kg (100 ppm) in soils and rocks, less than 10 mg/kg (10 ppm) in plant and animal tissues, or less than 1 mg/liter (1 ppm) in water. Elevated concentrations of inorganic trace elements in soils, waters, plants, and animals have been of concern for about a century, but they have been of greater concern in the past several decades. For instance, the discovery of human mercury poisoning in Japan in the 1950s and 1960s (Minamata and Itai-Itai diseases) triggered research into the environmental hazards of mercury in aquatic systems (Adriano, 1986). More recently, the discovery in the early 1980s of selenium poisoning of fish and waterfowl in the Kesterson National Wildlife Refuge has further heightened awareness of the potential hazards of naturally occurring trace elements in agricultural drainage waters (National Research Council, 1989b).

Deverel and Fujii (1990) state that the presence of elevated concentrations of trace elements in groundwaters and irrigated soils in the San Joaquin Valley's west side poses threats to agriculture as well as human and animal health. They found that the threat is reflected in three ways. First, trace elements can accumulate in plants to levels that cause phytotoxicity. Second, trace elements in plants can adversely affect humans and animals that consume those plants. Third, trace elements can migrate with seepage through the root zone into groundwaters, possibly reemerging with subsurface drainage in surface waters, thereby affecting wildlife, or with groundwater pumped for domestic use, thereby threatening human health.

The U.S. Environmental Protection Agency (EPA) (1986b) considers the following trace elements to be potentially harmful to human health if they appear in drinking water: arsenic, barium, cadmium, chromium, fluoride, lead, mercury, selenium, silver, copper, iron, manganese, and zinc. In addition to the trace elements listed above, other trace elements, including boron, nickel, uranium, tellurium, beryllium, and aluminum, have potential detrimental impacts on aquatic biota (San Joaquin Valley Drainage Program, 1990).

### Sources of Trace Elements

The primary sources of trace elements are earth materials and volcanic emanations. Trace elements are found in primary minerals in rocks and soils frequently as isomorphic substitution (isomorphic substitution is the substitution of one element for another in clay minerals that may result in a net change in the electrical charge of the mineral), for example, lead for potassium in feldspars, selenium for sulfur in pyrite, zinc for magnesium in olivines, and boron for silicon in micas (Sposito, 1989). Trace elements may also coprecipitate with secondary soil minerals, for example, molybdenum and arsenic with iron and aluminum oxides, cobalt and nickel with manganese oxides, vanadium and cadmium with calcium carbonates, titanium with vermiculites, and vanadium and copper with smectites. Numerous trace elements are also associated with soil organic matter, for example, aluminum, vanadium, chromium, manganese, iron, nickel, copper, zinc, cadmium, and lead (Sposito, 1989). These soil minerals serve as reservoirs for the trace elements, which are typically released at a slow rate into soil solutions and waters through a number of chemical weathering mechanisms.

Anthropogenic sources of trace elements (Adriano, 1986) include phosphatic fertilizers (zinc, copper, cadmium, chromium, nickel, and lead), liming materials (zinc, manganese, and copper), pesticides (mercury, arsenic, and lead), sewage sludges (cadmium, zinc, copper, lead, and nickel), animal wastes (copper, cobalt, and zinc), coal combustion residues (arsenic, cadmium, molybdenum, selenium, and zinc), mining and smelting residues (lead, copper, zinc, cadmium, cobalt, and manganese), and motor vehicle emissions (lead, zinc, cadmium, and nickel).

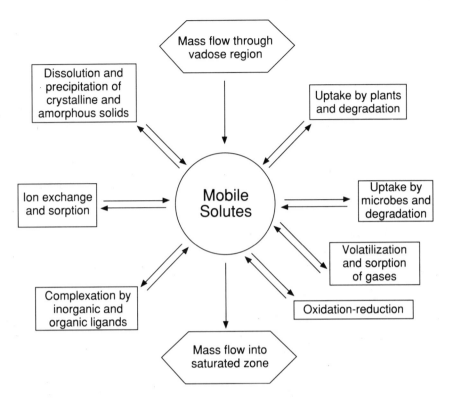

FIGURE 10-4 Possible abiotic and biotic processes affecting the reactivities and mobilities of trace elements. $EC_e$. electrical conductivity of the saturation extract (dS/m); $EC_w$, electrical conductivity of the irrigation water (dS/m). $EC_e = 1.5$ $EC_w$. Source: K. K. Tanji and L. Vallopi. 1989. Ground water contamination by trace elements. Agriculture, Ecosystems and the Environment 26:229–274. Reprinted with permission from © Elsevier Science Publishers, B.V.

### Reactivities and Mobilities of Trace Elements

The presence or accumulation of trace elements in agricultural drainage waters and groundwaters is influenced by a number of factors, including the nature and sources of trace elements, the particular trace element and its reactivity, and mobility and transport processes. The first item was addressed above.

Figure 10-4 represents some of the possible abiotic and biotic processes affecting the reactivities and mobilities of trace elements in soil-water systems (Tanji and Valoppi, 1989). The mobile forms of trace elements include various solutes and gaseous species that are subjected to numerous competing reactions. The rate and extent of these reactions are, in turn, influenced by a host of environmental conditions. Because of site-specific conditions and factors and the complexities of the reactivities of trace elements, only a few generalizations are possible.

For instance, the cationic trace elements (for example, heavy metals) such as copper, lead, zinc, cadmium, and nickel tend to be strongly retained by soils owing to ion exchange, sorption, and mineral precipitation. In contrast, anionic trace elements (for example, oxyanions) such as selenate ($SeO_4^{2-}$), chromate ($CrO_4^{2-}$), arsenate ($AsO_4^{2-}$), molybdate ($MoO_4^{2-}$), and vanadite ($VO_3^{2-}$) are subject to greater mobility, even though they may be retained to some extent by clays and sesquioxide surfaces. Differences in mobility, however, exist among the oxyanions of a given trace element. For example, selenite ($SeO_3^{2-}$) is more strongly sorbed by soils than selenate ($SeO_4^{2-}$), especially in the presence of elevated concentrations of $SO_4^{2-}$. Trace elements that tend to form complexes with inorganic and organic ligands have greater mobilities than do those that are not complexed. (Ligands are a group, ion, or molecule coordinated to an element forming a complex.)

The solubilities of minerals containing cationic trace elements typically increase as the pH decreases, whereas the mobilities of those containing anionic trace elements typically decrease as the pH decreases. The half-lives of reactions (Langmuir and Mahony, 1985) range from seconds (hydration, acid-base complexation, adsorption and desorption), minutes to hours (oxidation-reduction, gas solution, exsolution), weeks (precipitation, dissolution), months (polymerization and hydrolysis, isotopic exchange), and years (mineral crystallization).

## Effects of Trace Elements on Soils and Groundwaters

To provide examples of the effects of trace elements on soils and groundwaters, this section focuses on selenium, a naturally occurring trace element, and heavy metal accumulation in soils from sludge applications.

The distribution of selenium in the San Joaquin Valley in California is shown in Figure 10-5. Figure 10-5A shows the distribution of selenium in the top 30.5 cm (12 inches) of soils in the entire valley floor. Soils in the east side of the valley are extremely low in selenium except in localized sites. Animals that graze on soils on the east side of the valley frequently suffer from selenium deficiency since the selenium-poor soils that formed from sediments from the Sierra Nevada mountain range are mainly granitic. In contrast, soils that formed from sediments derived from the Coast Range mountains are rich in selenium.

Figure 10-5B shows the distribution of selenium found in shallow groundwaters on the west side of the San Joaquin Valley. The highest concentrations were in excess of 200 µg/liter (200 ppb) and the maximum was 3,800 µg/liter (3,800 ppb). Drainage water disposed into Kesterson Reservoir had an average selenium concentration of about 300 µg/liter (300 ppb). The soluble selenium fractions in the surface soil depths have been leached in profiles of soils that have been irrigated for decades. In contrast, investigators found elevated selenium levels in surface soil depths when the soils had not been irrigated. The annual precipitation of 100 to 300 mm (4 to 12 inches) is insufficient to leach selenium from surface soils. Based on the ratio of the isotopes, oxygen-18, and deuterium (hydrogen-2) in groundwaters, scientists from the U.S. Geological Survey (for example, Gilliom et al., 1989) have concluded that selenium and other dissolved mineral salts have been concentrated in the shallow groundwaters by evaporation.

Chang and colleagues (1984) studied the annual application of a composted sewage sludge and anaerobically digested liquid sludge for 6 years in cropped soils (Figure 10-6). They found that cadmium, nickel, chromium, lead, and copper accumulate almost entirely in the surface 15 cm of the soil. Findings similar to those presented in Figure 10-5 were obtained with the application of liquid sludge. These metals are comparatively immobile because of the strong soil retention mechanisms described above.

## Effects of Trace Elements on Plants

Plants differ in their abilities to absorb, accumulate, and tolerate trace elements. For a given species, concentrations of trace elements in various parts of the plant and among different plant cultivars also vary. Genetically controlled features of plants, morphological and anatomical differences between plants, and the physiology of a plant's ion transport mechanism may be responsible for these differences.

Plants can accumulate enough of certain trace elements such as cadmium, selenium, and molybdenum to cause acute toxicity or chronic metabolic imbalances in consumers of the plants. In some cases, plants do not absorb trace elements because they have a soil-plant barrier (Page et al., 1990); in those cases, the food chain is protected from accumulating harmful amounts of trace elements. Ingestion of contaminated soil or dust particles, however, may cause intake of toxic trace elements (Page et al., 1990).

The concentration of a trace element that results in toxicity to plants (phytotoxicity) may vary, and so ranges are usually reported. Table 10-1

A

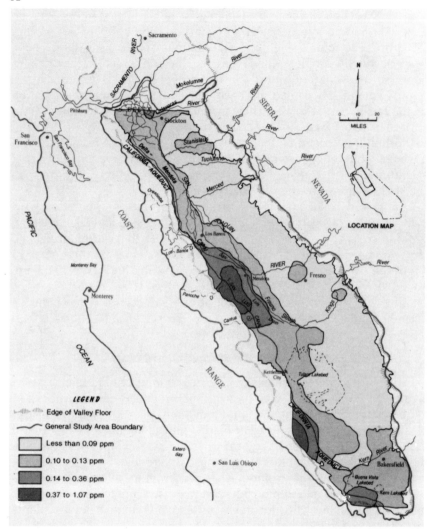

FIGURE 10-5   Total selenium concentrations in the top 30.5 cm (12 inches) of soil (A) and in shallow groundwater (B) from 1984 to 1989 in the San Joaquin Valley. Source: San Joaquin Valley Drainage Program. 1990. A Management Plan for

**B**

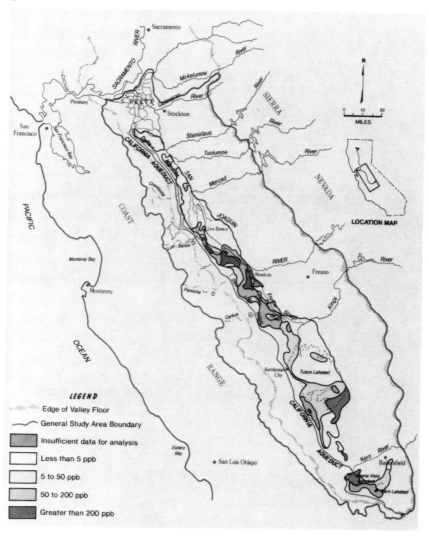

Agricultural Subsurface Drainage and Related Problems on the Westside San Joaquin Valley. Final Report. Sacramento, Calif.: San Joaquin Valley Drainage Program.

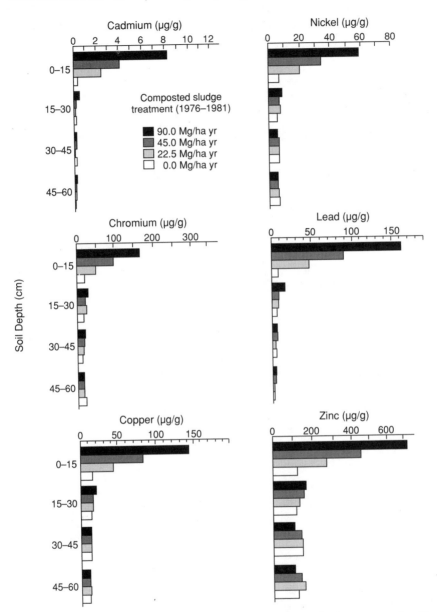

FIGURE 10-6  Heavy metal contents in Greenfield sandy loam treated with composted sludge from 1976 to 1981. Source: A. C. Chang, J. A. Warneke, A. L. Page, and L. J. Lund. 1984. Accumulation of heavy metals in sewage sludge-treated soils. Journal of Environmental Quality 13:87–91. Reprinted with permission from © American Society for Agronomy, Crop Science Society of America, and Soil Science Society of America.

TABLE 10-1    Concentration of Trace Elements
Commonly Observed in Forage Crops

| Element | Dry Weight (mg/kg) | |
| --- | --- | --- |
| | Typical | Phytotoxic |
| Arsenic | 0.01–1.0 | 3–10 |
| Boron | 7–75 | 75 |
| Cadmium | 0.10–1.0 | 5–700 |
| Cobalt | 0.01–0.3 | 25–100 |
| Copper | 3–20 | 25–40 |
| Molybdenum | 0.10–3.0 | 100 |
| Nickel | 0.10–5.0 | 50–100 |
| Selenium | 0.10–2.0 | 100 |
| Zinc | 15–150 | 500–1,000 |

SOURCE: A. L. Page, A. C. Chang, and D. C. Adriano. 1990. Deficiencies and toxicities of trace elements. Pp. 138–160 in Agricultural Salinity Assessment and Management, K. K. Tanji, ed. ASCE Manuals and Reports on Engineering Practice No. 71. New York: American Society of Civil Engineers. Reprinted with permission from © American Society of Civil Engineers.

summarizes the concentration ranges of trace elements associated with both normal and phytotoxic levels in forage crops (Page et al., 1990).

Table 10-2 presents the recommended maximum concentrations of 15 trace elements in irrigation waters for long-term protection of plants and animals (Pratt and Suarez, 1990). These concentrations should be considered as guidelines designed to protect the most sensitive crops and animals from receiving toxic amounts of trace elements.

Figure 10-7 gives the range of the selenium concentrations found in edible portions of crops and forage plants grown in selenium-impacted soils of the San Joaquin Valley's west side (Tanji, 1991b). These levels are not high enough to contribute amounts above normal dietary levels (50 to 200 μg/day [0.002 to 0.0007 ounces] for adults). However, some types of crops like the crucifers (for example, cabbage and mustard) are capable of assimilating high levels of selenium; this may be of concern in the future because agricultural and drainage practices will probably change in the west side of the San Joaquin Valley. Furthermore, public health officials give health advisories to rural residents who live where selenium levels are high. Those residents are advised to limit their consumption of home-grown foodstuffs.

Chang and colleagues (1984) applied soil sludge to winter barley and sorghum crops for 6 years (Figure 10-6). Table 10-3 summarizes the total amounts of cadmium and zinc removed by crops treated with the two

TABLE 10-2   Recommended Maximum Concentrations of 15 Trace Elements in Irrigation Waters for Long-Term Protection of Plants and Animals

| Element | Recommended Maximum Concentration (mg/liter)[a] | Comments |
|---|---|---|
| Arsenic | 0.10 | This guideline protects sensitive crops grown on sandy soils. Higher concentrations can be tolerated by some crops for short periods when they are grown in fine-textured soils. |
| Beryllium | 0.10 | Toxicities to plants have been reported at concentrations of as low as 0.5 mg/liter in nutrient solutions and at levels in the soil greater than 4 percent of the cation-exchange capacity. |
| Cadmium | 0.01 | Concentrations equal to or less than 0.01 mg/liter require 50 years or more to exceed the recommended maximum cadmium loading rate. Removal in crops and by leaching partially compensates and perhaps allows use of the water indefinitely. |
| Chromium | 0.10 | Toxicity in nutrient solutions has been observed at a concentration of 0.50 mg/liter and in soil cultures at a rate of 10 kg/ha. Toxicity depends on the form of chromium existing in the water and soil and on soil reactions. |
| Cobalt | 0.05 | A concentration of 0.10 mg/liter is near the toxic threshold for many plants grown in nutrient solution. Toxicity varies, depending on the type of crop and soil chemistry. |
| Copper | 0.20 | Concentrations of 0.1 to 1.0 mg/liter in nutrient solutions have been found to be toxic to plants, but soil reactions usually precipitate or adsorb copper, so that soluble copper does not readily accumulate. |
| Fluoride | 1.0 | This concentration is designed to protect crops grown in acidic soils. Neutral and alkaline soils usually inactivate fluoride, so higher concentrations can be tolerated. |
| Lead | 5.0 | Plants are relatively tolerant to lead, and soils effectively sorb or precipitate it. Toxicity to animals typically is caused not by lead adsorption from soils but by aerial deposition of lead on the foliage of pasture and forage plants. |
| Lithium | 2.5[b] | Most crops are tolerant to lithium up to 5 mg/liter in nutrient solutions. Citrus, however, is highly sensitive to lithium. Lithium is a highly mobile cation that leaches from soils over an extended period of time. |
| Manganese | 0.20 | Some crops show manganese toxicities at a fraction of an mg/liter in nutrient solution, but typical soil pH and oxidation-reduction potentials control manganese in soil solution, so that the manganese concentration in irrigation water is relatively unimportant. |

TABLE 10-2    (*Continued*)

| Element | Recommended Maximum Concentration (mg/liter)[a] | Comments |
|---|---|---|
| Molydenum | 0.01 | This concentration is below the phytotoxic level but is recommended to protect animals from molybdosis because of excess molybdenum in forages. |
| Nickel | 0.20 | Nickel is toxic to many plants at concentrations of 0.5 to 1.0 mg/liter. Toxicity from this element decreases with an increase in pH, so acidic soils are the most sensitive. |
| Selenium | 0.02 | This guideline protects livestock from selenosis because of selenium in forage. Selenium absorption by plants is greatly inhibited by sulfate, so the guideline for this element can be increased for gypsiferous soils and waters. |
| Vanadium | 0.10 | Toxicity to some plants has been recorded at vanadium concentrations above 0.5 mg/liter. |
| Zinc | 0.50 | Zinc is toxic to a number of plants at a concentration of 1 mg/liter in nutrient solution, but soils have a large capacity to precipitate this element. This guideline is designed to provide protection for acidic sandy soils. Neutral and alkaline soils can accept much greater concentrations without developing toxicities. |

[a]Loading rates in kg/ha-year can be calculated from the relationship that 1 mg/liter in the water gives 10 kg/ha-year when water is used at a rate of 10,000 $m^3$/ha-year.

[b]For citrus, the maximum recommended concentration is 0.075 mg/liter.

SOURCE: P. F. Pratt and D. L. Suarez. 1990. Irrigation water quality assessments. Pp. 220–236 in Agricultural Salinity Assessment and Management, K. K. Tanji, ed. ASCE Manuals and Reports on Engineering Practice No. 71. New York: American Society of Civil Engineers. Reprinted with permission from © American Society of Civil Engineers.

types of sludge applications. Winter barley and sorghum removed increasing amounts of cadmium and zinc with increasing sludge applications, with zinc uptake being far greater than cadmium uptake. The amount of metals taken up by the crops was insignificant (less than 1 percent of that applied). However, the metal contents in the crops were at phytotoxic levels (Table 10-1) and probably should not be consumed by animals including humans.

## ALTERNATIVE MANAGEMENT OPTIONS

Although water quality problems induced by irrigation tend to share some common features (National Research Council, 1989b), problems caused by site-specific conditions and processes prevail. Soils and

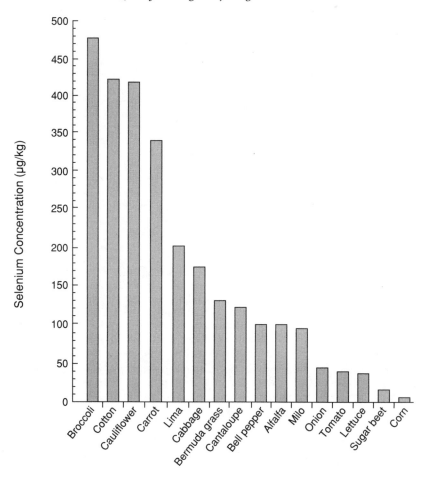

FIGURE 10-7 Concentrations of selenium in tissues of various edible crops. Source: K. K. Tanji. 1991. Principal Accomplishments 1985–90. Davis: University of California, Division of Agriculture and Natural Resources, UC Salinity/ Drainage Task Force.

cropping systems, irrigation and drainage systems, water rights, institutional infrastructure, and drainage water disposal practices differ. Nevertheless, it is possible to generically group alternative management options that can be used to control salinity and trace elements into (1) source control measures, (2) drainage water reuse, (3) drainage water treatment, (4) drainage water disposal, and (5) institutional changes.

TABLE 10-3  Total Removal by Crops of Cadmium and Zinc from Sludge-Treated Greenfield Sandy Loam Soils, 1976-1981[a]

| Cumulative Solids Applied in Sludge (metric ton/ha) | Concentration in Soil (g/ha) | |
|---|---|---|
| | Cadmium | Zinc |
| Composted sludge | | |
| 0 | 5.6 | 1,199 |
| 137 | 15.0 | 2,166 |
| 274 | 35.0 | 3,049 |
| 548 | 55.7 | 3,877 |
| Liquid sludge | | |
| 0 | 7.1 | 1,170 |
| 80 | 84.5 | 3,534 |
| 156 | 106.9 | 4,650 |
| 298 | 129.4 | 5,765 |

[a]The total amounts of cadmium and zinc were calculated as the sum of annual crop yields (grain and straw) multiplied by their corresponding metal contents.

SOURCE: A. C. Chang, J. A. Warneke, A. L. Page, and L. J. Lund. 1984. Accumulation of heavy metals in sewage sludge-treated soils. Journal of Environmental Quality 13:87–91. Reprinted with permission from © American Society for Agronomy, Crop Science Society of America, and Soil Science Society of America.

## Source Control Measures

The principal aim of source control is to use water and land resources efficiently with off-farm and on-farm measures that minimize salinity and trace element problems. The off-farm or irrigation project measures involve flexibility, reliability, and stream flow control in water delivery to the farm (Clemmons, 1987). A rotation schedule may range from fixed to seasonally varied water deliveries with regard to amounts and timing. In contrast, demand and arranged schedules allow farmers to have complete flexibility in the frequency, rate, and duration of water delivery. The use of modern technology that uses automated downstream or upstream canal regulation and centralized computer controls is desirable but costly. Substantial improvements in the timing, flow, and volume of water deliveries to farms may be achieved by well-trained ditch riders (personnel responsible for monitoring and adjusting water control gates in irrigation canals). Seepage of water from canals is a major problem, particularly in salt-affected areas, where seepage water picks up dissolved mineral salts from saline soils and geologic formations, for example, in the Grand Valley of Colorado.

### On-Farm Source Control Measures

The on-farm source control measures for salinity and trace elements involve the application of water more uniformly and efficiently to reduce surface as well as subsurface drainage. Irrigation scheduling to determine the timing and amount of irrigation water application is essential. Scheduling may be achieved by one of two general methods: (1) by monitoring soil and/or crop parameters and (2) by computing the soil-water balance. The first method involves measuring the soil water content or matrix potential (a measure of how tightly water molecules are bound to soil particles) and/or the leaf water potential of the crop. The second method requires an estimate of the storage capacity of water in soil, crop rooting depth, allowable soil water depletion, and crop evapotranspiration (Martin et al., 1990). Such water balance techniques range from simple "checkbook" accounting to computerized scheduling using real-time weather data like those provided by the California Irrigation Management Information System or the model of the Agricultural Research Service, U.S. Department of Agriculture.

### Water Application Systems

Water is applied to croplands by surface and pressurized irrigation application systems. The surface irrigation systems include furrow, border, and basin methods. The pressurized systems include high- and low-volume sprinklers and surface and subsurface drip or trickle irrigation systems. Each of these water application methods has its advantages and disadvantages, depending on site-specific conditions and agronomic practices. The performance characteristics of irrigation systems can be evaluated by measuring their application efficiency uniformities, deep percolation ratios, and tailwater ratios (Heerman et al., 1990). The potential attainable application efficiencies for these irrigation systems are nearly the same as those for properly designed and managed systems. However, the typical application efficiencies of surface methods tend to be lower (50 to 70 percent) than those of sprinklers and drip/trickle systems (75 to 90 percent). Improved furrow irrigation systems with shorter runs (for example, 200 m [218 yards]), modified set times, and tailwater return systems can achieve the application efficiencies of pressurized systems.

### Management of Salt-Affected, Waterlogged Croplands

Salt-affected, waterlogged croplands require additional considerations and special management practices. Reductions in subsurface drainage minimize off-farm environmental impacts. In addition to the distribution uniformity of water application, subsurface drainage is affected by the uniformity of water infiltration rates across irrigated fields. Deep-rooted, salt-tolerant crops may be able to use shallow groundwaters to meet a portion of their evapotranspiration needs. The salt balance in the crop root zone needs to be maintained to sustain crop production. When the source of soil salinity is the salts contained in the irrigation water, the guidelines of Ayers and Westcot (1985) can be used to establish a certain leaching fraction for salt control. If, however, the principal source of salts is naturally occurring salts in the soils, soil salinity needs to be monitored and water in excess of crop water needs to be applied periodically to control salinity in the root zone.

Because of the nonuniformity of both application of water and infiltration rates, subsurface drainage water and associated pollutants will be produced even if producers implement best-management practices. Moreover, salt accumulation in the root zone from either applied water or chemical weathering of soils needs to be controlled to sustain crop yields. Thus, irrigated agriculture inevitably results in the production of residuals (drainage water and pollutants) that need to be managed and/or disposed.

## Drainage Water Reuse

Source control can be viewed as the first line of action to reduce the off-site impacts of return flows of water from irrigation. A second management option is to reuse the subsurface drainage water until it is no longer usable. The San Joaquin Valley Drainage Program (1990) recommended a drainage water reuse strategy in waterlogged lands. High-quality irrigation water is used to irrigate the more salt-sensitive crops. The subsurface drainage from salt-sensitive crops is used to irrigate the more salt-tolerant crops and trees. The subsurface drainage from salt-tolerant plants is, in turn, reused to irrigate halophytes (plants that tolerate elevated salinities). In this strategy, the volume of drainage water is successively decreased, and concurrently, the salinity of the drainage water is successively increased so that further management of subsurface drainage waters could be carried out more efficiently, for example, disposal in off-site water bodies, salt harvesting in evaporation ponds, water treatment, or injection into deep wells.

Saline water as either fresh irrigation water or subsurface drainage water has been successfully used under certain conditions. For instance, irrigation water containing an average of 2,500 mg/L of total dissolved solids (2,500 ppm) has been used for decades in the Pecos Valley of

Texas (Moore and Hefner, 1977). In contrast, the Broadview Water District in the San Joaquin Valley's west side blended subsurface drainage water into fresh canal water for about 25 years, but it had to discontinue this practice. The blended water initially had an EC of 1.6 dS/m in 1956, but the EC rose to 3.2 dS/m by 1981, with a decline in the growth and yields of the more salt-sensitive crops. A major problem was the crusting of surface soils as a result of using waters with high sodium adsorption ratios (see above) and difficulties in seed germination and seedling growth.

Agroforestry is a new approach being tested in the San Joaquin Valley to lower high water tables through water extraction by tree roots as well as the reuse of saline drainage waters. Tanji and Karajeh (1991) have intensively monitored salt and water balances in a 9.4-ha (23.2-acre) eucalyptus plantation. The 6-year-old trees lowered the water table from about 0.6 to 2.2 m (2 to 7 feet) below the soil surface and are using irrigation drainage water with an EC of 10 dS/m from nearby croplands. However, a buildup of salinity to an average $EC_e$ of 25 dS/m has reduced the evapotranspiration rate by about 67 percent. The 16 percent leaching fraction has been increased to reduce soil salinity and improve the evapotranspiration rates of trees.

## Drainage Water Treatment

The California Department of Water Resources has investigated desalinization of agricultural drainage waters in the San Joaquin Valley. Reverse osmosis shows the greatest potential for achieving this (Lee, 1990). To successfully desalt drainage waters, however, the drainage waters must be pretreated to avoid scaling and biological fouling of the membranes used in the reverse osmosis process. Pretreatment involves removal of suspended solids, silica, and calcium and sulfate ions. Disinfection is also required. The estimated cost for reverse osmosis is $880/10^3$ m$^3$ ($1,090/acre-foot), excluding the cost of collecting and delivering the drainage water and disposing the treatment by-products (Lee, 1990).

### Biological and Physicochemical Processes

The San Joaquin Valley Drainage Program sponsored research on the removal of selenium and other trace elements from agricultural drainage waters, including biological and physicochemical processes. An anaerobic bacterial process used methanol as a source of carbon for microbes to reduce selenium, microfilters to remove fine suspended solids, and ion-exchange resins to polish the effluent. The 330 to 550 µg/liter of selenium of influent (330 to 550 ppb) was reduced to 16 to 50 µg/liter (16 to 50 ppb) in the biological reactor and to 10 to 40 µg/liter after microfiltration. Treatment costs ranged from \$118 to \$182/$10^3$ m$^3$ (\$145 to \$224/acre-foot) (Lee, 1990).

A second biological process that has been studied involved growth and harvesting of microalgae and bacteria, methane fermentation of the biomass, and ferric chloride treatment. In laboratory studies, the digested biomass reduced the selenium concentration in the influent from 367 to 20 µg/liter (367 to 20 ppb). In field studies, incorporation of a nitrate reduction process and treatment with ferric chloride further reduced the selenium level to less than 1 µg/liter (1 ppb). The treatment cost for this microalgal-bacterial process ranged from a conservative \$55 to \$83/$10^3$ m$^3$ (\$67 to \$102/acre-foot) (Lee, 1990).

A third biological process involves the volatilization of methylated selenides by several indigenous species of fungi. This volatilization process is applicable to both ponded waters and surface soils. Further research is needed to assess the volatilization of selenium from ponded waters (Lee, 1990).

The physicochemical treatment processes that have been investigated include chemical reduction and surface adsorption of selenium onto hydroxylated surfaces. Selenium can be reduced and precipitated from drainage waters by using heavy doses of ferrous hydroxide. The treatment costs for reducing selenium concentrations to 1 µg/liter (1 ppb) range from \$57 to \$125/$10^3$ m$^3$ (\$70 to \$154/acre-foot) (Lee, 1990).

Selenium can also be adsorbed to iron filings activated by oxygenation. Apparently, both surface adsorption to hydroxylated sites as well as chemical reactions aid in immobilizing selenium up to about 90 percent of the initial concentration. Under field conditions, serious problems of cementation occurred in the bed of iron filings. Depending on the life expectancy of the bed, the costs for reducing selenium concentrations ranged from \$57 to \$231/$10^3$ m$^3$ (\$70 to \$284/acre-foot) (Lee, 1990).

### Future Research Needs

The treatment studies described above were carried out in bench-scale models in the laboratory and in mini-pilot plants in the field. Although many of these processes show some promise in their ability to remove selenium, there remains a need for further research to understand the basic mechanisms by which selenium can be removed (Lee, 1990). The costs of these treatments are likely too expensive for irrigators to bear the entire costs. The anaerobic bacterial treatment process appears to show the greatest promise of removing selenium at a practical cost, and a pilot plant study is under way in the San Joaquin Valley (Lee, 1990).

## Drainage Water Disposal

Options for disposing of agricultural drainage waters include (1) deep percolation into the underlying groundwater basin; (2) discharge into surface waters, with the ultimate destination being the oceans or inland sinks; (3) disposal in agricultural evaporation ponds; and (4) deep well injection into permeable substrata. The first option was discussed earlier in this chapter.

### Discharge into Surface Waters

The practice of discharging collected surface irrigation return flow water into streams and lakes is widespread. However, increasing constraints are being placed on such discharges as more stringent water quality standards for receiving waters are being promulgated. Thus far, irrigation drainage water is considered a nonpoint source of pollution and is not regulated as much as point sources of discharge are. Increasing constraints will likely be placed on nonpoint sources of pollution.

### Disposal in Agricultural Evaporation Ponds

An example of disposal into agricultural evaporation ponds is the Kesterson Reservoir. Such a practice may pose risks to wildlife and groundwater. The 510-ha (1,260-acre) Kesterson Reservoir was constructed primarily to evaporate impounded saline drainage water and secondarily to maintain a habitat for waterfowl. (The impact of selenium accumulation in the aquatic food chain was discussed above.) A study done before reservoir construction indicated that about 40 percent of the impounded water would seep into the underlying aquifer (Benson et al., 1990). While Kesterson was under operation, a groundwater mound formed about 0.5 to 3.0 m (1.6 to 9.8 feet) above the regional groundwater level. The rate of lateral groundwater flow was about 4.6 m/year (15 feet/year). Much of the selenium present in the impounded water accumulated in the sediments and organic detritus. The selenium concentrations in shallow groundwater were low because of the transformation of oxidized forms of selenium to reduced forms (elemental selenium and selenides), which are immobile.

Elsewhere in the San Joaquin Valley's west side, agricultural evaporation ponds were installed between 1972 and 1985, mainly in the environs of Tulare Lake Basin, which is a hydrologically closed basin. Of the 28 ponds constructed, 5 are now inactive or closed. These ponds occupy a surface area of about 2,800 ha (about 7,000 acres) and vary from 4 to 720 ha (10 to 1,800 acres) (Tanji and Dahlgren, 1990). The evaporation ponds annually receive about 3,900 ha-m (31,600 acre-feet) of subsurface drainage from some 22,400 ha (55,350 acres) of tile-drained lands. The drainage waters discharged into the ponds annually contain about 0.72 million metric tons (0.8 million tons) of dissolved mineral salts, which is equivalent to about 25 percent of the annual salt accumulation in the west side of the San Joaquin Valley.

The selenium concentrations in the drainage waters disposed into these ponds vary from less than 1 to 610 µg/liter (1 to 610 ppb). The average concentration of selenium disposed into the Kesterson Reservoir was about 300 µg/liter (300 ppb). The evaporation pond facility receiving 610 µg/liter of selenium (610 ppb) was shut down in 1989 because the selenium concentrations in the evapoconcentrating water exceeded 2,000 µg/liter (2,000 ppb), and the water was judged to be hazardous liquid waste.

Selenium poisoning symptoms like those in the Kesterson Reservoir have been detected in several agricultural evaporation ponds that receive influent selenium concentrations at much lower levels (10 to 80 µg/liter [10 to 80 ppb]). The differences in toxicity threshold levels found between the Kesterson Reservoir and the agricultural evaporation ponds are attributed to site-specific conditions. For instance, the Kesterson Reservoir is located in an area surrounded by uncontaminated wetlands where some feeding by waterfowl took place. In contrast, the Tulare Lake Basin ponds are the only surface water bodies present and are the principal habitats and feeding grounds for waterfowl. Because of these new findings on the hazardous nature of evaporation ponds (Skorupa and Ohlendorf, 1991), this practice of drainage water disposal is expected to be severely curtailed for drainage waters containing potentially toxic amounts of trace elements.

## Deep-Well Injection

Deep-well injection, similar to the disposal of waste brine from oil fields, is another option for disposing agricultural drainage waters. A 2,400-m (7,900-feet)-deep well was constructed in the Westland Water District of the San Joaquin Valley to test this option. After drilling, injection tests indicated that the subsurface geologic formation had a substantially lower than expected permeability. A shallower formation with higher permeability is present, but the EPA has not approved this proposed deep-well injection. The service life of an injection well is governed largely by the buildup of water pressure in the aquifer as well as the physical and chemical properties of the aquifer (Lee, 1990). Particulate matter, chemical precipitation, and biological slime at the injection site can cause clogging. Thus, pretreatment appears to be a necessary process for deep-well injection of agricultural drainage waters. The estimated cost for this option is $132 to $172/10$^3$ m$^3$ ($164 to $213/acre-foot).

## Institutional Changes

Institutional changes that can be used to solve drainage water quality problems include (1) land retirement or idling, (2) tiered water pricing, (3) water marketing and transfers, and (4) the use of regional drainage management authorities.

### Land Retirement or Idling

One suggestion made by the San Joaquin Valley Drainage Program (1990) is to cease irrigation of areas where shallow groundwaters contain elevated levels of selenium and where it is difficult to drain the soils. Such lands could be permanently retired or idled until some future date. Conceptually, this option may be an attractive relative to other alternatives of source control, treatment, or disposal. However, specific criteria for land retirement have not yet been developed, and only preliminary economic analyses have been carried out.

### Tiered Water Pricing

Tiered water pricing or increasing block-rate prices for irrigation water may serve as a motivation to reduce the amount of drainage water. For instance, the Broadview Water District in the San Joaquin Valley implemented such a program in 1989 on a trial basis. Prior to this new rate structure, water was sold at $13/10$^3$ m$^3$ ($16/acre-foot). Crop-specific average water application rates were established, and any additional use above these basic rates were charged at $32/10$^3$ m$^3$ ($40/acre-foot). The district offered real-time weather data to growers so that they could better schedule irrigation of their crops. The estimated reduction in drainage water was about 23 percent (Wichelns, 1991). Part of this reduction may be attributed to a drought-related decrease in the water supply.

## Water Marketing and Transfer

Water marketing and transfer may provide, in some instances, an economic incentive to irrigators to consider off-farm uses of water. It requires a clear arrangement between the buyer and seller of the rights to transfer water for a limited period or longer. Such voluntary transfer of water, however, may create third-party impacts that have not been studied adequately. In California's fifth year of drought (1991), the California Department of Water Resources actively solicited water and was successful in arranging about 49,000 ha-m (400,000 acre-feet) of one-time water transfers to drought-stricken urban and agricultural water users. Water marketing for environmental benefits has been somewhat limited. For instance, 3,700 ha-m (30,000 acre-feet) of fresh water was purchased in 1989 by the California Department of Fish and Game and the Grasslands Water District from the U.S. Bureau of Reclamation to maintain wildlife and fish (Robert Potter, deputy director, California Department of Water Resources, personal communication, 1991). Existing state and federal laws relative to water rights may need to be reassessed to promote water marketing and transfers.

## Regional Drainage Management Authorities

Formation of regional drainage management authorities may aid in regulating drainage water production through, for instance, penalty costs on drainage water or subsidies for drainage water reductions. It is not unusual, however, to have an entity that delivers irrigation water and another entity that manages drainage water in a given region but with different boundaries. There is a need for joint planning and management of irrigation and drainage waters. Some efforts are being made to form drainage management authorities in the San Joaquin Valley.

# 11

## Manure and Nutrient Management

Concern about the impact of nutrient loadings on the environment, that is, the relation of livestock and poultry manure to groundwater contamination, is well documented both inside the United States (Brown et al., 1989; Frink, 1969; Harris, 1987; Lanyon and Beegle, 1989; Madison et al., 1986; Patni and Culley, 1989; Pinkowski et al., 1985; University of Wisconsin-Extension and Wisconsin Department of Agriculture, Trade and Consumer Protection, 1989; U.S. Congress, Office of Technology Assessment, 1990; Walter et al., 1987; Young et al., 1985) and outside the United States (Adams and McAllister, 1975; Phillips et al., 1982; Steenvoorden, 1986; Webster and Goulding, 1989).

### RESOURCE UTILIZATION OR WASTE DISPOSAL

A little more than 50 years ago, animal manures were considered a tremendous asset in providing fertility to U.S. soils. The 1938 yearbook of agriculture stated:

> One billion tons of manure, the annual product of livestock on American farms, is capable of producing $3,000,000,000 worth of increase in crops. The potential value of this agricultural resource is three times that of the Nation's wheat crop and equivalent to $440 for each of the country's 6,800,000 farm operators. The crop nutrients it contains would cost more than six times as much as was expended for commercial fertilizers in 1936. Its organic matter content is double the amount of soil humus annually destroyed in growing the Nation's grain and cotton crops (U.S. Department of Agriculture, 1938:445).

*399*

Today, animal excrements are largely referred to as wastes for disposal rather than as manures for utilization. The change in attitude toward manure has mainly been a result of two factors. First, livestock and poultry production has become concentrated in large-scale, confinement-type enterprises. These include dairy cow operations with hundreds of cows, beef and hog feedlots with thousands of animals, and poultry enterprises with many hundreds of thousands of birds. In 1987, for example, only 7.5 percent of all beef cow-calf producers had herds larger than 100 animals, but these producers produced 47.6 percent of all the calves produced. In 1990, cattle feedlots with capacities of 1,000 head or more represented only 4 percent of the total feedlots, but they fed 84 percent of the cattle that year (Krause, 1992). Such large concentrations of animals or birds have greatly magnified the problems related to handling of wastes, including hazards to human health and aesthetic nuisances. Second, marked improvements in the techniques for making fertilizer from atmospheric nitrogen were made in the period before World War II. During the war, the federal government built numerous plants for the manufacture of fixed nitrogen for munitions. At the war's end, these manufacturing plants became available for making farm fertilizers at relatively low prices. Equally spectacular achievements have been realized in the production of highly effective phosphorus fertilizers from rock phosphates.

## Benefits of Manure Application

Even though fertilizer prices have increased considerably in recent years as a result of increasing energy costs, in many instances they remain lower than the cost of handling animal manures. Therefore, if one looks only at the economic value of manure as a source of plant nutrients, particularly nitrogen, the use of manure may not be competitive. There are other benefits, however, from using animal manures in crop production systems. Continuous and judicious use of manure improves the physical and chemical properties of nearly all soils, particularly those that are shallow, coarse textured, or low in organic matter; and the potential for degradation of the quality of soil, air, and water resources is greatly reduced. More specifically, manure provides essential elements for crop growth. It adds organic matter, it improves soil structure and tilth, and it increases the soil's ability to hold water and nutrients as well as resist compaction and crusting (Madison et al., 1986).

The return of nutrients and organic matter to the soil by manure completes the ancient and natural cycle on which all life depends. Soil fertility—the ability of soil to provide nutrients for plant growth—is enhanced by such judicious returns of nutrients. The composition of manures, however, depends on the kind of animal or bird, the type of feed, storage and handling procedures, climate, and other factors. Nonetheless, soil chemical properties are generally improved by animal manures, unless the manures are added in excess.

### Supply of Manure

The American Society of Agricultural Engineers (1988) has established standard values for estimating the amounts and compositions of manures produced in the United States (Table 11-1). Cross and Byers (1990) used these values together with recent livestock and poultry statistics to estimate the total amount of manure and the economic value of the nutrients in manure produced in the United States (Table 11-2). Beef cattle in extensive (grazing) production systems contribute about seven-eighths of the total beef cattle manure, whereas beef cattle on feed contribute only about one-eighth. For dairy cattle manure, as much as one-half or two-thirds might be voided on pastures, although the trend

TABLE 11-1   Manure and Its Associated Nutrient Content

| Source | Millions of Metric Tons (dry weight) | | | |
|---|---|---|---|---|
| | Total Manure | Nitrogen | Phosphorus | Potassium |
| Cross and Byers (1990) | 143 | 5.9 | 1.8 | 3.7 |
| Van Dyne and Gilbertson (1978) | 102 | 3.7 | 0.9 | 2.2 |
| Committee on Long-Range Soil and Water Quality (1993; this report) | 124 | 5.1 | 1.4 | NR |

NOTE: NR, no data reported.

TABLE 11-2   Economic Value of Nitrogen, Phosphorus, and Potassium in Manures

| Source | Total Manure (millions of metric tons) | Nutrient Value (millions of dollars) | | |
|---|---|---|---|---|
| | | Nitrogen | Phosphorus | Potassium |
| Cross and Byers (1990) | 143 | 1,611 | 668 | 895 |
| Van Dyne and Gilbertson (1978) | 102 | 1,016 | 334 | 524 |
| Committee on Long-Range Soil and Water Quality (1993; this report) | 124 | 1,387 | 498 | NR |

NOTE: NR, no data reported.

to intensive (confinement) production systems is expected to continue. Essentially all poultry are produced in confinement. For intensive animal production systems, the predominant sources of voided manure in terms of solids and nitrogen appear to be dairy cattle, swine, beef cattle, broilers, turkeys, and laying hens.

Van Dyne and Gilbertson (1978) also estimated the amount of manure and nutrients contained in manure produced by livestock and poultry in 3,050 counties in the United States. Their estimates, 102 million, 3.7 million, 0.9 million, and 2.2 million metric tons (112 million, 4.1 million, 1 million, and 2.4 million tons) of manure, nitrogen, phosphorus, and potassium, respectively, on a dry-weight basis, were somewhat lower; but in most regards they were similar to the estimates of Cross and Byers (1990) given in Table 11-1. Van Dyne and Gilbertson (1978) also estimated that losses caused by volatilization, leaching, and runoff would reduce the dry weight by 10 percent, total nitrogen by 36 percent, phosphorus by 5 percent, and potassium by 4 percent. Of the 102 million metric tons (112 million tons) of dry manure produced, Van Dyne and Gilbertson (1978) estimated that 47 million metric tons (52 million tons) was economically recoverable.

The committee also estimated manure and nutrients produced by livestock and poultry as one element of estimated nitrogen (see Chapter 6) and phosphorus (see Chapter 7) mass balances (see the Appendix for a complete discussion of the methods used to estimate manure production) for each state and for the United States as a whole. The economic value of the nutrients available in manure shows the benefits that can be attained by using manure as a resource.

### Manure as Waste

The concentration and specialization of modern crop and livestock production systems has led to a breakdown of manure recycling as it occurred on livestock and crop farms in the past (Walter et al., 1987). Figure 11-1 shows the break in the livestock-crop nutrient cycle. Synthetic fertilizers replace manures as a nutrient source, and manure becomes a waste problem. Often, the least-cost solution to the manure waste problem is to apply it to the land (Walter et al., 1987).

Simply disposing of manure as a waste product can lead to serious degradation of both surface water and groundwater. Likely sources of surface water and groundwater contamination include runoffs and leaching from manure and wastewater applied to the land, open and unpaved feedlots, runoff holding ponds, manure treatment and storage lagoons, and manure stockpiles. Dead animal disposal and animal

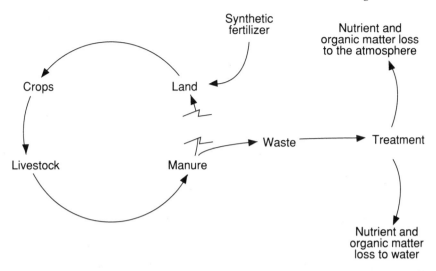

FIGURE 11-1 Schematic of livestock-crop system showing gap in traditional manure recycling system because of use of relatively inexpensive fertilizers. Source: M. F. Walter, T. L. Richard, P. D. Robillard, and R. Muck. 1987. Manure management with conservation tillage. Pp. 253–270 in Effects of Conservation Tillage on Groundwater Quality: Nitrates and Pesticides, T. J. Logan, J. M. Davidson, J. L. Baker, and M. R. Overcash, eds. Chelsea, Mich.: Lewis Publishers, a subsidiary of CRC Press, Boca Raton, Florida. With permission.

dipping vats may contribute to surface water and groundwater contamination. Manure accumulations around livestock watering locations, intermittent-use stock pens, and livestock grazing operations that occur on areas ranging from sparsely grazed rangelands to intensively grazed pastures may also influence surface water and groundwater quality (U.S. Congress, Office of Technology Assessment, 1990).

The presence of constituents such as pathogenic organisms, nitrate, and ammonia in livestock drinking water may adversely affect livestock health (U.S. Congress, Office of Technology Assessment, 1990). The presence of the same constituents in surface water or groundwater drink water supplies may adversely affect human health. Consequently, effective on-farm nutrient management is essential.

## IMPROVING MANURE MANAGEMENT

Technologies are available to improve manure management, and wider use of these technologies could reduce the water pollution caused by misuse of manures. Producers trying to improve the ways that they manage manures face particularly difficult problems that need to be overcome before new technologies can be successfully applied.

## Special Problems in Manure Management

Improvements in manure management can effectively capture the benefits of using manures as inputs to crop production systems and can reduce the water pollution associated with manure disposal. Special problems with the management of manures as production inputs must be overcome if manure management is to be improved. These problems include handling and application costs, estimating nutrient value, determining the amount of manure to be applied, balancing multiple nitrogen and phosphorus applications, concentration of livestock, and the need for storage and handling facilities.

### Handling and Application Costs

High labor requirements for manure handling, the increased travel distances required to spread manure, and reduced opportunities for using manure as a resource on farms have tended to increase the cost of using manure as a nutrient source. Investigators have developed management systems that reduce the amount of labor required to handle manure and that increase its nutrient value. These systems, however, require significant capital investments for the construction of storage and handling facilities and the purchase of application equipment. These capital costs can be an important constraint to the adoption of these systems.

### Difficulty Estimating Nutrient Value of Manures

Nitrogen is often the plant nutrient that first becomes limiting in crop production systems. Much of the nitrogen in manure is in organic form and must be mineralized before it can be used by plants. In contrast to nitrogen, the phosphorus in manures is generally conceded to be as effective as acid-treated forms of inorganic phosphorus, such as superphosphate (a common formulation used in commercial phosphorus fertilizers) (Azevedo and Stout, 1974). Potassium in manure is also considered readily available. Direct losses of nitrogen, phosphorus, and potassium via volatilization, leaching, and runoff are estimated to reduce the nutrient content of manure significantly. The nutrient contents of different manures can vary significantly, making estimation of application rates difficult. Furthermore, not all nutrients in manure are immediately available for crop growth. Improved storage, treatment, and application equipment can reduce these manure utilization problems. The uncertainty in estimating the quantity and availability of nutrients in manures can lead to their overapplication or to the use of supplemental nutrient sources when none are required for crop growth. Manure testing services to estimate the nutrient content of manures are available, however, and should be used to improve manure management.

### Reduced Need for Manure after Repeated Applications

Gilbertson and colleagues (1979) developed a technical guide (Table 11-3) that estimates the amount of manure that must be added to supply 112 kg of available nitrogen per hectare (100 lb/acre). This guide is based on the principles that the percentage of nitrogen in manure that is released in the first year increases with the amount of nitrogen in the manure and that it takes 3 or more years before most of the nitrogen present in manure is mineralized and available to plants. The important point is that when manure is applied to the same field year after year, each succeeding year requires less manure to maintain a supply of 112 kg of plant-available nitrogen per hectare (100 lb/acre).

TABLE 11-3  Quantity of Livestock or Poultry Manure Needed to Supply 100 kg of Nitrogen over the Cropping Year with Repeated Applications of Manure

| Number of Years Applied | Quantity (metric tons) Needed for Manures with the Following Percent Nitrogen | | | |
|---|---|---|---|---|
| | 0.25 | 1.0 | 2.0 | 4.0 |
| 1 | 154.1 | 22.2 | 7.0 | 1.4 |
| 2 | 79.3 | 15.6 | 5.8 | 1.4 |
| 3 | 53.8 | 12.7 | 5.1 | 1.4 |
| 4 | 40.9 | 11.0 | 4.7 | 1.3 |
| 5 | 33.0 | 9.8 | 4.4 | 1.3 |
| 10 | 17.0 | 6.9 | 3.7 | 1.3 |
| 15 | 11.5 | 5.6 | 3.3 | 1.2 |
| 20 | 8.7 | 4.8 | 3.0 | 1.2 |

SOURCE: Derived from J. S. Schepers and R. H. Fox. 1989. Estimation of N budgets for crops. Pp. 221–246 in Nitrogen Management and Ground Water Protection, R. F. Follet, ed. Developments in Agricultural and Managed-Forest Ecology 21. Amsterdam: Elsevier.

*Nitrogen-Phosphorus Trade-Off*

The ratio of nitrogen to phosphorus in fresh manure is generally 3 or 4. Since a significant amount of nitrogen is lost by volatilization, the nitrogen:phosphorus ratio of manure applied to the land is often less than 3. Therefore, when manure is applied at rates sufficient to supply adequate nitrogen for most cropping conditions, excess amounts of phosphorus and potassium are added. For example, Sharpley and colleagues (1984) found that 8 years of continuous manure usage resulted in large accumulations of available phosphorus. Christie (1987) also found significant increases in extractable phosphorus in soil that had been treated with cow or pig manure slurries.

The increased phosphorus contents of surface soil increase the potential for soluble and sediment-bound phosphorus to be transported in runoff and, therefore, has water quality implications, as discussed in Chapter 7. Sharpley and Smith (1989) recently developed an equation that predicts the soluble phosphorus concentration of runoff on the basis of the available phosphorus of the surface soil determined by a soil test.

Van Reimsdijk and colleagues (1987) also warned that spreading of animal manure on land in quantities exceeding the amount of phosphorus taken up by plants results in phosphorus accumulation. They stated that when the cumulative excess becomes large compared with the buffering capacity of the soil, phosphorus could leach to surface water and groundwater, causing eutrophication. (Eutrophication is the process by which a body of water becomes—either naturally or by pollution—rich in dissolved nutrients such as phosphates and, often, becomes seasonally deficient in dissolved oxygen.) They concluded, however, that these negative effects develop only after a relatively long period of large manure applications and that soils differ widely in their buffering capacities.

The potential phosphorus buildup from the use of manure poses a tremendous challenge for managing animal wastes. Historically, manure application rates have mainly been based on nitrogen loading rates, with little attention paid to phosphorus accumulation. However, with the growing environmental concern associated with phosphorus in surface water supplies, pressures are mounting for limiting or even banning phosphorus additions to soils that exceed a certain level of plant-available phosphorus on the basis of a soil test. Such criteria could result in significantly reduced manure application rates or even no manure applications. Michigan recently adopted guidelines indicating that manure additions should be restricted to rates adequate to replace the phosphorus removed by crops once the available phosphorus level determined by a soil test reaches a value of 160.

The use of phosphorus as the criterion for determining manure loading rates may be appropriate, particularly in regions containing surface waters where accelerated eutrophication can occur. In other areas, however, the benefits derived from using manures to enhance overall soil quality and as the primary source of nitrogen for supplementing plant growth may outweigh any potential negative effects associated with increased phosphorus levels. Therefore, risk assessments for specific regions rather than adoption of generic standards across regions should be made.

### Concentration of Livestock

The concentration of livestock production in large confinement feeding operations or regional concentrations of dairy, poultry, or other animal production systems has resulted in situations in which there is simply more manure being produced than can be used efficiently on nearby croplands. Nutrient flows in intensive livestock operations are directly related to outputs as animals and animal products. In contrast to the situation on cash grain farms, the proportion of nutrient input that exits in product output is usually much less for intensive livestock operations. This accumulation causes a net nutrient loading on the farm that may exceed the crop's nutrient needs (Lanyon and Beegle, 1989) and degrade water quality.

According to Young and colleagues (1985), southeastern Pennsylvania, and Lancaster County in particular, are among the most intensively farmed areas in the United States. In Lancaster County, between 1960 and 1986, beef cattle numbers increased by 55 percent, dairy cattle increased by 61 percent, hogs increased by 677 percent, poultry layers and pullets increased by 193 percent, and broilers increased by 504 percent (Lanyon and Beegle, 1989). This production intensity is linked to continuing development pressures on agricultural land coupled with excellent marketing conditions, in that one-third of all U.S. consumers live within 124 km (200 miles) of Lancaster County. This situation threatens local and regional environments because southeastern Pennsylvania agricultural land is a major source of the nutrients and pesticides that enter the Chesapeake Bay (Young et al., 1985).

The volume of manure produced in areas where livestock production is concentrated may well exceed the area of cropland available on which to apply manures at rates that minimize the potential for surface water or groundwater degradation. For example, 0.9 metric ton (1 ton) of solid dairy cattle manure contains approximately 4.5 kg (10 lb) of

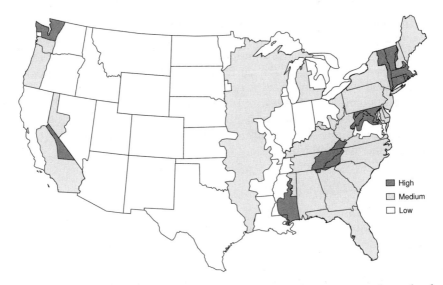

FIGURE 11-2 Ratio of amount of manure produced to amount of cropland available for manure application. Source: U.S. Department of Agriculture, Soil Conservation Service. 1989. Water Quality Indicator Guides: Surface Waters. Report No. SCS-TP-161. Washington, D.C.: U.S. Department of Agriculture.

nitrogen, 2.3 kg (5 lb) of phosphate ($P_2O_5$), and 4.5 kg (10 lb) of potash ($K_2O$) (Madison et al., 1986). If a 635-kg (1,400-lb) dairy cow, for example, produces 50 to 57 kg (110 to 125 lb) of manure and bedding daily, a 50-cow herd will produce about 454 metric tons (500 tons) of manure daily. At a maximum application rate of 56 metric tons/ha (25 tons/acre), 8 ha (20 acres) of cropland is required. Additional lands may be required to spread manure from calves, heifers, steers, or other livestock (Madison et al., 1986). Figure 11-2 illustrates the problem of manure production and land application.

Van Dyne and Gilbertson (1978) also estimated the ratio of manure production to land area in the United States. The average ratio of economically recoverable manure weight to cropland area and improved pasture averaged only 0.27 metric tons of manure per hectare (0.12 tons/acre) nationally across all 3,050 counties of the United States. The range was from less than 0.02 metric dry tons/ha (0.01 dry tons/acre) to a high of 4.0 metric dry tons/ha (1.8 dry tons/acre). Although there is adequate cropland in all areas of the United States to receive manure, it is often not in the immediate vicinity of the animal manure source or under the control of the livestock producer. As a result, these facilities often stockpile wastes or apply it to

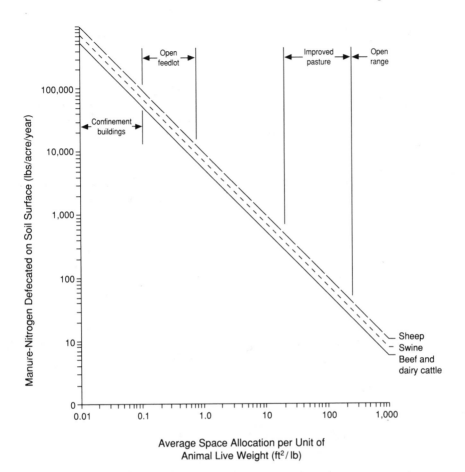

FIGURE 11-3 Average amount of manure nitrogen produced by animals per unit area in relation to animal spacing. Source: Adapted from U.S. Congress, Office of Technology Assessment. 1990. Technologies to improve nutrient and pest management. Pp. 81–167 in Beneath the Bottom Line: Agricultural Approaches to Reduce Agrichemical Contamination of Groundwater. Report No. OTA-F-418. Washington, D.C.: U.S. Government Printing Office.

croplands at rates in excess of those required for maintaining soil quality. Figure 11-3 shows manure production and nitrogen concentration (on an as-voided basis) within various intensive animal production systems versus extensive livestock production systems as a function of animal density and spacing per unit live weight (U.S. Congress, Office of Technology Assessment, 1990).

### Storage and Handling Facilities

Off-site utilization of manures is constrained because of the high costs associated with the facts that both semisolid and liquid manures are bulky material to handle, the use of manure in an environmentally acceptable manner limits the application time period, and sufficient hauling equipment must be available during the limited application time period (the equipment then goes unused for the remainder of the year) (Young et al., 1985).

Although some manure transactions occur now, there is not a flourishing market for manure, and the potential for marketing excess manure nutrients with a positive return to farmers is limited. Considering the volume of the excess manure on farms in southeastern Pennsylvania, Young and colleagues (1985) concluded there is no reason to anticipate improved circumstances for manure marketing with positive returns to farmers. The trend of increasing farm animal numbers adds to the dilemma.

Manure that is collectible is concentrated geographically. Gilbertson and colleagues (1979) mapped areas where manure from livestock and poultry can be collected and spread economically. There appear to be three geographic areas of special interest: (1) New York, Pennsylvania, and Vermont; (2) Wisconsin, Iowa, southern Minnesota, northern Illinois, eastern South Dakota, and eastern Nebraska; and (3) southern California and New Mexico (Walter et al., 1987).

## Opportunities for Improvement

For management purposes, animal wastes can be divided into point and nonpoint sources. Point sources, such as feedlots and other confinement facilities, are regulated by the U.S. Environmental Protection Agency (EPA). Point source animal waste management problems must be addressed in terms of existing situations as well as expanding and new facilities. It is essential that environmentally sound waste management plans be developed during the planning of all new and expanding confinement feeding operations. Improving existing facilities to meet regulatory standards presents a greater challenge.

Nonpoint sources of livestock manure are characterized by diffuse runoff from areas such as feeding and watering sites. In most cases, these sites are not regulated, do not produce collectable manure, and are manageable through proper site selection away from streams, soil erosion control, cover cropping, and use of vegetative filters to minimize transport of potential contaminants to streams (Sweeten, 1991). Manure-

Using a flush-clean system, cattle manure is washed from the confinement area to underground concrete tanks. After screening, the manure wash will be pumped onto fields with irrigation water. Credit: Agricultural Research Service, USDA.

fertilized pastures and croplands are also regarded as nonpoint sources of nitrogen and phosphorus pollution. With the increased national concern about the effects of nonpoint sources of pollution on water quality, increasing attention will be given to how and when animal wastes are applied to pastures and croplands and at what rates. Nutrient management plans, tailored for local conditions, are essential for safeguarding the soil and water resources over the long term.

### Point Source Control

The initial concern with confinement feeding operations was fish kills associated with the runoff that enters streams and lakes. Therefore, regulations and policies focused on controlling runoffs from those operations that were considered point sources.

Technologies or best-management practices for water quality protection from concentrated animal feeding operations have been well developed and are widely implemented (Sweeten, 1991). These point sources are directly regulated by the EPA and/or state agencies, with the basic requirement of no discharge, that is, containment and proper disposal of all manure, wastewater, and runoff for up to 25 years and for a 24-hour-duration storm event.

For purposes of water pollution control, intensive livestock production systems are defined by EPA regulations for feedlots as

> animal feeding operations [where animals are] stabled or confined and fed or maintained for a total of 45 days or more in any 2-month period, and . . . crops, vegetation, forage growth or post harvest residues are not sustained in the normal growing season over any portion of the lot or facility (U.S. Environmental Protection Agency, 1976:11,460, as cited by U.S. Congress, Office of Technology Assessment, 1990).

This definition covers many animal species, types of facilities, animal densities, climates, and soils. It uses a single, visually determined criterion—that is, the absence of vegetation. Under such conditions, manure production and animal traffic are of sufficient quantity and duration to prevent germination or growth of forage. This condition implies that runoff, volatilization, and leaching pathways may be proportionately larger from unvegetated surfaces than from vegetated surfaces (U.S. Congress, Office of Technology Assessment, 1990). Consequently, the U.S. Congress, Office of Technology Assessment (1990) has concluded that EPA regulations for confined livestock and poultry operations deal with surface water protection and do not include requirements for groundwater protection.

Several states and local entities do have groundwater protection requirements. For example, the Texas Water Commission regulation that governs confined, concentrated livestock and poultry feeding operations includes groundwater protection for lagoons and holding ponds. The regulation requires that all wastewater retention facilities be constructed of compacted, low-permeability soils (for example, a clay or clay loam) at a minimum thickness of 30 cm (12 inches) (U.S. Congress, Office of Technology Assessment, 1990).

Sweeten (1991) lists the following best-management practices as appropriate for achieving a no-discharge system:

- proper site selection;
- selection of appropriate types of facilities with respect to climate, topography, geology, soils, land resource base, land use, and proximity to surface water or groundwater or neighbors;
- reduced sizes of open feedlots, diversion of runoff outside the feedlot, covered manure storage facilities, and installation of roof gutters on feedlot buildings;
- use of frequent dry scraping, debris basins, or screen separators;

- use of lagoons, holding ponds, or storage tanks or pits for liquid manure and/or runoff;
- sealing or lining of earthen storage and treatment structures for groundwater protection, subject to permeability testing;
- adequate systems for manure and wastewater distribution on croplands or pasturelands;
- dead animal disposal in clay-lined dry pits or by composting and utilization; and
- land application at rates consistent with crop production and water quality goals.

*Nonpoint Source Control*

Control of nonpoint sources of animal wastes requires improvements in manure management in smaller and diverse livestock operations. Effective manure management strategies have been determined (University of Wisconsin-Extension and Wisconsin Department of Agriculture, Trade and Consumer Protection, 1989) and should include the following:

- awareness of the nutrient value of manure;
- manure analysis;
- proper crediting of nutrients;
- appropriate application methods, rates, and timing;
- site considerations;
- manure storage;
- effective and efficient disposal of animal wastes; and
- designated cattle lanes and fencing.

Although some technologies and best-management practices address all of these objectives, additional efforts are needed to promote the development and adoption of such practices. Of major importance is the need to develop and extend economic guidance for land application of manures, including soil and manure testing to define appropriate application rates and information about nutrient release rates to allow efficient and economically viable use of manures. These efforts must include quantification of the magnitude of nutrient losses from lagoons, storage tanks, and land application as a function of design, operation, and climatic variables to develop nutrient management plans and nutrient mass balance models.

Increases in the agronomic uses of manure might be fostered through joint efforts among states, cities, industry, and agriculture to promote manure processing and use on public and private lands. Development of incentives for manure use in cropping systems, particularly in areas with high levels of manure production, may offer an opportunity to enhance the agronomic use of this resource as opposed to treating it as a waste disposal problem.

Federal and state programs that include cost-sharing or other economic incentives that encourage livestock producers to adopt and implement water quality protection practices, particularly in areas where water is most vulnerable, could promote the adoption of such incentives. Technical assistance (provided by the Soil Conservation Service of the U.S. Department of Agriculture [USDA]), education (provided by the Cooperative Extension Service of USDA), and research (supported by the Agricultural Research Service of USDA) must be able to promote and support the adoption of water quality protection practices by producers. For example, demonstration livestock production operations in areas with high or low pollution potentials for groundwater could serve to disseminate information on appropriate best management practices that contain provisions for groundwater protection (U.S. Congress, Office of Technology Assessment, 1990).

Efforts to increase the adoption of improved manure management practices may face important economic and social obstacles. The effects of on-farm nutrient loss abatement practices on farm income may be great. Moderate nutrient loss reductions may be achieved with negligible effects on farm profits. Substantial reductions in nutrient losses, however, could have significant effects. Without taxpayer assistance, measures to reduce nutrient losses significantly could impose financial hardships on producers (Young et al., 1985). The economic farm model of Young and colleagues (1985) determined that nutrient losses could be reduced about 10 percent with negligible impacts on net farm economic returns by using manure storage, the more even application of manure on croplands, and changes in the intensity of crop rotation. Given the fact, however, that the additional income generated by adding a cow exceeds the additional expenses associated with disposing of the extra manure when society bears the pollution impact, overapplication of manure nutrients is economically rational (Young et al., 1985).

Facilitating change in management techniques can be a slow process. The rate of adoption of various practices is a function of many factors (Harris, 1987). Knowing that nutrient losses from farms must be reduced to achieve water quality improvements and that achieving such reductions is technically feasible is only the first step toward a sound nutrient management program. Social and economic information will be needed to determine what kinds of implementation policies are needed to obtain nutrient reductions in targeted watersheds in an efficient and equitable manner while giving due consideration to the taxpayers and the heritage of the area. These issues can be resolved only with additional social and economic research to complement and augment ongoing agronomic, biological, and hydrologic research (Young et al., 1985).

### Alternative Uses of Manure

There may be some promise of developing alternative products from animal manures. Livestock and poultry manures generated from concentrated and confined animal feeding facilities may be valuable sources of fertilizer, feedstuff, or fuel. Manure is widely used as an organic fertilizer in many areas. Certain types of manure also may receive limited use in specialized situations as animal feedstuffs, as a substrate for anaerobic digestion to produce methane gas, or as a fuel for combustion or gasification for electric power generation.

These alternatives are not without problems. Some alternative uses return all or part of the original manure fertilizer value as a residue that eventually is applied to land. Power generation requires economies of scale for it to be profitable, and refeeding of manure causes livestock health problems if it is done at high levels. Preliminary evaluations of these alternatives indicate that they would be expensive to implement at a scale sufficient to solve the excess nutrient problems in Lancaster County, for example. Moreover, care should be taken when evaluating alternatives for off-site manure disposal subsidized by public funds. Reducing manure disposal costs merely makes animal production more profitable. Thus, farmers are encouraged to expand their operations, further aggravating the problem. Animal numbers should not be increased unless the manure can be used in an environmentally acceptable manner (Young et al., 1985).

Alternative useful products can be produced from manure, and these products can be exported from the producing farm. However, increased development of manure management treatment and use technologies, particularly in relation to composting, methane gas generation, thermochemical conversion, fiber recovery, and marketing of such products, will be required to take advantage of these opportunities.

# 12

# A Landscape Approach to Agricultural Nonpoint Source Pollution

Most programs used to control agricultural nonpoint source pollution focus on in-field best-management practices, but there is a growing interest in the use of off-field control techniques (Clausen and Meals, 1989). The most commonly used off-field control practices are vegetative filter strips and riparian buffer zones. Vegetative filter strips are narrow strips of managed grassland situated directly adjacent to agricultural fields (Dillaha et al., 1989b). Riparian buffer zones are usually areas of natural forest vegetation situated between cropped areas and streams (Lowrance et al., 1984a). Interest in both of these practices has increased dramatically in recent years.

A focus on off-field controls changes the unit of analysis for physical and social science questions from the field to the landscape scale. Whereas nonpoint source pollution best-management practices and most agricultural policy instruments are directed toward activities that occur within agricultural fields, analysis of off-field nonpoint source pollution controls requires consideration of the interaction of crop fields with adjacent managed or unmanaged ecosystems and how those interactions affect water quality over an area larger than a specific crop field.

The emerging field of landscape ecology provides a conceptual basis for landscape analysis of agricultural nonpoint source pollutant problems (Forman and Godron, 1986). Landscape analysis considers the spatial juxtaposition and dynamic interactions between agricultural and adjacent ecosystems in the context of the water quality of the landscape as a single unit, for example, a watershed or groundwater recharge zone. This chapter describes the conceptual and practical bases for landscape analysis of agricultural nonpoint source pollution and discusses options for and obstacles to implementing this approach.

## NONPOINT SOURCE POLLUTANT ATTENUATION MECHANISMS

The basis for a landscape approach to agricultural nonpoint source pollutants is the use of particular areas as sinks for pollutants moving off agricultural fields. These sinks must be capable of intercepting the pollutants in either surface water runoffs and/or groundwater flows (Figure 12-1) and must support one or more of the processes that remove pollutants. These processes include plant and microbial uptake of nutrients and trace metals, microbial degradation of organic compounds, sediment trapping, microbial conversion of nitrate into nitrogen gas, and physical and chemical adsorption of metals and organic compounds.

Planning, implementation, and evaluation of the use of landscape sinks for nonpoint source pollutant control must consider two key factors: (1) the capability of a particular area to intercept surface water- and/or groundwater-borne pollutants and (2) the activities of different pollutant removal processes. Analysis of these factors is relevant in several contexts including field-scale development of specific off-field control practices such as grass vegetative filter strips, farm-scale analysis of where off-field controls should be established, and watershed-scale analysis of the effectiveness of existing sink areas such as riparian areas or wetlands.

### Sediment Trapping

Extensive research has demonstrated that grass vegetative filter strips have high sediment-trapping efficiencies if the flow is shallow and the vegetative filter strips are not filled with sediment. Trapping efficiency has been found to decrease dramatically at high runoff rates (Barfield et al., 1979; Schwer and Clausen, 1989). Several short-term experimental studies have reported the effectiveness of grass filter strips in reducing the amounts of sediments in runoffs (Dillaha et al., 1988; Magette et al., 1987; Young et al., 1980). These short-term studies found that grass filter strips are effective at removing sediments and sediment-bound pollutants at trapping efficiencies that exceed 50 percent if the flow is shallow (shallow flow refers to water flowing in sheets across a field or grass filter strip). Grass filter strip

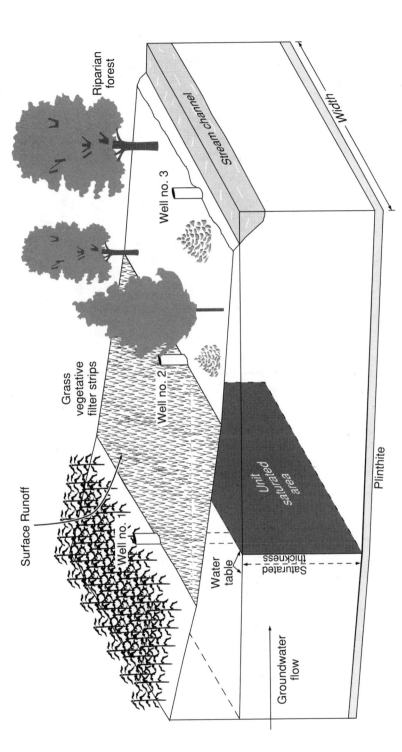

FIGURE 12-1 Conceptual diagram of a landscape showing potential for grass vegetative filter strips and riparian buffer zones to intercept nonpoint source pollutants transported by surface water runoff and groundwater flow. Source: Adapted from R. R. Lowrance, R. L. Todd, and L. E. Rasmussen. 1984b. Nutrient cycling in an agricultural watershed. I. Phreatic movement. Journal of Environmental Quality 13:22–27. Reprinted with permission from © American Society for Agronomy, Crop Science Society of America, and Soil Science Society of America.

Switchgrass in a laboratory test channel gives agricultural engineer and agronomist a chance to measure the grass's sediment trapping capability in a controlled environment. Credit: Agricultural Research Service, USDA.

plots with concentrated flows (concentrated flow refers to water flowing in channels across fields or grass filter strips) similar to those expected under field conditions were reported to be much less effective than the experimental plots with shallow flows used in most vegetative filter strip research (Dillaha et al., 1989b).

Dillaha and colleagues (1989a) studied existing grass filter strips on 18 farms in Virginia and found them to be extremely variable in their effectiveness at removing sediments. Most grass filter strips in hilly areas were ineffective because runoff usually crossed the strip as a concentrated flow. In flatter regions, grass filter strips were more effective because slopes were more uniform and more runoff entered the strip as a shallow flow. Several 1- to 3-year-old vegetative filter strips were observed to have trapped so much sediment that they produced more sediments than adjacent upland fields. In these cases, runoff flowed parallel to the vegetative filter strip until it reached a low point, where it crossed the vegetative filter strip as a concentrated flow. These vegetative filter strips needed maintenance to regain their sediment-trapping abilities.

Several models have been developed for vegetative filter strip design and evaluation. GRASSF is an event-based model developed for design-ing vegetative filter strips with respect to sediment removal (Barfield et al., 1979; Hayes et al., 1979). The model was evaluated by using plot data from multiple events; and predicted values were in good agreement with observed values, even though the model does not consider deposition in the water ponded upslope of the grass strip, which is where most deposition occurs in grass filter strips (Hayes and Hairston, 1983). By neglecting the deposition of sediment in the upslope ponded water, a model tends to underpredict the trapping capability of the strip. GRAPH (Lee et al., 1989), a derivative of the GRASSF model, simulates nutrient transport in vegetative filter strips. GRAPH considers the effects of advection (transport by water flows) as well as adsorption and desorption processes. The model also considers the effects of changes in sediment size distribution and the chemical transport processes in vegetative filter strips.

The chemicals, runoff, and erosion from agricultural management systems (CREAMS) model can also be used to evaluate the trapping of sediment by grass filter strips from overland and concentrated flow (Williams and Nicks, 1988) and from deposition where the upper edge of a vegetative filter strip has redirected runoff from overland to concen-trated flow. Flanagan et al. (1989) derived simplified equations to design grass filter strips and compared estimates from their equations and from the CREAMS model with observed data. In both cases, they found good agreement. An important advantage of the CREAMS model is the ability to compute the trapping of sediment in the ponded water created by a filter strip placed across an area where concentrated flow is occurring. RUSLE, the revised universal soil loss equation, also computes the effect of grass strips on erosion and sediment yield.

If grass filter strips are so narrow that the strips completely fill with deposited sediment, CREAMS overestimates the trapping of sediment because the model does not account for sediment deposited in the grass strip. However, most filter strips used where overflow is occurring are usually wide enough that the width of the grass strip is not a factor in the amount of sediment that is trapped. The critical factor is how well the model reflects the reduced transport capacity that is created when the grass strip reduces the velocity of the runoff water.

Riparian forest buffer zones have the ability to absorb as many or more sediments than grass filter strips (Cooper et al., 1987; Lowrance et al., 1988). As with grass filter strips, concentrated flow, sediment accumu-lation, and buffer zone disturbances can reduce the sediment-trapping ability or cause the accumulated sediments to be released from riparian forest buffer zones.

## Plant Uptake

Plants can effectively take up nutrient elements such as nitrate and phosphate and can also absorb many heavy metals such as lead, cadmium, copper, and zinc. The importance of plants as a nonpoint source pollutant sink depends on their ability to absorb the nutrients moving in either surface or subsurface water flows. Pollutants moving in groundwater are accessible to plants only when the water table is high in the soil profile, such as in wetlands. In these situations, plants can be an important sink for groundwater-borne pollutants (Ehrenfield, 1987). Plant roots may not be able to take up nutrients if water flow is too rapid. The nutrients in water moving across the soil surface in a concentrated flow or percolating rapidly through soil macropores may not be susceptible to plant uptake. Large rainfall events—which often transport a very high percentage of nonpoint source pollutants—readily produce concentrated surface flows and macropore-dominated percolation.

Plant species differ in their abilities to take up different pollutants and in the rate at which uptake occurs. There is ample opportunity to manage the plant community in vegetative filter strips, such as through selection of appropriate grass species and harvesting, which removes the accumulated pollutants (Brown and Thomas, 1978). In riparian forests, management of the plant community is more difficult, but it can be accomplished through selective cutting and replanting. Actively managing riparian forests could substantially increase their effectiveness in preventing water pollution. Selection of the optimal grass species for vegetative filter strips and development of management plans for riparian forests are major topics of research at several locations in the United States.

### Seasonal Dynamics

Although some plants are able to take up pollutants, they may not be reliable long-term sinks for pollutants in the landscape. Uptake of pollutants by plants necessarily declines or stops during the winter, which is often when most movement of surface water and groundwater from upland areas toward water bodies occurs. The use of a mixture of plant materials (for example, cool and warm season grasses) can extend the period of plant activity, but in many areas, a significant dormant season is inevitable.

A major concern with uptake of pollutants by plants is that the nutrients trapped in plant tissues can later be released back into the soil solution as these tissues decompose. Storage of pollutants in the structural tissues of trees represents a relatively long-term attenuation, but it still does not result in removal of pollutants from the ecosystem. The nutrients released from decomposing plant tissues may be attenuated by microbial, physical, or chemical pollutant mechanisms in surface soils. Release of pollutants by decomposition may be beneficial if the vegetation removed the nutrients from groundwater, where the potential for attenuation is often quite low.

### Temporal Dynamics

In addition to seasonal dynamics, longer-term temporal dynamics affect the ability of plants to act as pollutant sinks. Over time, plants in a riparian buffer zone or vegetative filter strip can become "saturated" in their ability to absorb nutrients, resulting in a decline in their absorption capacity (Aber et al., 1989). Although most plants show marked growth responses to nitrogen and phosphorus, after a period of high input, other nutrients become limiting and nitrogen and phosphorus are no longer absorbed. Unless there is some type of nutrient removal through harvesting, saturation will likely occur at some time.

## Microbial Processes

Several microbial processes can serve to attenuate nonpoint source pollutants in different components of the landscape. Like plants, microbes can take up or "immobilize" nutrients and metals in their tissues to support growth. As in plants, this immobilization is reversible, and the accumulated pollutants can be released upon microbial death and decomposition. Since microbial turnover is quite rapid (Paul and Clark, 1989), immobilization in microbes is likely not a significant long-term sink for pollutants in the landscape.

Microorganisms have the ability to degrade organic compounds such as pesticides and many pesticides are highly susceptible to microbial degradation. Landscape areas that support high levels of organic matter and microbial populations (for example, vegetative filter strips or riparian forests and wetlands) may be important sites for degradation of the compounds that leave agricultural fields.

Denitrification is a microbial process that converts nitrate, the most common form of nitrogen leaving agricultural fields, into nitrogen gas. This process occurs under anaerobic conditions and has been found to be an important nitrate attenuation mechanism in the surface soils of wet riparian forests (Ambus and Lowrance, 1991; Jacobs and Gilliam,

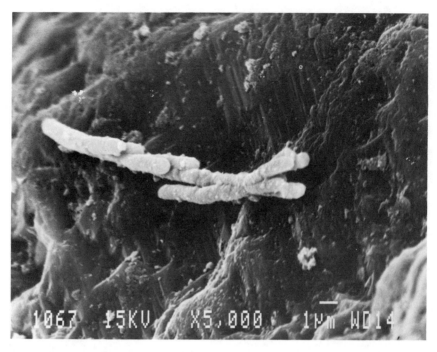

Pesticide-degrading bacteria on the surface of a grain of sand. Credit: Agricultural Research Service, USDA.

1985; Lowrance et al., 1984b; Peterjohn and Correll, 1984) and, to a lesser extent, in vegetative filter strips (Groffman et al., 1991).

Investigators have expressed great interest in assessing the potential for denitrification in groundwater. Although some studies have found significant potential for denitrification in groundwater (Slater and Capone, 1987; Smith and Duff, 1988; Trudell et al., 1986), others have found little or no denitrification activity (Parkin and Meisinger, 1989). A key question relates to the availability of carbon to support microbial activity in the subsurface (Obenhuber and Lowrance, 1991). Groundwater-borne nitrate beneath wet riparian zones may be subject to attenuation more than the nitrate in groundwater beneath more upland areas is, since water tables are high in wetlands, allowing groundwater-borne nitrates to interact with the biologically active zone of the soil with high levels of carbon.

A major mechanism of nitrogen loss in the surface soils of riparian wetlands may be denitrification of the nitrate removed from groundwater by plants (Groffman et al., 1992; Lowrance, 1992b). Uptake of nitrate from groundwater by plants can lead to increases in the amount of nitrogen in plant litter (Lowrance et al., 1984a). High nitrogen levels in plant litter can lead to increases in nitrogen mineralization and the amount of available nitrate in surface soils. This nitrate is then subject to denitrification.

## Adsorption

Several physical and chemical adsorption processes in soil attenuate pollutants in vegetative filter strips, riparian buffer zones, and other landscape sink areas. These processes include the attraction of cations to negatively charged sites on clay and organic matter, chemical binding of organic compounds on clay and organic matter, and physical fixation of the ions within clay minerals. Adsorption processes are controlled by the amount of clay and organic matter in the soil.

## PROCESS-PLACE INTERACTIONS

Two key factors control the effectiveness of landscape-scale nonpoint source pollutant sinks: (1) the capability of a particular area to intercept surface water- or groundwater-borne pollutants and (2) the ability of a particular area to support different pollutant removal processes. By considering the discussion of attenuation mechanisms presented above, it becomes possible to identify those key landscape components that may be effective as nutrient sinks.

Riparian forests, especially those dominated by wetlands, are major potential pollutant sinks. Because of their physical position in the landscape, they can intercept a high percentage of the surface runoff and groundwater flow that moves from upland areas before it reaches streams. Wetland areas have a unique ability to interact with groundwater because the water tables in these areas are close to the soil surface, allowing for the interaction of roots and microorganisms with groundwater-borne pollutants. In addition, riparian areas have the potential to attenuate many pollutants. In most naturally vegetated areas, plant uptake is vigorous and soil organic matter levels are high, which increases the potential for microbial attenuation and chemical adsorption processes. For these reasons, riparian areas have justifiably been the focus of much research on landscape-scale sinks of agricultural nonpoint source pollutants.

A considerable body of evidence confirms that riparian forests can be effective sinks for agriculturally derived nonpoint source pollutants. Several studies have found that strips of riparian forest vegetation are important for maintaining stream water quality in areas where uplands are formed intensively (Groffman et al., 1992; Jacobs and Gilliam, 1985; Karr and Schlosser, 1978; Lowrance et al., 1984b; Peterjohn and Correll, 1984; Simmons et al., 1992). These studies have included comparisons of watersheds with and without riparian vegetation as well as process-level studies of pollutant removal from riparian soils and vegetation.

Investigators have several uncertainties about the performance of riparian forests as nonpoint source pollutant filters. Regional variations in their effectiveness may be important. Although a relatively large body of research suggests that these areas are effective in the Southeast (Jacobs and Gilliam, 1985; Lowrance et al., 1984b; Peterjohn and Correll, 1984), fewer data have been collected for other parts of the United States. Several studies have suggested that riparian zones are effective in the Corn Belt (Huggins et al., 1990; Kovacic et al., 1991; Schlosser and Karr, 1978) and Northeast (Simmons et al., 1992), but more data for these regions need to be collected.

A major concern with riparian zones is their long-term effectiveness. Over time, the ability of these areas to absorb sediments and nutrients may decline. Sites for sediment trapping may become filled, and the capacity of plants and microorganisms to take up nutrients may become saturated. The accumulated pollutants may be released if riparian areas are disturbed by logging, fire, windstorms, or flooding. Investigators now need to research riparian zones that have been absorbing pollutants from uplands for many years.

Given what investigators know about pollutant attenuation processes, predicting the value of upland grass filter strips as pollutant sinks is more problematic than predicting that of riparian forests. The plant communities in filter strips and riparian areas potentially could be managed intensively, but plants may be the least reliable pollutant attenuation mechanism. Upland areas generally have relatively low organic matter levels, so microbial sinks in upland areas are usually less vigorous than they are in riparian zones. There is little potential for upland grass filter strips to interact with groundwater. The main advantage of vegetative filter strips is their proximity to agricultural fields and the potential for aggressive management of the plant community and the physical condition of the vegetative filter strip.

## IMPLEMENTING A LANDSCAPE APPROACH

Making full use of off-field nonpoint source pollutant control mechanisms presents distinct challenges to both physical and social scientists. For example, determining which areas must be placed into vegetative filter strips or riparian buffer zones and how these areas should be managed is a challenge to physical scientists. Development of policy instruments that can be used to implement these changes is a challenge to social scientists. Evaluation of the performance of new policies and practices is a challenge to both groups.

For field-scale applications, the U.S. Forest Service of the U.S. Department of Agriculture (USDA) has produced specific guidelines for riparian buffer zone planning, design, and maintenance (Welsch, 1991). The guidelines call for three zones on which different management practices are used. Zone 1, nearest the stream, consists of undisturbed forest, mostly for stream habitat protection. Zone 2 consists of managed forest and is the site of most pollutant removal activity. Zone 3 is a grassed area, with management that differs with upland land use. If the upland land use is row crops, zone 3 may contain water bars (small dams placed to slow and redirect runoff), spreaders, or other devices to convert concentrated runoff into a dispersed flow. If the upland land use is pasture, management of zone 3 is less intense, with controlled grazing permitted under certain conditions.

At the landscape scale, determination of areas to be used for vegetative filter strips or riparian buffer zones must be based on scientific evaluations of how effective these areas are likely to be for pollutant control. Moreover, unless off-field control practices are instituted in a systematic way across a given landscape unit, overall landscape water quality improvements will be minimal (Phillips, 1989). The challenge, then, is to identify key areas that need to be managed for their pollutant control value and to develop policy instruments to facilitate land use changes.

Approaches such as USDA's Conservation Reserve Program will have to be substantially restructured to increase the production of riparian areas. Although land suitable for buffer zones is eligible for enrollment in the Conservation Reserve Program, most producers prefer to enroll whole fields. Figure 12-2 points out both the problems with land retirement on a field-by-field basis and the potential for more effective use of programs such as the Conservation Reserve Program. Programs like the Conservation Reserve Program will not produce the necessary land use changes unless water quality problems and riparian areas are clearly stated priorities (Dillaha et al., 1989a).

Conversion of different types of land into different types of pollutant sinks will require a diverse set of policy instruments. Implementation of field-scale vegetative filter strips is the most conceptually straightforward approach. For example, if investigators determine that all fields that exceed a certain length or slope criterion must have a vegetative

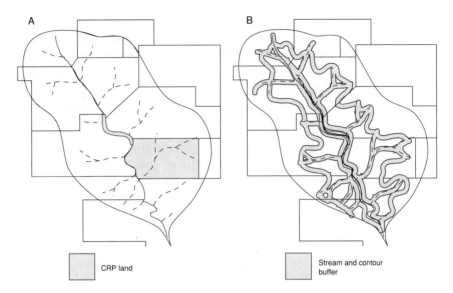

FIGURE 12-2   Conceptual diagram comparing (A) cropland enrolled by field in the Conservation Reserve Program (CRP) with (B) the same area of land set aside in riparian buffer zones.

filter strip at the field's edge, then voluntary or mandatory instruments can be devised to motivate the land manager to install the vegetative filter strip. This type of implementation may require the loss of productive land.

Implementation of riparian buffer zones on a landscape scale will be more challenging. Since these areas are necessarily linked to certain water bodies, designation of the areas to be maintained as buffer zones will be based on watershed delineations that likely are not consistent with land ownership patterns or governmental jurisdictions. A major problem will arise from the fact that not all lands in a watershed will have riparian forests. It is important to consider situations in which one land user owns lands with riparian forest that filters pollutants that run off the lands of many upland land users. A system could be worked out whereby producers in areas without riparian compensate producers that establish or protect riparian areas for the loss of land necessary to meet watershed-wide water quality goals.

# APPENDIX

# Nitrogen and Phosphorus Mass Balances: Methods and Interpretation

The flux of nutrients through an agroecosystem is an important determinant of the productivity of the farming system and the potential for water pollution from losses of nitrogen and phosphorus. Mass balances can be used to assess the transformations and transfers that occur in and between components of the farming system and to assess the efficiencies of nutrient use in the system. Often such assessments point out where further study or quantification of transformations or transfers is needed.

In most natural ecosystems, nutrient inputs and outputs are limited and most nutrients are cycled and recycled through the system. External inputs and overall exports or losses are minimal, and inputs and outputs remain in relative balance. In modern agricultural systems, however, the amount of external inputs, such as fertilizer, added to the system is very large to enable large outputs of the crops that are harvested for food and fiber.

Nutrient budgets and mass balances have been approached on various scales, ranging from experimental plots to estimates of global balances (i.e., Follett et al., 1987; Hauck and Tanji, 1982; Meisinger and Randall, 1991; Power, 1981; Thomas and Gilliam, 1978). Nutrient budgets can be approached at various levels of detail and with varying degrees of completeness. Some budgets even attempt to characterize the inputs, storage, and processing by insects and soil and plant microbiota (Stinner et al., 1984).

The general form of a nutrient budget can be viewed, in simple fashion, as a simple equation: the nutrient *inputs* to the ecosystem *minus* the nutrient *outputs must equal* the change in nutrient *storage* within the ecosystem. This is deceptively simple because quantifying all of the inputs and outputs is difficult, and even defining the "system" in space and time is problematic. Simplifying assumptions and partial budgets still provide important insights, depending on how the balances are to be used. Partial nutrient balances are used, though often only implicitly, in establishing nutrient-crop yield response models and fertilizer recommendations for crop producers.

In farming systems, nutrient budgets can be used to review the balance of major inputs and outputs to assess where the opportunities lie for improvements in efficiency. Although the nature and amount of nutrient inputs and outputs vary among farming systems, regions, and even among fields, the mass balance concept provides a framework that can be applied systematically across a diversity of farming systems, as the field and farm scale balances discussed in Chapters 6 and 7 illustrate. The point of presenting partial balances here is to illustrate that this approach can be used even at the state and regional levels, as part of the analysis of farming systems, to guide program development and targeting to improve input use efficiency.

As summarized by Meisinger and Randall (1991), agricultural watersheds with the largest nitrate losses are associated with excess nitrogen inputs—that is, fertilizer, manure, and legume nitrogen inputs greatly exceed the nitrogen that is taken up by the crop. As Meisinger and Randall (1991:85) note: "These are also the sites where improved N[nitrogen]-management practices will have the greatest chance of improving groundwater quality."

## ESTIMATION OF STATE, REGIONAL, AND NATIONAL BUDGETS

The committee did not attempt to estimate complete mass balances of the nitrogen and phosphorus flux in U.S. agriculture. The committee's intent was to present a partial balance for harvested cropland, focusing on the manageable nitrogen and phosphorus inputs and the resulting outputs of nitrogen and phosphorus in the harvested crops. The purpose of the committee's estimates is to illustrate (1) the opportunities that are available for improving the efficiency with which nitrogen and phosphorus are used in farming systems and (2) large-scale approaches that can be used to target and evaluate programs to improve nutrient use efficiencies across a diversity of farming systems.

The committee's estimates do not include inputs or outputs from range-or pastureland or various set-aside or idled lands. Hence, these partial balances take on a different formula than the general form presented above: major *inputs* of nitrogen and phosphorus to croplands *minus* the major *outputs* of nitrogen and phosphorus in harvested crops *equals* the *balance* or residual nitrogen and phosphorus. The balance, therefore, is an estimate of the amount of the nitrogen and phosphorus inputs that cannot be accounted for in the harvested crops. These nitrogen and phosphorus balances may represent (1) storage in the soil or (2) losses from the farming system into the environment. The magnitude of the balance, and the relative magnitude of the inputs, provide insights into the opportunities to improve the management of nutrients.

### Data Used

The committee estimated nitrogen and phosphorus balances for 1987 using data on crop and livestock production from the latest available Census of Agriculture (U.S. Department of Commerce, Bureau of the Census, 1989). The other primary data source was total fertilizer nutrient data by state, compiled by the Tennessee Valley Authority, National Fertilizer Research Center (Hargett and Berry, 1988). In perspective, 1987 is probably a representative, and perhaps a conservative year for calculating nutrient balances. Fertilizer use was down from peak usage and crop acreage for some commodities was below average because of annual set-asides. Enrollment of cropland into the Conservation Reserve Program was just expanding, and while 1987 was the beginning of a drought period in the midwest, crop yields were still above average in the midwest and nationally. The methods used to estimate nutrient inputs and outputs are outlined below.

### Estimation of Inputs

The nitrogen and phosphorus inputs estimated by the committee include only the major, primary inputs of nutrients to cropland including the nitrogen and phosphorus in commercial fertilizer, the nitrogen and phosphorus in manure, dinitrogen fixation of nitrogen (and/or nitrogen accumulation) by legumes, and the nitrogen and phosphorus content of crop residues. Estimates of inputs were limited to primary sources since these nitrogen and phosphorus inputs can be directly affected by management and since it was desirable to limit the amount of computation required.

There are other input sources, such as nutrients in precipitation and dry deposition, crop seed, foliar absorption, and nonsymbiotic fixation of nitrogen. These are minor or secondary inputs and they are not typically manageable, seldom measured, and are explicitly ignored in most studies and management systems. There are two other inputs that are important but highly variable, spatially and temporally. The nitrogen in irrigation water, generally available as nitrate, can be an important management consideration locally (see Chapter 6, and Schepers and Mosier, 1991). It was not possible to characterize irrigation water nitrogen inputs for this analysis. Other important inputs are the nitrogen and phosphorus contributed from mineralization from the soil. Although important, this factor is difficult to estimate at the scale of this analysis and has become a relatively small component in many farming systems. Of greater importance, given the need to manage the annual variability of nutrient availability to improve efficiencies, is an assessment of the nutrients available from mineralization *plus* the nutrients available as residual from inputs in previous years. The nutrients in the balance unaccounted for in the harvested crop (output term) may carry over as inputs in following crop years. The buildup of soil test phosphorus levels, presented in Chapter 7, is an example of the accumulation of phosphorus in the farming system from this balance term over time.

### Fertilizer Nutrients

Estimates of fertilizer inputs are the most reliable of the estimates of inputs and outputs calculated for nitrogen and phosphorus balances. The nutrient inputs from commercial fertilizers were estimated directly from the sales and tonnage records of state agricultural agencies compiled by the Tennessee Valley Authority, National Fertilizer Research Center (Hargett and Berry, 1988, and unpublished data). These data were provided to the committee by the U.S. Geological Survey and had been adjusted for estimates of sales that crossed county and state lines (Fletcher, 1991: personal communication). The values in the adjusted data set generally vary by less than 2 percent from the unadjusted Tennessee Valley Authority compilations.

### Manure Nutrients

The amount of nitrogen and phosphorus in manure applied to cropland was estimated using standard assumptions and values for the nitrogen and phosphorus content of livestock manures developed by the American Society of Agricultural Engineers (see Chapter 11, and American Society of Agricultural Engineers, 1988; Midwest Planning Service, 1985). First, the total amount of nitrogen and phosphorus voided by livestock was estimated by using the following calculation:

TABLE A-1   Factors Used to Estimate Total Nitrogen and Phosphorus Voided in Manures

| Farm Animal | Ratio of Animals Sold/ Breeding Herd | Nutrient Production (kg/day/454-kg animal) | | Mean Animal Size (kg) | Production Period (days) |
|---|---|---|---|---|---|
| | | Nitrogen | Phosphorus | | |
| Livestock | | | | | |
| Beef cattle | 1/1 | 0.15 | 0.042 | 363 | 180 |
| Dairy cattle | NA | 0.20 | 0.043 | 635 | 365 |
| Hogs and pigs | 10/1 | 0.24 | 0.082 | 61 | 90 |
| Sheep and lambs | 5/1 | 0.19 | 0.039 | 27 | 365 |
| Poultry | | | | | |
| Broilers and meat chickens | NA | 0.50 | 0.136 | 0.9 | 42 |
| Hens and pullets | NA | 0.38 | 0.136 | 1.8 | 365 |
| Turkeys | NA | 0.28 | 0.104 | 6.8 | 72 |

NOTE: NA, not applicable.

(total number of animals) * (estimated average weight of the livestock or poultry species/454 kg [1,000 lbs]) * (production, or residence period of the livestock or poultry species, in days) * (kg of nitrogen or phosphorus voided/day/454 kg of animal weight).

The factors used to estimate the amount of nitrogen and phosphorus voided in manures are given in Table A-1.

The production period for dairy cattle and hens or pullets was assumed to be 365 days, and the year-end inventory data in the 1987 Census of Agriculture were used to estimate the number of dairy cattle and hens or pullets that were present during 1987. Annual sales data for beef cattle, swine, sheep, and poultry were used rather than inventory numbers. The sales data were adjusted using standard ratios of breeding stock to sales for animal numbers; and standard estimates of the number of days in production were used to estimate the number of beef cattle, swine, and sheep that were present in 1987.

Not all manure voided by livestock and poultry are economically recoverable for application to croplands. Manure voided on pasture or rangeland, for example, cannot be collected for use. The committee included estimates of the nitrogen and phosphorus in only those manures that are collectable and recoverable. The proportion of the total nitrogen and phosphorus voided in manures that can be economically recovered varies depending on collection, storage, and application methods. Therefore, the manure nutrient production of each state was adjusted by

TABLE A-2    Nitrogen Voided in Recoverable Manures

| State | Percent Recoverable Manure-N | State | Percent Recoverable Manure-N |
|---|---|---|---|
| Alabama | 34 | Nebraska | 31 |
| Alaska | 0 | Nevada | 14 |
| Arizona | 40 | New Hampshire | 50 |
| Arkansas | 41 | New Jersey | 57 |
| California | 36 | New Mexico | 22 |
| Colorado | 35 | New York | 64 |
| Connecticut | 57 | North Carolina | 39 |
| Delaware | 50 | North Dakota | 22 |
| Florida | 21 | Ohio | 40 |
| Georgia | 37 | Oklahoma | 17 |
| Hawaii | 17 | Oregon | 23 |
| Idaho | 31 | Pennsylvania | 57 |
| Illinois | 32 | Rhode Island | 0 |
| Indiana | 33 | South Carolina | 24 |
| Iowa | 32 | South Dakota | 26 |
| Kansas | 33 | Tennessee | 27 |
| Kentucky | 28 | Texas | 25 |
| Louisiana | 18 | Utah | 32 |
| Maine | 69 | Vermont | 67 |
| Maryland | 65 | Virginia | 36 |
| Massachusetts | 67 | Washington | 33 |
| Michigan | 49 | West Virginia | 33 |
| Minnesota | 43 | Wisconsin | 57 |
| Mississippi | 26 | Wyoming | 20 |
| Missouri | 26 | | |
| Montana | 15 | United States | 34 |

SOURCE: D. L. Van Dyne and C. B. Gilbertson. 1978. Estimating U.S. Livestock and Poultry Manure Production. Report ESCS-12. Washington, D.C.: U.S. Department of Agriculture, Economics, Statistics, and Cooperative Service.

estimates of the proportion of manure that is recoverable for cropland use, and for nitrogen the value was further reduced for storage and handling losses. The state totals were derived from the analysis of Van Dyne and Gilbertson (1978) and are given in Tables A-2 and A-3. Only about one-third of the total nitrogen voided in manures was estimated as recoverable, while about one-half of the total phosphorus voided in manures was estimated as recoverable. Though developed in the 1970s, Van Dyne and Gilbertson's review is the most thorough of its kind, and the basic assumptions have not changed. Using their 1978 values for recoverable manure provided a conservative estimate for 1987 because livestock production had, by then, generally become more concentrated,

TABLE A-3  Phosphorus Voided in Recoverable Manures

| State | Percent Recoverable Manure-P | State | Percent Recoverable Manure-P |
|---|---|---|---|
| Alabama | 45 | Nebraska | 46 |
| Alaska | 0 | Nevada | 25 |
| Arizona | 45 | New Hampshire | 100 |
| Arkansas | 54 | New Jersey | 50 |
| California | 54 | New Mexico | 30 |
| Colorado | 42 | New York | 87 |
| Connecticut | 100 | North Carolina | 61 |
| Delaware | 100 | North Dakota | 25 |
| Florida | 33 | Ohio | 64 |
| Georgia | 57 | Oklahoma | 19 |
| Hawaii | 50 | Oregon | 27 |
| Idaho | 42 | Pennsylvania | 79 |
| Illinois | 57 | Rhode Island | 0 |
| Indiana | 58 | South Carolina | 43 |
| Iowa | 56 | South Dakota | 36 |
| Kansas | 48 | Tennessee | 35 |
| Kentucky | 38 | Texas | 31 |
| Louisiana | 20 | Utah | 43 |
| Maine | 100 | Vermont | 100 |
| Maryland | 83 | Virginia | 47 |
| Massachusetts | 100 | Washington | 44 |
| Michigan | 69 | West Virginia | 50 |
| Minnesota | 64 | Wisconsin | 81 |
| Mississippi | 39 | Wyoming | 27 |
| Missouri | 42 | | |
| Montana | 20 | United States | 49 |

SOURCE: D. L. Van Dyne and C. B. Gilbertson. 1978. Estimating U.S. Livestock and Poultry Manure Production. Report ESCS-12. Washington, D.C.: U.S. Department of Agriculture, Economics, Statistics, and Cooperative Service.

with more cattle and swine in confinement operations and less on pasture and range. Hence, a greater proportion of manure would have been collectable in 1987 than in 1974.

### Legume Nitrogen

The importance of symbiotic dinitrogen fixation has been known, and that knowledge used to enhance crop production, since ancient times. Yet estimates of rates of fixation by legumes (legume-N) vary widely, depending on the species, the age, density, and vigor of the crop, the amount of nitrogen in the soil, and the number of years the legume

TABLE A-4   Estimates of Nitrogen Fixation by
Legumes

| | Nitrogen Fixation Rate (kg/ha/yr) | |
|---|---|---|
| Legume | Low Estimate | High Estimate |
| Alfalfa | 70 | 600 |
| Soybeans | 15 | 310 |
| Midwest | 55 | 95 |
| Southeast | 70 | 220 |
| Dry Beans | 2 | 215 |
| Peanuts | 40 | 60 |
| Cowpeas | 80 | 100 |
| Chickpeas | 25 | 80 |
| Clover (various) | 100 | 200 |
| Sweet Clover | 4 | 130 |
| Fava Bean | 175 | 200 |
| Lentils | 165 | 190 |
| Lupins | 150 | 215 |
| Peas | 55 | 195 |
| Vetch | 90 | 120 |

SOURCE: Data derived from Evans and Barber, 1977; Follet et al.,
1987; Heichel, 1987; Meisinger and Randall, 1991; Peterson and
Russelle, 1991; Schepers and Fox, 1989; Schepers and Mosier, 1991;
Thurlow and Hiltbold, 1985; Tisdale and Nelson, 1966.

stand remains in the field before being turned under. A summary of
estimates for various legumes is given in Table A-4.

The lower values for fixation by perennial legumes given in Table A-4
generally reflect fixation during the first year of growth. The higher
values reflect fixation that has occurred in stands of legumes that have
been in place for 2 or more years. The estimates given in Table A-4 are
estimates of the total fixed legume-N and are greater than values often
estimated as available to crops that are planted after the legumes are
harvested. The total fixation value, however, includes the amount fixed
and taken up in the legume biomass, most of which is subsequently
harvested and unavailable to succeeding crops.

Estimates of the amount of nitrogen actually fixed by a particular
species of legume are problematic because there are no unequivocal
methods for measurement. LaRue and Patterson (1981) summarized
published research and concluded there is not a single legume crop for
which valid estimates of the nitrogen fixed in agricultural production
were available. They did report consistent ranges for some legumes that
emerged from various research.

Part of the measurement problem occurs because legumes use nitrogen in the soil as well as fix nitrogen from the atmosphere. Generally, legumes will fix nitrogen only after taking up much of the nitrogen that is available in the soil (see, e.g., Phillips and DeJong, 1984). In fertile soils, with substantial soil organic matter and soil nitrogen, they will not fix as much as in soils of low fertility. For example, in Table A-4, the different values for fixation by soybeans under midwestern and southeastern conditions reflect the general differences in soils between those regions.

Some studies suggest that some legumes (particularly annual grain legumes) may remove more nitrogen from the soil than they fix and, hence, the legumes may represent a net loss of nitrogen (Schepers and Mosier, 1991; Follett et al., 1987). Although unequivocal estimates of the nitrogen input from symbiotic nitrogen fixation are still open to discussion, crop rotation with legumes consistently produces a yield benefit to the succeeding crop, with reduced nitrogen inputs. This undoubtedly reflects various rotation effects as well as any nitrogen residuals supplied from true fixation or accumulation from other sources.

To minimize environmental losses of nitrogen and to optimize crop yields, some estimate of the legume contribution to crop rotations *must* be made. To account for the combined effects of rotation and fixation, legume benefits are often estimated as a fertilizer nitrogen replacement value or a fertilizer nitrogen equivalence. Based on consistent results from many experiments throughout the United States, Schepers and Fox (1989) summarize that the fertilizer nitrogen equivalence of a 2- to 4-year old "good" alfalfa stand is at least 100 to 150 kg/ha for the first succeeding crop and 30 to 50 kg/ha for the second crop; and nitrogen fertilizer applications to crops following soybeans should be reduced by approximately 15 to 17 kg/ha per Mg/ha of soybean yield (~1 lb/acre/bu soybeans).

In the committee's estimates of nitrogen balances, the input values for legume-N are balanced by estimates of the output of nitrogen in harvested alfalfa and soybean crops. The result is an estimate of the total nitrogen that may accumulate and remain as a nitrogen replacement value. In some cases, this replacement value may also reflect rotation benefits other than legume-N.

Another problem in estimating the nitrogen supplied by legumes is that some legume-N, particularly from perennial forages such as alfalfa, may be available for succeeding crops several years after the forage crop is harvested and plowed down. For the first crop year the nitrogen replacement value of the legume may equal 150 to 200 kg/ha. Calculated over a 2- to 5-year period following plow down, some estimates of the nitrogen replacement value range as high as 450 kg/ha (Peterson and Russelle, 1991; Schepers and Mosier, 1991). The committee used only estimates of the hectares of alfalfa hay and soybeans harvested in 1987 and the tons of

TABLE A-5  Estimated Rates of Nitrogen Accumulation and Nitrogen Replacement Value for Alfalfa and Soybeans in Low-, Medium-, and High-Fixation Scenarios

| Legume | Estimate Scenario | Total Nitrogen Fixed (kg/ha) | Nitrogen Harvested[a] (kg/ha) | Nitrogen Replacement Value[b] (kg/ha) |
|--------|-------------------|------------------------------|-------------------------------|----------------------------------------|
| Alfalfa | Low | 230 | 185 | 45 |
| | Medium | 250 | 185 | 65 |
| | High | 380 | 185 | 195 |
| Soybean | Low | 175 | 165 | 10 |
| | Medium | 200 | 165 | 35 |
| | High | 220 | 165 | 55 |

[a]Includes nitrogen in the harvested portion of the legume and in crop residue.
[b]Includes the amount of fixed nitrogen available to a succeeding crop and the reduced need for supplemental nitrogen that may be a result of rotation effects.

alfalfa hay or soybeans harvested in 1987 to estimate a replacement value for legume-N in 1987. Alfalfa and soybeans are the two major legume crops, but many others (see Table A-4) also contribute nitrogen inputs. Alfalfa and other legumes grown in pasture were not considered in the committee's estimates. Data on the land area planted to legumes and their yields were derived from the 1987 Census of Agriculture.

Because of the uncertainty in estimates of fixation and nitrogen replacement value, the committee generated three scenarios using a low, medium, and high estimate of the nitrogen supplied by alfalfa and soybeans in 1987. The input value used was considered to be an estimate of total fixation and accumulation of nitrogen by these legumes. The amounts of nitrogen harvested in alfalfa hay and in soybean grain and residues were entered on the output side of the balance. The difference between the total alfalfa or soybean nitrogen input and the alfalfa or soybean nitrogen output was used as an estimate of the residual nitrogen replacement value potentially available to a succeeding crop. The magnitude of the estimates of this nitrogen replacement value was compared to standard estimates of the nitrogen replacement value of alfalfa and soybeans derived from other research. The nitrogen fixation rates, estimates of nitrogen harvested in alfalfa hay and soybeans, and the resulting replacement values used in the committee's estimates of nitrogen balances are summarized on a per-hectare basis in Table A-5.

The fixation values used in the committee's estimates agree well with those used by other scientists. The medium nitrogen fixation value for alfalfa (250 kg/ha/yr) is approximately the same as that used by Peterson and Russelle (1991) (252 kg/ha/yr). The maximum value (380 kg/ha/yr) results in an estimated replacement value to a subsequent grain crop of 195 kg/ha, slightly higher than the ranges suggested by Schepers and Fox (1989). However, only about 30 percent of alfalfa acres are typically rotated to another crop in most years. The low value for alfalfa nitrogen fixation produces an estimated resultant value for the total alfalfa hectares harvested in 1987 that approximates the replacement value expected if 30 percent of the alfalfa acres in 1987 were rotated to another crop in 1988.

For soybeans, the estimates of fixation rates and harvest values used result in nitrogen replacement values equivalent to 5, 16, and 25 kg/ha/yr per Mg/ha of soybeans harvested (or about 0.3, 0.9, and 1.5 lb/acre/per bu soybeans). The replacement values, which are of most interest here, are in line with most recent estimates (see Meisinger and Randall, 1991; Schepers and Mosier, 1991; Schepers and Fox, 1989), with the medium to low estimates providing the best estimates for an annual balance.

### Crop Residues

*Crop residues* are the mass of plant matter that remains in the field after harvest. The committee estimated the volume of crop residues using published estimates of the amount (ratio) of residue produced related to the amount of harvested grain (Larson et al., 1978). The phosphorus and nitrogen content of crop residues was derived from the United States-Canadian Tables of Feed Composition (National Research Council, 1982). The nitrogen content of residues was calculated assuming crude protein as 16 percent nitrogen. The grain-to-residue ratios used to estimate crop residue values the percentages of nitrogen and phosphorus in residues are given in Table A-6. (Crop residues are further discussed below.)

In an operational system, to use such balances for analysis, the legume residual from one year would become the input term for the next: that is, to assess opportunities for improved input management for 1987, the legume residual for 1986 should actually be used. In the committee's illustrative analysis, the 1987 legume residuals are used, for simplicity, but this also provides a conservative estimate; the area of soybeans and alfalfa was greater in 1986 than 1987, while the yields were greater in 1987.

### Estimation of Outputs

Only the primary desirable nutrient outputs were estimated by the committee, that is, the nitrogen and phosphorus taken up in the harvested crops and in crop residues. Other outputs include undesirable losses into the environment through ammonia volatilization, denitrification, soil erosion and runoff losses, and leaching losses. As discussed

TABLE A-6  Factors Used to Estimate Nitrogen and
Phosphorus in Crop Residues

| Crop | Ratio of Residue to Grain | Percent in Nitrogen Residue | Percent in Phosphorus Residue |
|---|---|---|---|
| Alfalfa hay[a] | NA | NA | NA |
| Barley | 1.5 | 0.64 | 0.09 |
| Corn | 1.0 | 0.89 | 0.09 |
| Cotton | 1.0 | 0.59 | 0.05 |
| Dry beans[b] | 1.5 | 0.73 | 0.05 |
| Hay[a] | NA | NA | NA |
| Oats | 2.0 | 0.65 | 0.06 |
| Peanuts[c] | 0.14 | 1.11 | 0.14 |
| Potatoes[c] | 0.15 | 0.37 | 0.03 |
| Rice | 1.5 | 0.48 | 0.07 |
| Sorghum | 1.0 | 0.74 | 0.12 |
| Soybeans | 1.5 | 0.74 | 0.05 |
| Tobacco[a] | NA | NA | NA |
| Wheat | 1.7 | 0.51 | 0.04 |

NOTE: Conversion factor coefficients for bushels, bales, and hundredweight were obtained from the U.S. Department of Agriculture (1988). NA, not applicable.

[a]Total plant harvest was assumed for alfalfa hay, hay, and tobacco.

[b]The soybean residue/grain rate was used to estimate dry edible bean residue.

[c]The peanut and potato ratio of residue to grain were obtained from J. W. Gilliam, North Carolina State University, personal communication, 1991.

above, the committee made no attempt to estimate the magnitude of these losses. The unaccounted for balance includes both storage and undesired losses that may be subject to improved management.

Other nutrient outputs from farming systems may include other gaseous losses such as $N_2O$ evolution during nitrification, decomposition of nitrous acid, or losses directly from maturing or senescent crops (see Bremner et al., 1981; Meisinger and Randall, 1991; Nelson, 1982). Some nutrients are taken up by weeds or immobilized by microbes and enter the storage pool. These outputs are small relative to other outputs and typically have been implicitly included in nutrient-crop yield response models.

### Harvested Crops

The desired nutrient output from an agricultural ecosystem is the nutrient taken up in the harvested crop. The nitrogen and phosphorus

TABLE A-7  Nitrogen and Phosphorus Content of
Harvested Crops

| Crop | Percent Nitrogen | Percent Phosphorus |
|---|---|---|
| Alfalfa hay | 2.8 | 0.17 |
| Barley | 1.9 | 0.34 |
| Corn | 1.5 | 0.26 |
| Cotton | 1.6 | 0.11 |
| Dry beans[a] | 3.6 | 0.52 |
| Hay[b] | 1.9 | 0.27 |
| Oats | 1.9 | 0.33 |
| Peanuts | 4.4 | 0.30 |
| Potatoes | 0.35 | 0.06 |
| Rice | 1.3 | 0.28 |
| Sorghum | 1.8 | 0.29 |
| Soybeans | 6.3 | 0.60 |
| Tobacco[c] | 2.7 | 0.29 |
| Wheat | 2.3 | 0.37 |

NOTE: Conversion factor coefficients for bushels, bales, and hundredweights were obtained from the U.S. Department of Agriculture (1988).

[a]Percent nitrogen for navy beans was used for the dry edible bean.

[b]To estimate percent nitrogen for hay, the average of five crops (Kentucky bluegrass, brome, fescue, oats [sun-cured hay], and timothy hay) was used.

[c]Percent nitrogen for tobacco was obtained from J. S. Schepers and R. H. Fox. 1989. Estimation of N budgets for crops. Pp. 221–246 in Nitrogen Management and Ground Water Protection, R. F. Follett, ed. Developments in Agricultural and Managed-Forest Ecology 21. Amsterdam: Elsevier.

in the harvested portion of the crop are effectively removed from the farming system, unless they are fed to livestock. Estimates of the yield of harvested crops in the 1987 Census of Agriculture were used by the committee to estimate phosphorus and nitrogen outputs. The phosphorus and nitrogen content of the harvested crops was derived from the United States-Canadian Tables of Feed Composition (National Research Council, 1982); nitrogen was calculated assuming crude protein as 16 percent nitrogen. The estimates of the nitrogen and phosphorus content of the harvested crops used to estimate nutrient outputs are given in Table A-7.

Only the crops listed in Table A-7 were considered by the committee in estimating outputs of nitrogen and phosphorus from croplands. Analysis of national production statistics for all crops show that the crops included here account for approximately 90 percent of the total mass of harvested crops and more than 97 percent of the harvested nitrogen and phosphorus.

### Crop Residues

The nitrogen and phosphorus in crop residues were estimated as described above. For the committee's purposes, crop residues were considered as both inputs and outputs and hence offset each other in the calculation of balances. In reality, however, residues from the previous year should be estimated as the input and the residues of the current year in the output term. Over time, however, most of the residues remain in the system and the nitrogen and phosphorus in those residues would appear alternately as inputs and outputs. Crop residues are often not counted as inputs in cropland balances.

## Estimation of Balances

The committee's estimates, as discussed earlier, are partial nutrient balances for harvested cropland. The balance, or residual term represents an estimate of the amount of nitrogen and phosphorus inputs that (1) may go into storage or (2) may potentially be lost from the system (outputs) into the environment. The magnitude of the balance, then, provides insights into the potential for water pollution that may be created by nutrient fluxes through farming systems aggregated at the state, regional, and national levels. The magnitude of the balance and the relative importance of various inputs also provide insights into the opportunities to improve the management of nitrogen and phosphorus. These implications and opportunities are discussed in more detail in other portions of the report, as are comparisons with prior published balances (see Chapters 2, 3, 6, and 7). A summary of the committee's balance estimates is given in Table A-8.

## BALANCE ESTIMATES IN PERSPECTIVE

The methods and values chosen for any nutrient balance, at the scale of this analysis, are clearly equivocal and subject to error. Despite the assumptions the committee was forced to make to estimate nitrogen and phosphorus balances, the committee's estimates are similar to other past published budgets (see Chapter 6). For the most part, the choices of methods and values made by the committee would be expected to produce a conservative estimate of the unaccounted for balance of

TABLE A-8 Inputs and Outputs of Nitrogen and Phosphorus on Croplands in the United States, 1987

| Inputs and Outputs | Nitrogen | | | | | | Phosphorus | |
| | Low Scenario | | Medium Scenario | | High Scenario | | | |
| | Metric Tons | Percent of Total Inputs[a] | Metric Tons | Percent of Total Inputs | Metric Tons | Percent of Total Inputs | Metric Tons | Percent of Total Inputs |
| --- | --- | --- | --- | --- | --- | --- | --- | --- |
| Inputs | | | | | | | | |
| Fertilizer | 9,390,000 | 47 | 9,390,000 | 45 | 9,390,000 | 42 | 3,570,000 | 79 |
| Manure | 1,730,000 | 9 | 1,730,000 | 8 | 1,730,000 | 8 | 655,000 | 15 |
| Legumes | 6,120,000 | 30 | 6,870,000 | 33 | 8,560,000 | 38 | NA | NA |
| Crop residues | 2,890,000 | 14 | 2,890,000 | 14 | 2,890,000 | 13 | 272,000 | 6 |
| Total | 20,100,000 | 100 | 20,900,000 | 100 | 22,600,000 | 100 | 4,500,000 | 100 |
| Outputs | | | | | | | | |
| Harvested crops | 10,600,000 | 53 | 10,600,000 | 51 | 10,600,000 | 47 | 1,320,000 | 29 |
| Crop residues | 2,890,000 | 14 | 2,890,000 | 14 | 2,890,000 | 13 | 272,000 | 6 |
| Total | 13,500,000 | 67 | 13,500,000 | 64 | 13,500,000 | 60 | 1,600,000 | 36 |
| Balance | 6,670,000 | 33 | 7,420,000 | 36 | 9,110,000 | 40 | 2,900,000 | 63 |

NOTE: NA, not applicable.

[a]Input or output as a percent of the total mass of inputs.

nitrogen and phosphorus. Some perspectives on the effect of the assumptions the committee made are worthy of review.

Although data for only 1 year were used, as noted, 1987 appears to have been representative for the committee's purposes. It would be better, of course, if annual data on inputs and outputs were used to account for year-to-year variation in the crops planted, inputs used, and crop yields. The estimate of balances over time would be a particularly useful way to assess progress in improving input use efficiency and would be particularly useful since crop outputs can vary tremendously from year to year. Also, some inputs and outputs are not in steady state and should be factored over a period of years. Some of the nitrogen fixed by legumes, for example, is available for more than 1 year. Also, the nitrogen in manures is released over time, and multiyear decay constants are sometimes used to estimate the annual contribution of nutrients from manures (Schepers and Mosier, 1991). The committee's estimates of inputs are probably low, particularly the estimates for phosphorus and the estimates in the low and medium nitrogen scenarios, since no effort was made to account for the multiyear contributions or buildup of nitrogen and phosphorus over time.

Standard assumptions were used by the committee to estimate nitrogen and phosphorus inputs from manures. Some of the manure produced undoubtedly would have been applied to pastures and other land, resulting in an overestimate of that applied to cropland. As discussed, however, the values used to estimate recoverable manure were conservative; they were derived for 1970s conditions, and by 1987 livestock production had become more concentrated, with more cattle and swine in confinement operations and less on pasture and range. Hence, a greater proportion of manure would be collectable in 1987 than 1974.

Given the uncertainties for legume-N inputs, the assumptions made by the committee have been conservative. On the input side, only estimates for accumulation/fixation by alfalfa and soybeans were used. However, on the output side of this balance, the nitrogen harvested in other important legumes is included, notably dry beans and peanuts (Table A-6). Legumes account for more than 35 percent of the total nitrogen harvested in crops. Balances estimated on a crop-by-crop basis, then, would result in much higher ratios of inputs to outputs for some crops than is estimated by aggregating all crops together by state, region, or for the United States as a whole. The committee, however, did not estimate the nitrogen and phosphorus harvested in citrus and vegetables, undoubtedly resulting in too low an output value for some states, such as California, Florida, and Texas.

The most important factor leading to conservative estimates of the difference between inputs and outputs was the committee's decision not to correct estimates of the harvested crop for moisture content. The committee did not correct for moisture content because of uncertainties in how the yields of several crops were reported. Correcting for the moisture contents of harvested crops would reduce the harvested nutrient output by about 15 percent on average across the crops considered.

## IMPLICATIONS OF BALANCES

Even with the relatively conservative assumptions used, subtracting harvested crop outputs from inputs results in a residual balance equal to more than 60 percent of the phosphorus inputs and from 33 (low scenario) to 40 percent (high scenario) of the nitrogen inputs. The proportion of the residual nitrogen balance contributed by legumes was about 6, 9, and 16 percent less than that of the low, medium, and high scenarios, respectively.

Although the nutrient balance after harvest may go into storage or environmental losses, as discussed in Chapters 6 and 7, many studies have indicated that the magnitude of such balances are directly related to the magnitude of environmental losses. Phosphorus is relatively immobile and may build up in the soil over time, and the concentrations in the surface soil, in particular, are directly related to runoff losses. Nitrogen is much more mobile in the environment, and few farming systems show any increase in soil nitrogen over time, even where large balances are found.

There are several facets of the nitrogen budget and balances that need further discussion. As many large-scale and multicrop nitrogen balances would note, the nitrogen removed in the harvested crop is slightly more than the fertilizer nitrogen input. However, more than 35 percent of the total nitrogen in harvested crops is accounted for by legumes, which receive very little nitrogen fertilizer. Major commodities such as corn, cotton, potatoes, rice, and wheat account for more than 80 percent of the nitrogen applied in fertilizers, but the nitrogen harvested with these crops accounts for only 57 percent of the fertilizer nitrogen input. If all legume inputs and outputs are taken out of the estimated balances, the nitrogen harvested in the remaining crops is only about 35 to 40 percent of the nitrogen fertilizer and manure inputs. Nutrient balances constructed on a crop-by-crop basis, and analysis by input category can provide important guidance to programs seeking to improve the efficiency of nutrient use in farming systems.

# References

Abdalla, C. W. 1990. Agriculture and groundwater quality: Emerging issues and policies. Pp. 1–16 in Proceedings of the Philadelphia Society for the Promotion of Agriculture. Philadelphia, Pa.: The Society.

Aber, J. D., K. J. Nadelhoffer, P. Steudler, and J. M. Melillo. 1989. Nitrogen saturation in northern forest ecosystems. BioScience 39:378–386.

Abler, D. A., and J. S. Shortle. 1991. The political economy of water quality protection from agricultural chemicals. Northeastern Journal of Agricultural and Resource Economics 20:53–60.

Adams, R. S., and J. S. V. McAllister. 1975. Nutrient cycles involving phosphorus and potassium on livestock farms in Northern Ireland. Journal of Agricultural Science 85:345–349.

Adams, R. T., and F. M. Kurisu. 1976. Simulation of Pesticide Movement on Small Agricultural Watersheds. Publication No. EPA-600/3-76-066. Sunnyvale, Calif.: Environmental Systems Laboratory.

Addiscott, T., and D. Powlson. 1989. Laying the ground rules for nitrate. New Scientist 122(1662):28–29.

Addiscott, T. M., and R. J. Darby. 1991. Relating the nitrogen fertilizer needs of winter wheat crops to the soil's mineral nitrogen. Influence the downward movement of nitrate during winter and spring. Journal of Agricultural Science 117(part 2):241–249.

Addiscott, T. M., and R. J. Wagenet. 1985. Concepts of solute leaching in soils: A review of modeling approaches. Journal of Soil Science 36:411–424.

Adriano, D. C. 1986. Trace Elements in the Terrestrial Environment. New York: Springer-Verlag.

Alberts, E. E., W. H. Neibling, and W. C. Moldenhauer. 1981. Transport of sediment nitrogen and phosphorus in runoff through cornstalk residue strips. Soil Science Society of America Journal 45:1177–1184.

Alberts, E. E., and R. G. Spomer. 1985. Dissolved nitrogen and phosphorus in runoff from watersheds in conservation and conventional tillage. Journal of Soil and Water Conservation 40:153–157.

*449*

Alexander, E. B., and J. C. McLaughlin. 1992. Soil porosity as an indication of forest and rangeland soil condition (compaction) and relative productivity. Pp. 52–61 in Proceedings of the Soil Quality Standards Symposium. Report No. W0-WSA-2. Washington, D.C.: U.S. Department of Agriculture, Forest Service.

Allmaras, R. R., R. E. Burwell, and R. F. Holt. 1967. Plow-layer porosity and surface roughness from tillage as affected by initial porosity and soil moisture at tillage time. Soil Science Society of America Proceedings 31:550–556.

Allmaras, R. R., G. W. Langdale, P. W. Unger, R. H. Dowdy, and D. M. VanDoren. 1991. Adoption of conservation tillage and associated planting systems. Pp. 53–84 in Soil Management for Sustainability, R. Lal and F. J. Pierce, eds. Ankeny, Iowa: Soil and Water Conservation Society.

Alonzo, C. V., and F. D. Theurer. 1988. Environmental degradation of salmon spawning gravels in Tucannon River. Pp. 411–416 in Proceedings of the ASCE National Conference on Hydraulic Engineering. New York: American Society of Civil Engineers.

Ambus, P., and R. Lowrance. 1991. Comparison of denitrification in two riparian soils. Soil Science Society of America Journal 55:994–997.

American Society of Agricultural Engineers. 1977. Soil Erosion and Sediment. Proceedings of a National Symposium: Soil Erosion and Sedimentation by Water. ASAE Publication 4–77. St. Joseph, Mich.: American Society of Agricultural Engineers.

American Society of Agricultural Engineers. 1988. Manure Production and Characteristics. ASAE Data D384. St. Joseph, Mich.: American Society of Agricultural Engineers.

Anderson, H. W. 1975. Sedimentation and turbidity hazards in wildlands. Pp. 347–376 in Watershed Management Symposium Proceedings. Irrigation and Drainage Division. New York: American Society of Civil Engineers.

Anderson, R. L., and L. A. Nelson. 1975. A family of models involving intersecting straight lines and concomitant experimental designs useful in evaluating response to fertilizer nutrients. Biometrics 31:303–318.

Andraski, B. J., D. H. Mueller, and T. C. Daniel. 1985. Phosphorus losses in runoff as affected by tillage. Soil Science Society of America Journal 49:1523–1527.

Angle, J. S., G. McClung, M. S. McIntosh, P. M. Thomas, and D. C. Wolf. 1984. Nutrient losses in runoff from conventional and no-till corn watersheds. Journal of Environmental Quality 13:431–435.

Arnold, R. W., I. Zaboles, and V. C. Targulian, eds. 1990. Global Soil Change. Report of IIASA-ISSS. U.N. Environment Program Task Force on the Role of Soil in Global Change. Laxenburg, Austria: International Institute for Applied Systems Analysis.

Arshad, M. A., and G. M. Coen. 1992. Characterization of soil quality: Physical and chemical criteria. American Journal of Alternative Agriculture 7:25–32.

Atkinson, S. E., and T. H. Tietenberg. 1982. The empirical properties of two classes of designs for transferable discharge permit markets. Journal of Environmental Economics and Management 9:101–121.

Ayers, R. S., and D. W. Westcot. 1985. Water Quality for Agriculture. FAO Irrigation and Drainage Paper 29, Rev. 1. Rome: Food and Agriculture Organization of the United Nations.

Azevedo, J., and P. R. Stout. 1974. Farm Animal Manures: An Overview of Their Role in the Agricultural Environment. California Agricultural Experiment Station and Extension Service Manual 44. Berkeley: California Agricultural Experiment Station and Extension Service.

Backlund, V. L., and R. R. Hoppes. 1984. Status of salinity in California. California Agriculture (October):8–9.

Baker, D. B. 1985. Regional water quality impacts of intensive row-crop agriculture: A Lake Erie Basin case study. Journal of Soil and Water Conservation 40:125–132.

Baker, J. L., H. P. Johnson, M. A. Borcherding, and W. R. Payne. 1978. Nutrient and pesticide movement from field to stream: A field study. Pp. 213–245 in Best Management Practices for Agriculture and Silviculture, R. C. Loehr, D. A. Haith, M. F. Walter, and C. S. Martin, eds. Proceedings of the 1978 Cornell Agricultural Waste Conference. Ann Arbor, Mich.: Ann Arbor Science.

Baker, J. L., and J. M. Laflen. 1983. Water quality consequences of conservation tillage. Journal of Soil and Water Conservation 38:186–193.

Baker, J. L., and S. W. Melvin. 1992. Chemical management: Status and Findings II, continued use of agricultural drainage wells. Pp. 17–45 in Agricultural Drainage Well Research and Demonstration Project. Annual Report 1992. Des Moines, Iowa: Iowa Department of Agriculture and Land Stewardship, and Iowa State University.

Barfield, B. J., E. W. Tollner, and J. C. Hayes. 1979. Filtration of sediment by simulated vegetation. I. Steady-state flow with homogeneous sediment. Transactions of the American Society of Agricultural Engineers 22:540.

Barisas, S. G., J. L. Baker, H. P. Johnson, and J. M. Laflen. 1978. Effect of tillage systems on runoff losses of nutrients: A rainfall simulation study. Transactions of the American Society of Agricultural Engineers 21:893–897.

Barkema, A., and M. L. Cook. 1993. The changing U.S. pork industry: A dilemma for public policy. Economic Review, 2nd Quarter:49–65.

Barkema, A., M. Drabenstott, and K. Welch. 1991. The quiet revolution in the U.S. food market. Economic Review, May/June:Q5–41.

Barrows, H. L., and V. J. Kilmer. 1963. Plant nutrient losses from soils by water erosion. Advances in Agronomy 15:303–316.

Bartfeld, E. 1992. Point/Nonpoint Source Trading: Looking Beyond Potential Cost Saving. M.S. thesis. University of Michigan, School of Natural Resources, Ann Arbor, Michigan.

Batie, S. B. 1983. Soil Erosion: Crisis in America's Croplands? Washington, D.C.: The Conservation Foundation.

Batie, S. B. 1985. Soil conservation in the 1980s: A historical perspective. Agricultural History 59:107–123.

Beasley, D. B., L. F. Huggins, and E. J. Monhe. 1980. ANSWERS: A model for watershed planning. Transactions of the American Society of Agricultural Engineers 23:938–944.

Bennett, H. H., and W. R. Chapline. 1928. Soil Erosion a National Menace. U.S. Department of Agriculture, Circular No. 33. Washington, D.C.: U.S. Government Printing Office.

Benson, S. A., M. Delamore, and S. Hoffman. 1990. Kesterson crisis: Sorting out the facts. Pp. 453–460 in Proceedings, 1990 National Conference on Irrigation and Drainage, S. C. Harris, ed. New York: American Society of Civil Engineers.

Bharati, M. P., D. K. Wigham, and R. D. Voss. 1986. Soybean response to tillage and nitrogen phosphorus and potassium fertilization. Agronomy Journal 78:947–950.

Binford, G. D., and A. M. Blackmer. 1993. Visually rating the nitrogen status of corn. Journal of Production Agriculture 6:41–46.

Binford, G. D., A. M. Blackmer, and M. E. Cerrato. 1992. Relationships between corn yields and soil nitrate in late spring. Agronomy Journal 84:53–59.

Bjork, S. 1972. Swedish lake restoration program gets results. Ambio 1:153–165.

Black, A. L., A. D. Halvorson, and F. H. Siddoway. 1981. Dryland cropping strategies for efficient water use to control saline seeps in the Northern Great Plains, U.S.A. Agricultural Water Management 4:295–311.

Blackmer, A. M. 1984. Losses of fertilizer N from soils. Report No. CE-2081. Ames, Iowa: Iowa State University, Cooperative Extension Service.

Blackmer, A. M. 1986. Potential yield response of corn to treatments that conserve fertilizer nitrogen in soils. Agronomy Journal 78:571–575.

Blackmer, A. M., and T. Morris. 1992. Selecting nitrogen fertilizer rates for corn: New options. Pp. 19–24 in Building Bridges: Cooperative Research and Education for Iowa Agriculture. Ames, Iowa: Leopold Center for Sustainable Agriculture, Iowa State University.

Blackmer, A. M., D. Pottker, M. E. Cerrato, and J. Webb. 1989. Correlations between soil nitrate concentrations in late spring and corn yields in Iowa. Journal of Production Agriculture 2:103–109.

Blake, G. R., W. W. Nelson, and R. R. Allmaras. 1976. Persistence of subsoil compaction in a mollisol. Soil Science Society of America Journal 40:943–948.

Bloom, P. R. 1981. Phosphorus adsorption by an aluminum-peat complex. Soil Science Society of America Journal 45:267–272.

Bock, B. R., and G. W. Hergert. 1991. Fertilizer nitrogen management. Pp. 140–164 in Managing Nitrogen for Groundwater Quality and Farm Profitability, R. F. Follet, D. R. Keeney, and R. M. Cruse, eds. Madison, Wis.: Soil Science Society of America.

Bock, B. R., and F. J. Sikora. 1990. Modified-quadratic/plateau model for describing plant responses to fertilizer. Soil Science Society of America Journal 54:1784–1789.

Boone, R. D. 1990. Soil organic matter as a potential net nitrogen sink in a fertilized corn field, South Deerfield, Massachusetts, USA. Plant and Soil 128:191–198.

Bormann, F. H., and G. E. Likens. 1979. Pattern and Process in a Forest Ecosystem: Disturbance, Development, and The Steady State Based on the Hubbard Brook Ecosystem Study. New York: Springer-Verlag.

Bortleson, G. C., and G. F. Lee. 1974. Phosphorus, iron, and manganese distribution in sediment cores of Wisconsin lakes. Limnology and Oceanography 19:794–801.

Bosch, D. J., J. W. Pease, S. S. Batie, and V. O. Shanholtz. 1992. Crop selection, tillage practices, and chemical and nutrient applications in two regions of the Chesapeake Bay watershed. Water Resources Research Center Bulletin No. 176. Blacksburg: Virginia Polytechnic Institute and State University, Virginia Water Resources Research Center.

Boschwitz, R. 1987. Decouple supports first; then target benefits. Choices 2(1):34–35.

Bouldin, D. R., S. D. Klausner, and W. S. Reid. 1984. Use of nitrogen from manure. Pp. 221–248 in Nitrogen in Crop Production, R. D. Hauck, ed. Madison, Wis.: American Society of Agronomy, Crop Science Society of America, and Soil Science Society of America.

Bouma, J. 1989. Using soil survey data for quantitative land evaluation. Advances in Soil Science 9:177–213.

Bower, B. T. 1980. Implementation incentives in phosphorus management strategies for the Great Lakes. Pp. 123–157 in Phosphorus Management Strategies for Lakes, R. C. Loehr, C. S. Martin, and W. Rast, eds. Ann Arbor, Mich.: Ann Arbor Science Publishers.

Bowman, R. A., J. D. Reeder, and R. W. Lober. 1990. Changes in soil properties in a central plains rangeland soil after 3, 20, and 60 years of cultivation. Soil Science 150:851–857.

Breeusma, A., J. H. M. Wosten, J. J. Vleeshouwer, A. M. Slobbe, and J. Bouma. 1986. Derivation of land qualities to assess environmental problems from soil surveys. Soil Science Society of America Journal 50:186–190.

Bremner, J. M., G. A. Breitenback, and A. M. Blackmer. 1981. Effect of anhydrous ammonia fertilizer on emission of nitrous oxide from soils. Journal of Environmental Quality 10:77–80.

Bromley, D. W. 1990. Property rights and environmental policy: Is agriculture paying its way? Paper presented at the symposium Managing Agriculture for Environmental Goals, May 16–18, 1990, Washington, D.C.

Brown, K. W., and J. C. Thomas. 1978. Uptake of N by grass from septic fields in three soils. Agronomy Journal 70:1037–1040.

Brown, L. R., A. Durning, C. Flavin, H. French, J. Jacobson, M. Lower, S. Postel, M. Renner, L. Starke, and J. Young. 1990. State of the World, 1990. New York: Worldwatch Institute and W. W. Norton.

Brown, M. P., P. Longabucco, M. R. Rafferty, P. D. Robillard, M. F. Walter, and D. A. Haith. 1989. Effects of animal waste control practices on nonpoint source phosphorus loading in the west branch of the Delaware River watershed. Journal of Soil and Water Conservation 44:67–70.

Bruce, R. R., L. A. Harper, R. A. Leonard, W. M. Snyder, and A. W. Thomas. 1975. A model for runoff of pesticides from small upland watersheds. Journal of Environmental Quality 4:541–548.

Bruce, R. R., A. W. White, Jr., A. W. Thomas, W. M. Snyder, G. W. Langdale, and H. F. Perkins. 1988. Characterization of soil-crop yield relations over a range of erosion on a landscape. Geoderma 43:99–116.

Buckman, H. O., and N. C. Brady. 1969. The Nature and Properties of Soils, 7th Ed. London: MacMillan.

Bundy, L. G., and E. S. Malone. 1988. Effect of residual profile nitrate on corn response to applied nitrogen. Soil Science Society of America Journal 52:1377–1383.

Buringh, P. 1981. An Assessment of Losses and Degradation of Productive Agricultural Land in the World. Working Group on Soils Policy. Rome: Food and Agriculture Organization of the United Nations.

Burwell, R. E., D. R. Timmons, and R. F. Holt. 1975. Nutrient transport in surface runoff as influenced by soil cover and seasonal periods. Soil Science Society of America Proceedings 39:523–528.

Cahill, T. H., R. W. Pierson, Jr., and B. Cohen. 1978. The evaluation of best management practices for the reduction of diffuse pollutants in an agricultural watershed. Pp. 465–490 in Best Management Practices for Agriculture and Silviculture, R. C. Loehr, D. A. Haith, M. F. Walter, and C. S. Martin, eds. Proceedings of the 1978 Cornell Agricultural Waste Management Conference. Ann Arbor, Mich.: Ann Arbor Science.

Capalbo, S., and T. Phipps. 1990. Designing in environmental quality: Possibilities in U.S. agriculture. Paper presented at the American Enterprise Institute for Public Policy Research Conference, June 11–12, 1990, Washington, D. C.

Carey, A. E. 1991. Agriculture, agricultural chemicals, and water quality. Pp. 78–85 in Agriculture and the Environment: The 1991 Yearbook of Agriculture. Washington, D.C.: U.S. Government Printing Office.

Carignan, R., and J. Kalff. 1980. Phosphorus sources for aquatic weeds: Water or sediments? Science 207:987–988.

Carlson, G. A., and S. Shui. 1991. Farm programs and pesticide demand. Paper presented at American Agricultural Economics Association Meetings August 4–7, 1991, Manhattan, Kansas.

Carpentier, L. 1993. A GIS Approach to Point/Nonpoint Source Trading in Agriculture. Unpublished Paper. Blacksburg: Virginia Polytechnic and State University, Department of Agricultural Economics.

Carr, P. M., G. R. Carlson, J. S. Jacobsen, G. A. Nielsen, and E. O. Skogley. 1991. Farming soils, not fields: A strategy for increasing fertilizer profitability. Journal of Production Agriculture 4:57–61.

Carsel, R. G., C. N. Smith, L. A. Mulkey, J. D. Dean, and P. P. Jowise. 1984. User's Manual for Pesticide Root Zone Model (PRZM): Release 1. EPA-600/3–84–109. Athens, Ga.: U.S. Environmental Protection Agency.

Carson, R. 1962. Silent Spring. Boston: Houghton Mifflin.

Carter, D. L., J. A. Bondurant, and C. W. Robbins. 1971. Water-soluble $NO_3$-nitrogen, $PO_4$-phosphorus, and total salt balances on a large irrigation tract. Soil Science Society of America Proceedings 35:331–335.

Cartwright, N. L., Clark, and P. Bird. 1991. The impact of agriculture on water quality. Outlook on Agriculture 20:145–152.

Central Platte Natural Resources District. 1992. Central Platte NRD's Groundwater Quality Management Program: Groundwater Quality Improvement for the Central Platte Valley. Grand Island, Neb.: Central Platte Natural Resources District.

Cerrato, M. E., and A. M. Blackmer. 1990. Comparison of models for describing corn yield response to nitrogen fertilizer. Agronomy Journal 82:138–143.

Cerrato, M. E., and A. M. Blackmer. 1991. Relationship between leaf nitrogen concentrations and the nitrogen status of corn. Journal of Production Agriculture 4:525–531.

Chang, A. C., J. A. Warneke, A. L. Page, and L. J. Lund. 1984. Accumulation of heavy metals in sewage sludge-treated soils. Journal of Environmental Quality 13:87–91.

Charudattan, R. 1991. The mycoherbicide approach with plant pathogens. Pp. 24–57 in Microbial Control of Weeds, D. D. TeBeest, ed. London: Chapman and Hall.

Cheng, H. H., ed. 1990. Pesticides in the Soil Environment: Processes, Impacts, and Modeling. Soil Science Society of America Book Series No. 2. Madison, Wis.: Soil Science Society of America.

Childs, E. C., and N. Collis-George. 1950. The permeability of porous materials. Proceedings of the Royal Society (London) Series A 201:392–405.

Christie, P. 1987. Long term effects of slurry on grassland. Pp. 301–304 in Animal Manure on Grassland and Fodder Crops: Fertilizer or Waste?, H. G. Van Der Meer, R. J. Unwin, T. A. Van Kijk, and G. C. Enik, eds. The Hague, The Netherlands: Martinus Nijhoff.

Clark, E. H., II, J. A. Haverkamp, and W. Chapman. 1985. Eroding Soils: The Off-Farm Impacts. Washington, D.C.: The Conservation Foundation.

Clausen, J. C., and D. W. Meals, Jr. 1989. Water quality achievable with agricultural best management practices. Journal of Soil and Water Conservation 44:593–596.

Clemmons, A. J. 1987. Delivery System Schedules and Required Capacities. In Planning, Operation, Rehabilitation and Automation of Irrigation Water Delivery Systems, Proceedings of an ASCE Irrigation and Drainage Division Symposium, D. D. Zimbelman, ed. New York: American Society of Civil Engineers.

Cochrane, W. W. 1986. A new sheet of music: How Kennedy's farm advisor has changed his tune about commodity policy and why. Choices 1:11–15.

Cochrane, W. W., and C. F. Runge. 1993. Reforming Farm Policy: Toward a National Agenda. Ames: Iowa State University Press.

Cooper, J. R., J. W. Gilliam, R. B. Daniels, and W. P. Robarge. 1987. Riparian areas as filters for agricultural sediment. Soil Science Society of America Journal 51:416–420.

Coote, D. R., E. M. MacDonald, and R. DeHaan. 1978. Relationships between agricultural land and water quality. Pp. 79–92 in Best Management Practices for Agriculture and Silviculture, R. C. Loehr, D. A. Haith, M. F. Walter, and C. S. Martin, eds. Proceedings of the 1978 Cornell Agricultural Waste Conference. Ann Arbor, Mich.: Ann Arbor Science.

Cope, J. T., Jr. 1981. Effects of 50 years of fertilization with phosphorus and potassium on soil test levels and yields at six locations. Soil Science Society of America Journal 45:342–347.

Cox, F. R., E. J. Kamprath, and R. E. McCollum. 1981. A descriptive model of soil test nutrient levels following fertilization. Journal of the Soil Science Society of America 45:529–532.

Creason, J. R., and C. F. Runge. 1990. Agricultural Competitiveness and Environmental Quality: What Mix of Policies Will Accomplish Both Goals? St. Paul: University of Minnesota, Center for International Food and Agricultural Policy.

Cross, H. R., and F. M. Byers. 1990. Current Issues in Food Production: A Perspective on Beef as a Component in Diets for Americans. College Station: Texas A&M University, and Englewood, Colo.: National Cattlemens Association.

Crosson, P. 1985. Impact of erosion on land productivity and water quality in the United States. Pp. 217–236 in Soil Erosion and Conservation, S. A. El-Swaify, W. C. Moldenhauer, and A. Lo eds. Ankeny, Iowa: Soil Conservation Society of America.

Crosson, P. R., P. Dyke, J. Miranowski, and D. Walker. 1985. A framework for analyzing the productivity costs of soil erosion in the United States. Pp. 481–503 in Soil Erosion and Crop Productivity, R. F. Follett, B. A. Stewart, eds. Madison, Wis.: American Society of Agronomy.

Crosson, P. R., and A. J. Stout. 1983. Productivity Effects of Cropland Erosion in the United States. Washington, D.C.: Resources for the Future.

Crosswhite, W. M., and C. L. Sandretto. 1991. Trends in resource protection policies in agriculture. Pp. 42–49 in Agricultural Resources: Cropland, Water and Conservation Situation and Outlook, Report No. AR-23. Washington, D.C.: U.S. Department of Agriculture, Economic Research Service, Resources and Technology Division.

Dalton, F. N., W. N. Herkelrath, D. S. Rawlins, and J. D. Rhoades. 1984. Time-domain reflectometry: Simultaneous measurement of sink water content and electrical conductivity with a single probe. Science 224:989–990.

Daniel, T. C., R. C. Wendt, P. E. McGuire, and D. Stoffel. 1982. Nonpoint source loading rates from selected land uses. Water Resources Bulletin 18:117–120.

Deason, J. P. 1989. Irrigation-induced contamination: How real a problem? Journal of Irrigation and Drainage Engineering 115:9–20.

DeBach, P., and D. Rosen. 1991. Biological Control by Natural Enemies. Second Edition. Cambridge: Cambridge University Press.

Dendy, F. E., and G. C. Bolton. 1976. Sediment yield-runoff drainage area relationships in the United States. Journal of Soil and Water Conservation 31:264–266.

Deverel, J., and R. Fujii. 1990. Chemistry of trace elements in soils and ground water. Pp. 64–90 in Agricultural Salinity Assessment and Management, K. K. Tanji, ed. ASCE Manuals and Reports on Engineering Practice No. 71. New York: American Society of Civil Engineers.

Dillaha, T. A., R. B. Reneau, S. Mostaghimi, and D. Lee. 1989b. Vegetative filter strips for agricultural nonpoint source pollution control. Transactions of the American Society of Agricultural Engineers 32:513–519.

Dillaha, T. A., J. H. Sherrard, and D. Lee. 1989a. Long-term effectiveness of vegetative filter strips. Water, Environment and Technology 1:419–421.

Dillaha, T. A., J. H. Sherrard, D. Lee, S. Mostaghimi, and V. O. Shanholtz. 1988. Evaluation of vegetative filter strips as a best management practice for feed lots. Journal of the Water Pollution Control Federation 60:1231–1238.

Dillon, P. J., and F. H. Rigler. 1974. The phosphorus-chlorophyll relationship in lakes. Limnology and Oceanography 19:767–773.

Dixon, O., P. Dixon, and J. Miranowski. 1973. Insecticide requirement in an efficient agricultural sector. The Review of Economics and Statistics 55:423–432.

Dobbs, T. L., D. C. Taylor, and J. D. Smolik. 1992. Farm, Rural Economy, and Policy Implications of Sustainable Agriculture in South Dakota. Agricultural Experiment Station Bulletin 713. Brookings: South Dakota State University.

Doering, E. J., and F. M. Sandoval. 1976. Hydrology of saline seeps in the Northern Great Plains. Transactions of the American Society of Civil Engineers 19:856–861, 865.

Doering, O. 1991. Federal policies as incentives or disincentives to ecologically sustainable agricultural systems. Paper presented to Environmental Protection Agency Sustainable Agricultural Workshop, July 22, 1991.

Doering, O., and D. Ervin. 1990. Policy options to help the environment. Pp. 21–22 in Agricultural Outlook, AO-167. Rockville, Md.: U.S. Department of Agriculture.

Dolan, M. S., R. H. Dowdy, W. B. Voorhees, J. F. Johnson, and A. M. Bidwell-Schrader. 1992. Corn phosphorus and potassium uptake in response to soil compaction. Agronomy Journal 84:639–642.

Doneen, L. D., A. Quek, and K. K. Tanji. 1960. Water quality in relation to soil and crop production, Lower San Joaquin Valley. Pp. C-1-C-125 in Lower San Joaquin Valley Water Quality Investigation. Bulletin No. 89. Sacramento: California Department of Water Resources.

Donigian, A. S., Jr., D. C. Berzerlein, H. H. Davis, Jr., and N. H. Crawford. 1977. Agricultural Runoff Management (ARM) Model—Version II. Refinement and Testing. Publication No. EPA-600/3-77-098. Palo Alto, Calif.: Hydrocomp.

Donigian, A. S., Jr., and N. H. Crawford. 1976. Modeling Pesticides and Nutrients on Agricultural Lands. Publication No. EPA-600/23-76-043. Palo Alto, Calif.: Hydrocomp.

Doran, J. W., and M. S. Smith. 1991. Role of cover crops in nitrogen cycling. Pp. 85–90 in Cover Crops for Clean Water, W. L. Hargrove, ed. Ankeny, Iowa: Soil and Water Conservation Society.

Dorich, R. A., D. W. Nelson, and L. E. Sommers. 1980. Algal availability of sediment phosphorus in drainage water of the Black Creek watershed. Journal of Environmental Quality 9:557–563.

Dorich, R. A., D. W. Nelson, and L. E. Sommers. 1985. Estimation of algal available phosphorus in suspended sediments by chemical extraction. Journal of Environmental Quality 14:400–405.

Draycott, A. P., R. Hull, A. B. Messem, and J. Webb. 1970. Effects of soil compaction on yield and fertilizer requirement of sugar beet. Journal of Agricultural Science (Cambridge) 75:533–537.

Dudal, R. 1981. An evaluation of conservation needs. Pp. 3–12 in Soil Conservation, Problems and Prospects, R. P. C. Morgan, ed. Chichester, U.K.: Wiley.

Dudal, R. 1982. Land degradation in a world perspective. Journal of Soil and Water Conservation 37:345–349.

Duffy, M., and L. Thompson. 1991. The Extent and Nature of Iowa Crop Production Practices, 1989. Report No. FM 1839. Ames: Iowa State University, Cooperative Extension Service.

Duxbury, J. M., D. Bouldin, R. E. Terry, and R. L. Tate, III. 1982. Emissions of nitrous oxide from soils. Nature 298:462–464.

Duxbury, J. M., and J. H. Peverly. 1978. Nitrogen and phosphorus losses from organic soils. Journal of Environmental Quality 7:566–570.

Edwards, R. C., and R. E. Ford. 1992. Integrated pest management in the corn/soybean agroecosystem. Pp. 13–56 in Food, Crop Pests, and the Environment, F. G. Zalom and W. E. Fry, eds. St. Paul, Minn.: APS Press.

Edwards, W. M. 1991. Soil structure: Processes and management. Pp. 7–14 in Soil Management for Sustainability, R. Lal and F. J. Pierce, eds. Ankeny, Iowa: Soil and Water Conservation Society.

Eheart, J. W., E. D. Brill, Jr., and R. M. Lyon. 1983. Transferable discharge permits for control of BOD: An overview. Pp. 163–195 in Buying a Better Environment: Cost-Effective Regulation Through Permit Trading, E. F. Joeres and M. H. David, eds. Madison: University of Wisconsin Press.

Eheart, J. W., E. Wayland, D. Brill, Jr., B. J. Lence, J. D. Kilgore, and J. G. Uber. 1987. Cost efficiency of time-varying discharge permit programs for water quality management. Water Resources Research 23:245–251.

Ehrenfield, J. G. 1987. The role of woody vegetation in preventing ground water pollution by nitrogen from septic tank leachate. Water Resources Research 21:605–614.

Eichner, M. J. 1990. Nitrous oxide emissions from fertilized soils: Summary of available data. Journal of Environmental Quality 19:272–280.

El-Hout, N. M., and A. M. Blackmer. 1990. Nitrogen status of corn after alfalfa in 29 Iowa fields. Journal of Soil and Water Conservation 45:115–117.

Enache, A., and R. D. Ilnicki. 1990. Weed control by subterranean clover (*Trifolium subterraneum*) used as a living mulch. Weed Technology 4:534–538.

Engberg, R. A., M. A. Sylvester, and H. R. Fettz. 1991. Effects of drainage on water, sediment and biota. Pp. 801–807 in Proceedings of the National Conference on Irrigation and Drainage, N. F. Ritter, ed. New York: American Society of Civil Engineers.

Eradat Oskoui, K., and W. B. Voorhees. 1990. Economic consequences of soil compaction. ASAE Paper No. 90–1089. St. Joseph, Mich.: American Society of Agricultural Engineers.

Esseks, J. D., S. E. Kraft, and L. K. Vinis. 1990. Agriculture and the Environment: A Study of Farmer's Practices and Perceptions. Washington, D.C.: American Farmland Trust.

Evans, H. J., and L. E. Barber. 1977. Biological nitrogen fixation for food and fiber production. Science 197:332–339.

Exner, M. E. 1985. Concentration of nitrate-nitrogen in groundwater, Central Platte region, Nebraska, 1984. Lincoln: Conservation and Survey Division, University of Nebraska.

Exner, M. E., and R. F. Spaulding. 1976. Groundwater quality of the Central Platte region, 1974. Resource Atlas No. 2. Lincoln: Conservation and Survey Division, University of Nebraska.

Exner, M. E., and R. F. Spalding. 1990. Occurrence of pesticides and nitrate in Nebraska's groundwater. Lincoln: Water Center, Institute of Agriculture and Natural Resources, The University of Nebraska.

Faeth, P., Repetto, R., Q. Din, and G. Helmers. 1991. Paying the Farm Bill: U.S. Agricultural Policy and the Transition to Sustainable Agriculture. Washington, D.C.: World Resources Institute.

Falth, P. 1993. Evaluating Agricultural Policy and the Sustainability of Production Systems: An Economic Framework. Journal of Soil and Water Conservation 48(2):94–99.

Fan, S.-S. 1988. Twelve Selected Computer Sedimentation Models Developed in the United States. Washington, D.C.: Federal Energy Regulatory Commission, Federal Interagency Advisory Committee on Water Data.

Fausey, N. R., and A. S. Dylla. 1984. Effects of wheel traffic along one side of corn and soybean rows. Soil Tillage Research 4:147–154.

Ferguson, R. B., C. A. Shapiro, G. W. Hergert, W. L. Kranz, N. L. Klocke, and D. H. Krull. 1991. Nitrogen and irrigation management practices to minimize nitrate leaching from irrigated corn. Journal of Production Agriculture 4:186–192.

Flanagan, D. C., W. H. Neibling, and J. P. Burt. 1989. Simplified equations for filter strip design. Transactions of the American Society of Agricultural Engineers 32:2001–2007.

Flaxman, E. M. 1972. Predicting sediment yield in Western United States. Journal of the Hydraulic Division, American Society of Civil Engineers 98:2073–2085.

Fleming, M. H. 1987. Agricultural chemicals in ground water: Preventing contamination by removing barriers against low-input farm management. American Journal of Alternative Agriculture 2:124–130.

Fletcher, J. J., and T. T. Phipps. 1991. Data needs to assess environmental quality issues related to agriculture and rural areas. American Journal of Agricultural Economics 3:926–932.

Flint, M. L. 1989. Annual Report, Statewide IPM Project. Focus: Reducing Pesticide Use. Davis: University of California.

Follett, R. F., ed. 1989. Nitrogen Management and Ground Water Protection. Developments in Agricultural and Managed-Forest Ecology 21. Amsterdam: Elsevier.

Follett, R. F., S. C. Gupta, and P. G. Hunt. 1987. Conservation practices: relation to the management of plant nutrients for crop production. Pp. 19–51 in Soil Fertility and Organic Matter as Critical Components of Production Systems, Soil Science Society of America, Special Publication 19, R. F. Follett, et al., eds. Madison, Wis.: American Society of Agronomy.

Follett, R. F., D. R. Keeney, and R. M. Cruse, eds. 1991. Managing Nitrogen for Groundwater Quality and Farm Profitability. Madison, Wis.: Soil Science Society of America.

Follett, R. F., and D. S. Schimel. 1989. Effect of tillage practices on microbial biomass dynamics. Soil Science Society of America Journal 53:1091–1096.

Food and Agriculture Organization of the United Nations. 1983. Guidelines: Land Evaluation for Rainfed Agriculture. Soils Bulletin No. 52. Rome: Food and Agriculture Organization of the United Nations.

Foran, J. A., P. Butler, L. B. Cleckner, and J. W. Bulkley. 1991. Regulating nonpoint source pollution in surface waters: A proposal. Water Resources Bulletin 27:479–484.

Forman, R. T. T., and M. Godron. 1986. Landscape Ecology. New York: Wiley.

Foster, G. R., and L. J. Lane. 1987. User Requirements. USDA-Water Erosion Prediction Project (WEPP). NSERL Report No. 1. West Lafayette, Ind.: U.S. Department of Agriculture.

Foster, G. R., L. J. Lane, J. D. Nowlin, J. M. Laflen, and R. A. Young. 1981. Estimating erosion and sediment yield on field-sized areas. Transactions of the American Society of Agricultural Engineers 24:1253–1262.

Fox, R. H., G. W. Roth, K. V. Iversen, and W. P. Piekielek. 1989. Soil and tissue test compared for predicting soil nitrogen availability to corn. Agronomy Journal 81:971–974.

Fox, R. L., and E. J. Kamprath. 1971. Adsorption and leaching of P in acid organic soils and high organic matter sand. Soil Science Society of America Proceedings 35:154–156.

Fox, T. R., N. B. Comerford, and W. W. McFee. 1990a. Kinetics of phosphorus release from Spodosols: Effects of oxalate and formate. Soil Science Society of America Journal 54:1441–1447.

Fox, T. R., N. B. Comerford, and W. W. McFee. 1990b. Phosphorus and aluminum release from a spodic horizon mediated by organic acids. Soil Science Society of America Journal 54:1763–1767.

Fraczek, W. 1988. Assessment of the Effects of Changes in Agricultural Practices on the Magnitude of Floods in Coon Creek Watershed Using Hydrograph Analysis and Airphoto Interpretation. M.S. thesis. University of Wisconsin, Madison.

Franzmeier, D. P. 1990. Soil landscapes and erosion processes. Pp. 81–104 in Proceedings of Soil Erosion and Productivity Workshop, W. E. Larson, G. R. Foster, R. R. Allmaras, and C. M. Smith, eds. St. Paul: University of Minnesota.

Frere, M. H. 1978. Models for predicting water pollution from agricultural watersheds. Pp. 501–509 in Conference on Modeling and Simulation of Land, Air and Water Resources Systems. Ghent, Belgium: International Federation of Information Process.

Frink, C. R. 1969. Water pollution potential estimated from farm nutrient budgets. Agronomy Journal 61:550–553.

Frye, W. W. 1987. The effects of soil erosion on crop productivity. Pp. 151–172 in Agricultural Soil Loss: Processes, Policies, and Prospects, J. M. Harlin and G. M. Berardi, eds. Boulder, Colo.: Westview Press.

Frye, W. W., J. J. Varco, R. L. Blevins, M. S. Smith, and S. J. Corak. 1988. Role of annual legume cover crops in efficient use of water and nitrogen. Pp. 129–154 in Cropping Strategies for Efficient Use of Water and Nitrogen, W. L. Hargrove, ed. ASA Special Publication No. 51. Madison, Wis.: American Society of Agronomy, Crop Science Society of America, and Soil Science Society of America.

Garner, W. A., R. C. Honeycutt, and H. N. Nigg. 1986. Evaluation of Pesticides in Groundwater. ACS Symposium Series 315. Washington, D.C.: American Chemical Society.

Gianessi, L. P., H. M. Peskin, P. Crosson, and C. Pheffer. 1986. Nonpoint source pollution: Are agplane controls the answer? Journal of Soil and Water Conservation 41:215–219.

Gilbertson, C. B., F. A. Nordstadt, A. C. Mathers, R. F. Holt, A. P. Barnett, T. M. McCalla, C. A. Onstad, and R. A. Young. 1979. Animal waste utilization on cropland and pastureland: A manual for evaluating agronomic and environmental effects. Pp. 32–34 in USDA Report No. URR-6. Hyattsville, Md.: U.S. Department of Agriculture, Science and Education Administration.

Gilliam, J. W., T. J. Logan, and F. E. Broadbent. 1985. Fertilizer use in relation to the environment. Pp. 561–588 in Fertilizer Technology and Use, O. P. Engelstad, ed. Madison, Wis.: Soil Science Society of America.

Gilliom, R. J. 1989. Preliminary Assessment of Sources, Distribution and Mobility of Selenium in the San Joaquin Valley, California. USGS Water Resources Investigations Report No. 88–4186. Regional Aquifer System Analysis. Sacramento, Calif.: U.S. Geological Survey.

Glymph, L. M. 1956. Importance of sheet erosion as a source of sediment. Transactions of the American Geophysical Union 38:903–907.

Gold, A. J., W. R. DeRagon, W. M. Sullivan, and J. L. Lemunyon. 1990. Nitrate-nitrogen losses to groundwater from rural and suburban land uses. Journal of Soil and Water Conservation 45:305–310.

Goodroad, L. L., D. R. Keeney, and L. A. Peterson. 1984. Nitrous oxide emissions from agricultural soils in Wisconsin. Journal of Environmental Quality 13:557–561.

Granatstein, D. 1991. "Living soil" depends on microbial management. Sustainable Farming Quarterly 3:1–4.

Granatstein, D., and D. F. Bezdicek. 1992. The need for a soil quality index: Local and regional perspectives. American Journal of Alternative Agriculture 7:12–16.

Grattan, S. R., and C. M. Grieve. 1992. Mineral element acquisition and growth response to plants grown in saline environments. Agriculture, Ecosystems and the Environment 38:275–300.

Green, R. E., C. C. K. Liu, and N. Tamraker. 1986. Modeling pesticide movement in the unsaturated zone of Hawaiian soils under agricultural use. Pp. 366–383 in Evaluation of Pesticides in Groundwater, W. A. Garner, R. C. Honeycutt, and H. N. Nigg, eds. ACS Symposium Series 315. Washington, D.C.: American Chemical Society.

Griffith, R. W., C. Goudey, and R. Poff. 1992. Current application of soil quality standards. Pp. 1–5 in Proceedings of the Soil Quality Standards Symposium. Report No. W0-WSA-2. Washington, D.C.: U.S. Department of Agriculture, Forest Service.

Grissinger E. H., A. J. Bowie, and J. B. Murphey. 1991. Goodwin Creek bank instability and sediment yield. Pp. 532–539 in Proceedings of the Fifth Federal Interagency Sedimentation Conference.

Groffman, P. M., E. A. Axelrod, J. L. Lemunyon, and W. M. Sullivan. 1991. Denitrification in vegetated filter strips. Journal of Environmental Quality 20:671–674.

Groffman, P. M., A. J. Gold, and R. C. Simmons. 1992. Nitrate dynamics in riparian forests: Microbial studies. Journal of Environmental Quality 21:666–671.

Groszyk, W. S. 1978. Nonpoint source pollution control strategy. Pp. 3–10 in Best Management Practices for Agriculture and Silviculture, R. C. Loehr, D. A. Haith, M. F. Walter, and C. S. Martin, eds. Proceedings of the 1978 Cornell Agricultural Waste Conference. Ann Arbor, Mich.: Ann Arbor Science.

Grove, J. H., W. O. Thom, L. W. Murdock, and J. H. Herbeck. 1987. Soybean response to available potassium in three silt loam soils. Soil Science Society of America Journal 51:1231–1238.

Grunig, J. E., C. L. Nelson, S. J. Richburg, and T. J. White. 1988. Communication by agricultural publics: Internal and external orientations. Journalism Quarterly 65:26–38.

Gupta, S. C., and W. E. Larson. 1979a. Estimating soil water retention characteristics from particle-size distribution, organic matter percent and bulk density. Water Resources Research 15:1633–1635.

Gupta, S. C., and W. E. Larson. 1979b. A model for predicting packing density of soils using particle-size distribution. Soil Science Society of America Journal 44:758–764.

Gupta, S. C., J. F. Moncrief, and R. P. Ewing. 1991. Soil crusting in the midwestern United States. In International Symposium on Soil Crusting: Chemical and Physical Processes. Gengia, USA: University of Athens.

Haberen, J. 1992. Coming full circle: The new emphasis on soil quality. American Journal of Alternative Agriculture 7:3–4.

Hagen, L. J. 1988. Wind erosion prediction system: An overview. Paper 88-2554 presented at the winter meeting of the American Society of Agricultural Engineers, December 13–16, 1988, Chicago, Ill.

Hagen, L. J., and P. T. Dyke. 1980. Merging data from disparate sources. Agricultural Economics Research 32(4):45–49.

Hagen, L. J., T. Zobeck, and D. W. Fryrear. 1988. Concepts for Modeling Wind Erosion. Proceedings of International Conference on Dryland Farming, Amarillo, Texas. Washington, D.C.: U.S. Department of Agriculture: Agriculture Research Service.

Hahne, H. C. H., W. Kroontje, J. A. Lutz, Jr. 1977. Nitrogen fertilization. I. Nitrate accumulation and losses under continuous corn cropping. Journal of Soil Science Society of America 41:562–567.

Haith, D. A. 1980. A mathematical model for estimating pesticide losses in runoff. Journal of Environmental Quality 9:428–433.

Haith, D. A. 1986. Simulated regional variations in pesticide runoff. Journal of Environmental Quality 15:5–8.

Hakansson, I. 1985. Swedish experiments on subsoil compaction by vehicles with high axle load. Soil Use Management 1:113–116.

Hall, D. W. 1992. Effects of Nutrient Management on Nitrate levels in ground water near Ephrata, Pennsylvania. Ground Water 30:720–730.

Hallberg, G. R. 1987. Nitrates in groundwater in Iowa. Pp. 23–68 in Rural Groundwater Contamination, F. M. D'Itri and L. G. Wolfson, eds. Chelsea, Mich.: Lewis Publishing.

Hallberg, G. R. 1989a. Pesticide pollution of groundwater in the humid United States. Agriculture Ecosystems and Environment 26:299–368.

Hallberg, G. R. 1989b. Nitrate in ground water in the United States. Pp. 35–74 in Nitrogen Management and Ground Water Protection, Developments in Agricultural and Managed-Forest Ecology 21, R. F. Follet, ed. Amsterdam: Elsevier.

Hallberg, G. R., C. K. Contant, C. A. Chase, G. A. Miller, M. D. Duffy, R. J. Killorn, R. D. Voss, A. M. Blackmer, S. C. Padgett, J. R. DeWitt, J. B. Guilliford, D. A. Lindquist, L. W. Asell, D. R. Keeney, R. D. Libra, and K. D. Rex. 1991. A Progress Review of Iowa's Agricultural-Energy-Environmental Initiatives: Nitrogen Management in Iowa. Technical Information Series 22. Des Moines: Iowa Department of Natural Resources.

Halvorson, A. D. 1990. Management of dryland saline seeps. Pp. 372–392 in Agricultural Salinity Assessment and Management, K. K. Tanji, ed. ASCE Manuals and Reports on Engineering Practice No. 71. New York: American Society of Civil Engineers.

Hamilton, N. D. 1993. Feed our future: Six philosophical issues shaping agricultural law. Nebraska Law Review 701:717–720.

Hamlett, J. M., D. A. Miller, R. L. Day, G. W. Petersen, G. M. Baumer, and J. Russo. 1992. Statewide GIS-based ranking of watersheds for agricultural pollution prevention. Journal of Soil and Water Conservation 47:399–404.

Hanks, J., and J. T. Ritchie, eds. 1992. Modeling Plant and Soils Systems. Agronomy Monographs No. 31. Madison, Wis.: American Society of Agronomy.

Hanway, J. J., S. A. Barber, R. H. Bray, A. C. Caldwell, M. Fried, L. T. Kurtz, K. Lawton, J. T. Pesek, K. Pretty, M. Reed, and F. W. Smith. 1962. North Central Regional Potassium Studies. III. Field Studies with Corn, Research Bulletin 503. Des Moines, Iowa: Iowa Agricultural Home Economics Experiment Station.

Hanway, J. J., and J. M. Laflen. 1974. Plant nutrient losses from tile-outlet terraces. Journal of Environmental Quality 3:351–356.

Hargett, N. L., and J. T. Berry. 1988. Commercial Fertilizers, 1987. Muscle Shoals, Ala.: Tennessee Valley Authority.

Hargrove, W. L., ed. 1988. Cropping Strategies for Efficient Use of Water and Nitrogen. ASA Special Publication No. 51. Madison, Wis.: American Society of Agronomy, Crop Science Society of America, and Soil Science Society of America.

Hargrove, W. L., ed. 1991. Cover Crops for Clean Water. Ankeny, Iowa: Soil and Water Conservation Society.

Hargrove, W. L., A. L. Black, and J. V. Mannering. 1988. Cropping strategies for efficient use of water and nitrogen: Introduction. Pp. 1–6 in Cropping Strategies for Efficient Use of Water and Nitrogen, W. L. Hargrove, ed. ASA Special Publication No. 51. Madison, Wis.: American Society of Agronomy, Crop Science Society of America, and Soil Science Society of America.

Hargrove, W. L., G. W. Langdale, and A. W. Thomas. 1984. Role of legume cover crops in conservation tillage production systems. Paper of the American Society of Agricultural Engineers Microfiche Collection (fiche no. 84–2038). St. Joseph, Mich.: The Society.

Harrington, D. H., and O. C. Doering, III. 1993. Agricultural policy reform: A proposal. Choices (1st quarter):14–17 and 40–41.

Harris, J. H. 1987. The Delaware solution to poultry manure management and groundwater quality. Pp. 145–157 in Chautaugua groundwater workshop for extension agents: Proceedings of the Workshop held at the Chautaugua Institution, A. Rudd, ed. University Park, Pa.: Northeast Regional Center for Rural Development.

Harwood, R. R. 1990. A history of sustainable agriculture. Pp. 3–19 in Sustainable Agricultural Systems, C. A. Edwards, R. Lal, P. Madden, R. H. Miller, and G. House, eds. Ankeny, Iowa: Soil and Water Conservation Society.

Harwood, R. R. 1993. Managing the Living Soil for Human Well-being in an Increasing Populous and Interdependent World. Presentation made at Winrock International, May 5, 1993, Morrilton, AR.

Hauck, R. D., and K. K. Tanji. 1982. Nitrogen transfers and mass balances. Pp. 891–925 in Nitrogen in Agricultural Soils. Agronomy Monographs No. 22, F. J. Stevenson, ed. Madison, Wis.: American Society of Agronomy.

Hayes, J. C., B. J. Barfield, and R. I. Barnhisel. 1979. Filtration of sediment by simulated vegetation. Unsteady flow with non-homogeneous sediment. Transactions of the American Society of Agricultural Engineers 22:1063.

Hayes, J. C., and J. E. Hairson. 1983. Modeling the long-term effectiveness of vegetative filters on on-site sediment controls. ASAE Paper No. 83–2081. St. Joseph, Mich.: American Society of Agricultural Engineers.

Heady, E. O., J. T. Pesek, and W. G. Brown. 1955. Crop response surfaces and economic optima in fertilizer use. Iowa Experiment Station Research Bulletin 424. Ames, Iowa: Iowa State University.

Heerman, D. F., W. W. Wallender, and M. G. Bos. 1990. Irrigation efficiency and uniformity. Pp. 125–149 in Management of Farm Irrigation Systems, G. J. Hoffman, T. A. Howell, and K. H. Solomon, eds. St. Joseph, Mich.: American Society of Agricultural Engineers.

Heichel, G. H. 1987. Legumes as a source of nitrogen in conservation tillage systems. Pp. 29–35 in The Role of Legumes in Conservation Tillage Systems, J. F. Power, ed. Ankeny, Iowa: Soil Conservation Society of America.

Hertel, T. W., M.E. Tsigas, and P. V. Preckel. 1990. Unfreezing program payment yields: Consequences and alternatives. Choices 5(2):32–33.

Hillel, D. J. 1991. Out of the Earth: Civilization and the Life of the Soil. New York: The Free Press.

Hjemfelt, A. T. Jr., and L. A. Kramer. 1988. Unit hydrograph variability for a small agricultural watershed. Pp. 357–366 in Modeling Agricultural, Forest, and Rangeland Hydrology. Proceedings of the 1988 International Symposium, American Society of Agricultural Engineers, December 12–13, 1988, Chicago.

Holden, L. R., J. A. Graham, R. W. Whitmore, W. J. Alexander, R. W. Pratt, S. K. Liddle, and L. L. Piper. 1992. Results of the National Alachlor Well Water Survey. Environmental Science and Technology 26:935–943.

Holden, P. W. 1986. Pesticides and Groundwater Quality. Issues and Problems in Four States. Washington, D.C.: National Academy Press.

Hornsby, A. G. 1988. Applications of results: Use of interpreted results to develop management strategies for preventing groundwater contamination. Pp. 120–128 in Methods for Groundwater Quality Studies, D. W. Nelson and R. H. Dowdy, eds. Lincoln: University of Nebraska.

Hornsby, A. G. 1992. Site-specific pesticide recommendations: The final step in environmental impact prevention. Weed Technology 6:736–754.

Hornsby, A. G., and R. B. Brown. 1992. Soil parameters significant to pesticide fate. Pp. 62–71 in Proceedings of the Soil Quality Standards Symposium. Report No. W0-WSA-2. Washington, D.C.: U.S. Department of Agriculture, Forest Service.

Hortensius, D., and S. Nortcliff. 1991. International standardization of soil quality measurement procedures for the purpose of soil protection. Soil Use and Management 7:163–166.

Hrubovcak, J., M. LeBlanc, and J. Miranowski. 1990. Limitations in evaluating environmental and agricultural coordination benefits. American Economic Review 80:208–212.

Hubbard, R. K., A. E. Erickson, B. G. Ellis, and A. R. Wolcott. 1982. Movement of diffuse source pollutants in small agricultural watersheds of the Great Lakes basin. Journal of Environmental Quality 11:117–123.

Huggins, D. G., M. L. Johnson, P. M. Liechti, T. M. Anderson, S. Meador, and J. L. Whistler. 1990. Establishment of empirical relationships between land use/land cover and nonpoint source pollution stream effects within an ecoregion. MPS analysis, Project Report 3, Region VII, Kansas City, Kans. Washington, D.C.: U.S. Environmental Protection Agency.

Inman, R., D. Olson, and D. King. 1984. Grand Valley salt pickup calculations. Pp. 157–167 in Salinity in Watercourses and Reservoirs, R. H. French, ed. Boston: Butterworth.

Jackson, W., and J. Piper. 1989. The necessary marriage between ecology and agriculture. Bulletin of the Ecological Society of America 70:1591–1593.

Jacobs, J. J., and G. L. Casler. 1979. Internalizing externalities of phosphorus discharges from crop production to surface water: Effluent taxes versus uniform reductions. American Journal of Agricultural Economics 61:309–312.

Jacobs, T. C., and J. W. Gilliam. 1985. Riparian losses of nitrate from agricultural drainage water. Journal of Environmental Quality 14:272–278.

Jayman, T. C. Z., and S. Sivasubramaniam. 1975. Release of bound iron and aluminum from soils by the root exudates of tea (*Camellia sinensis*) plants. Journal of Science and Food Agriculture 26:1895–1898.

Johnson, J. F., S. D. Logsdon, W. B. Voorhees, and G. W. Randall. In press. Soil and plant responses to high axle load wheel traffic and subsoil tillage.

Johnson, J. F., W. B. Voorhees, W. W. Nelson, and G. W. Randall. 1990. Soybean growth and yield as affected by surface and subsoil compaction. Soil Science Society of America Journal 82:973–979.

Johnson, J. S. 1978. The role of conservation practices as best management practices. Pp. 69–78 in Best Management Practices for Agriculture and Silviculture, R. C. Loehr, D. A. Haith, M. F. Walter, and C. S. Martin, eds. Proceedings of the 1978 Cornell Agricultural Waste Conference. Ann Arbor, Mich.: Ann Arbor Science.

Johnson, L. C. 1987. Soil loss tolerance: Fact or myth? Journal of Soil and Water Conservation 42:155–160.

Johnson, M. G., D. A. Lammers, C. P. Andersen, P. T. Rygiewcz, and J. S. Kern. 1992. Sustaining soil quality by protecting the soil resource. Pp. 72–80 in Proceedings of the Soil Quality Standards Symposium, San Antonio, Texas, October 23, 1990. Watershed and Air Management Report No. WO-WSA-2. Washington, D.C.: U.S. Department of Agriculture, U.S. Forest Service.

Jokela, W. E. 1992. Nitrogen fertilizer and dairy manure effects on corn yield and soil nitrate. Soil Science Society of America Journal 56:148–154.

Jokela, W. E., and G. W. Randall. 1989. Corn yield and residual soil nitrate as affected by time and rate of nitrogen application. Agronomy Journal 81:720–726.

Jones, A. R. 1984. Controlling salinity in the Colorado River basin, the arid west. Pp. 337–347 in Salinity in Watercourses and Reservoirs, R. H. French, ed. Boston: Butterworth.

Judson, S. 1981. What's happening to our continents? Pp. 12–139 in B. J. Skinner, ed. Use and Misuse of Earth's Surface, Los Altos, Calif.: William Kaufman.

Jury, W. A., W. R. Gardner, and W. H. Gardner. 1991. Soil Physics, 5th Ed. New York: Wiley.

Jury, W. A., W. F. Spencer, and W. J. Farmer. 1984. Behavior assessment model for trace organics in soils. III. Application of screening model. Journal of Environmental Quality 13:573–579.

Jury, W. A., G. Sposito, and R. E. White. 1986. A transfer function model of solute transport through soil. 1. Fundamental concepts. Water Resources Research 22:243–247.

Jury, W. A., A. Ul-Hassan, and G. Butters. 1988. The difficulty in predicting ground water Contamination. Pp. 111–120 in Proceedings of the Sixteenth Biennial Conference on Groundwater. Report No. 66. Riverside: University of California, Water Resources Center.

Kafkafi, U., B. Bar-Yosef, R. Rosenberg, and G. Sposito. 1988. Phosphorus adsorption by kaolinite and montmorillonite. II. Organic anion competition. Soil Science Society of America Journal 52:1585–1589.

Kamprath, E. J. 1967. Residual effect of large applications of phosphorus on high phosphorus fixing soils. Agronomy Journal 59:25–27.

Kamprath, E. J. 1989. Effect of starter fertilizer on early soybean growth and grain yield on Coastal Plain soils. Journal of Production Agriculture 2:318–320.

Kamprath, E. J., and M. E. Watson. 1980. Conventional soil and tissue tests for assessing the phosphorus status of soils. Pp. 433–469 in The Role of Phosphorus in Agriculture, F. E. Khasawneh, E. C. Sample, and E. J. Kamprath, eds. Madison, Wis.: American Society of Agronomy.

Karim, M. I., and W. A. Adams. 1984. Relationships between sesquioxides, kaolinite, and phosphate sorption in a catena of Oxisols in Malawi. Soil Science Society of America Journal 48:406–409.

Karr, J. R., and I. J. Schlosser. 1978. Water resources and the land-water interface. Science 201:229–234.

Kawai, K. 1980. The relationship of phosphorus adsorption to amorphous aluminum for characterizing Andosols. Soil Science 129:186–190.

Kay, B. D. 1989. Rates of change of soil structure under different cropping systems. Advances in Soil Science 12:1–52.

Keeney, D. R. 1982. Nitrogen management for maximum efficiency and minimum pollution: Farmed soils, fertilizer, agro-ecosystems. Pp. 605–649 in Nitrogen in Agricultural Soils. Agronomy Monographs No. 22, F. J. Stevenson, ed. Madison, Wis.: American Society of Agronomy.

Keeney, D. R. 1986a. Sources of nitrate to ground water. Critical Reviews in Environment Control 16:257–304.

Keeney, D. R. 1986b. Nitrate in ground water: Agricultural contribution and control. Pp. 329–351 in Agricultural Impacts on Groundwater. Worthington, Ohio: National Water Well Association.

Kelly, H. W. 1984. Communicating conservation. Journal of Soil and Water Conservation 39:23–25.

Khalid, R. A., W. H. Patrick, Jr., and R. D. DeLaune. 1977. Phosphorus sorption characteristics of flooded soils. Soil Science Society of America Journal 41:305–310.

Killorn, R., and D. Zourarakis. 1992. Nitrogen fertilizer management effect on corn grain yield and nitrogen uptake. Journal of Production Agriculture 5:142–148.

Kimes, S. C., J. L. Baker, and H. P. Johnson. 1979. Sediment Transport from Field to Stream: Particle Size and Yield. American Society of Agricultural Engineers Paper 79–2529. St. Joseph, Mich.: American Society of Agricultural Engineers.

Kiniry, L. N., C. L. Scrivner, and M. E. Keener. 1983. A soil productivity index based upon predicted water depletion and root growth. Research Bulletin 1051. Columbia: Missouri Agricultural Experiment Station.

Kirschenmann, F. 1991. Fundamental fallacies of building agricultural sustainability. Journal of Soil and Water Conservation 46:165–168.

Knisel, W. G. 1980. CREAMS: A Field Scale Model for Chemicals, Runoff, and Erosion from Agricultural Management Systems. Conservation Research Report No. 26. Washington, D.C.: Agricultural Research Service, U.S. Department of Agriculture.

Knox, J. C. 1977. Human impacts on Wisconsin stream channels. Annals of the Association of American Geographers 67:323–342.

Kovacic, D. A., L. L. Osborne, and B. C. Dickson. 1991. Buffer Strips and Nonpoint Pollution. Illinois Natural History Survey Reports No. 304.

Kramer, R. A., and S. S. Batie. 1985. The cross-compliance concept in agricultural programs: The New Deal to the present. Agricultural History 59(4):307–319.

Krause, K. R. 1992. The Beef Cow-Calf Industry, 1964–87: Location and Size. Agricultural Economic Report No. 659. Washington, D.C.: U.S. Department of Agriculture, Economic Research Service, Commodity Economics Division.

Kross, B. C., G. R. Hallberg, D. R. Bruner, R. D. Libra, and K. D. Rex. 1990. The Iowa State-Wide Rural Well-Water Survey, Water Quality Data: Initial Analysis. Technical Information Series 19. Des Moines: Iowa Department of Natural Resources, Geological Survey Bureau.

Krupnick, A. J. 1989. Tradeable Nutrients Permits and the Chesapeake Bay Compact. Discussion Paper QE89–07. Washington, D.C.: Resources for the Future.

Kuo, S. 1990. Phosphate sorption implications on phosphate soil tests and uptake by corn. Soil Science Society of America Journal 54:131–135.

Ladd, J. N., M. Amato, R. B. Jackson, and J. H. A. Butler. 1983. Utilization by wheat crops of nitrogen from legume residues decomposing in the field. Soil Biology and Biochemistry 18:417–425.

Laflen, J. M., W. J. Elliot, J. R. Simanton, C. S. Holzhey, and K. D. Kohl. 1991b. WEPP, soil erodibility experiments for rangeland and cropland soils. Journal of Soil and Water Conservation 46:39–44.

Laflen, J. M., R. Lal, and S. A. El-Swaify. 1990. Soil erosion and a sustainable agriculture. Pp. 569–581 in Sustainable Agricultural Systems, C. A. Edwards, R. Lal, P. Madden, R. H. Miller, and G. House, eds. Ankeny, Iowa: Soil and Water Conservation Society.

Laflen, J. M., L. J. Lane, and G. R. Foster. 1991a. WEPP, a new generation of erosion prediction technology. Journal of Soil and Water Conservation 46:34–38.

Lake, J. E. 1983. Sharing conservation tillage information. Journal of Soil and Water Conservation. Ankeny, Iowa: Soil Conservation Society of America 38:158–159.

Lal, R. 1987. Effects of soil erosion on crop production. CRC Critiques and Reviews in Plant Science 5:303–367.

Lal, R. 1988. Soil erosion by wind and water: Problems and prospects. Pp. 1–6 in Soil Erosion Research Methods, R. Lal, ed. Ankeny, Iowa: Soil and Water Conservation Society.

Lal, R. 1990. Soil erosion and land degradation.: The global risks. Advances in Soil Science 11:129–172.

Lal, R., and F. J. Pierce. 1991. The vanishing resource. Pp. 1–5 in Soil Management for Sustainability, R. Lal and F. J. Pierce, eds. Ankeny, Iowa: Soil and Water Conservation Society of America.

Lal, R., E. Regnier, D. J. Eckert, W. M. Edwards, and R. Hammond. 1991. Expectations of cover crops for sustainable agriculture. Pp. 1–11 in Cover Crops for Clean Water, W. L. Hargrove, ed. Ankeny, Iowa: Soil and Water Conservation Society.

Lal, R., and B. A. Stewart. 1990a. Need for action: Research and development priorities. Advances in Soil Science 11:331–336.

Lal, R., and B. A. Stewart. 1990b. Soil degradation: A global threat. Advances in Soil Science, 11:13–17.

Lamp, J. 1986. Minimum data sets and basic procedures for global assessments. Pp. 238–245 in Transactions XIII Congress of the International Society of Soil Science, Vol. V. Hamburg: International Society of Soil Science.

Lane, L. J., and M. A. Nearing. 1989. Profile Model Documentation, USDA-Water Erosion Prediction Project (WEPP). NSERL Report No. 2. West Lafayette, Ind.: National Soil Erosion Research Laboratory, Agricultural Research Service, U.S. Department of Agriculture.

Langdale, G. W., R. L. Blevins, D. L. Karlen, D. K. McCool, M. A. Nearing, E. L. Skidmore, A. W. Thomas, D. D. Tyler, and J. R. Williams. 1991. Cover crop effects on soil erosion by wind and water. Pp. 15–21 in Cover Crops for Clean Water, W. L. Hargrove, ed. Ankeny, Iowa: Soil and Water Conservation Society.

Langdale, G. W., R. A. Leonard, and A. W. Thomas. 1985. Conservation practice effects on phosphorus losses from Southern Piedmont watersheds. Journal of Soil and Water Conservation. Soil Conservation Society of America 40:157–161.

Langdale, G. W., W. C. Mills, and A. W. Thomas. 1992a. Use of conservation tillage to retard erosive effects of large storms. Journal of Soil and Water Conservation 47:257–260.

Langdale, G. W., L. T. West, R. R. Bruce, W. P. Miller, and A. W. Thomas. 1992b. Restoration of eroded soil with conservation tillage. Catena 5:81–90.

Langmuir, D., and J. Mahony. 1985. Chemical equilibrium and kinetics of geochemical processes in ground water studies. Pp. 69–95 in First Canadian/American Conference on Hydrogeology: Practical Application of Ground Water Geochemistry: Banff, Alberta, Canada, June 22–26, 1984. Worthington, Ohio: National Well Water Association.

Lanyon, L. E., and D. B. Beegle. 1989. The role of on-farm nutrient balance assessments in an integrated approach to nutrient management. Journal of Soil and Water Conservation 44:164–168.

Larsen, J. E., R. Langston, and G. F. Warren. 1958. Studies on the leaching of applied labeled phosphorus in organic soils. Soil Science Society of America Proceedings 22:558–560.

Larson, W. E., C. E. Clapp, W. H. Pierre, and Y. B. Morachan. 1972. Effects of increasing amounts of organic residues on continuous corn. II. Organic carbon, nitrogen, phosphorus, and sulfur. Agronomy Journal 64:204–208.

Larson, W. E., G. R. Foster, R. R. Allmaras, and C. M. Smith, eds. 1990. Proceedings of the Soil Erosion and Productivity Workshop. St. Paul, Minnesota: University of Minnesota.

Larson, W. E., R. F. Holt, and C. W. Carlson. 1978. Residues for soil conservation. Pp. 1–15 in Crop Residue Management Systems. Madison, Wisc.: Agronomy Society of America, Crop Science Society of America, Soil Science Society of America.

Larson, W. E., W. G. Lovely, J. T. Pesek, and R. E. Burwell. 1960. Effects of subsoiling and deep fertilizer placement on yields of corn in Iowa and Illinois. Agronomy Journal 54:185–189.

Larson, W. E., and F. J. Pierce. 1991. Conservation and the enhancement of soil quality. Pp. 175–203 in Evaluation for Sustainable Land Management in the Developing World, Vol. 2: Technical Papers. Bangkok, Thailand: International Board for Soil Research and Management.

Larson, W. E., and F. J. Pierce. 1994. The dynamics of soil quality as a measure of sustainable management. In Defining Soil Quality for a Sustainable Environment. Special Publication 33. Madison, Wis.: Soil Science Society of America.

Larson, W. E., F. J. Pierce, and R. H. Dowdy. 1983. The threat of soil erosion to long-term crop production. Science 219:458–465.

Larson, W. E., F. J. Pierce, and R. H. Dowdy. 1985. Loss in long-term productivity from soil erosion in the United States. Pp. 262–217 in Soil Erosion and Conservation, S. A. El-Swaify, W. C. Moldenhauer, and A. Lo, eds. Ankeny, Iowa: Soil Conservation Society of America.

Larson, W. E., and P. C. Robert. 1991. Farming by soil. Pp. 103–112 in Soil Management for Sustainability, R. Lal and F. J. Pierce, eds. Ankeny, Iowa: Soil and Water Conservation Society.

Larson, W. E., and B. A. Stewart. 1992. Thresholds for soil removal for maintaining cropland productivity. Pp. 6–14 in Proceedings of the Soil Quality Standards Symposium, San Antonio, Texas, October 23, 1990. Watershed and Air Management Report No. WO-WSA-2. Washington, D.C.: U.S. Department of Agriculture, U.S. Forest Service.

LaRue, T. A., and T. G. Patterson. 1981. How much nitrogen do legumes fix? Advances in Agronomy, 34:15–38.

Lee, D. L., T. A. Dillaha, and J. H. Sherrard. 1989. Modeling phosphorus transport in grass buffer strips. Journal of Environmental Engineering 115:409–427.

Lee, E. W. 1990. Drainage water treatment and disposal options. Pp. 432–449 in Agricultural Salinity Assessment and Management, K. K. Tanji, ed. ASCE Manuals and Reports on Engineering Practice No. 71. New York: American Society of Civil Engineers.

Leedy, J. B. 1979. Observations on the Sources of Sediment in Illinois Streams. Illinois Water Information System Report No. 18.

Legg, J. O., and J. J. Meisinger. 1982. Soil nitrogen budgets: Nitrogen balance, and cycle. Pp. 503–566 in Nitrogen in Agricultural Soils. Agronomy Monographs No. 22, F. J. Stevenson, ed. Madison, Wis.: American Society of Agronomy.

Legg, T. D., J. J. Fletcher, and K. W. Easter. 1989. Nitrogen budgets and economic efficiency: A case study of southeastern Minnesota. Journal of Production Agriculture 2:110–116.

Leonard, R. A., W. G. Knisel, and D. A. Still. 1987. GLEAMS: Groundwater loading effects of agricultural management systems. Transactions of the American Society of Agricultural Engineers 30:1403–1418.

Leonard, R. A., and W. D. Wauchope. 1980. The pesticide submodel. Pp. 88–112 in CREAMS: A Field-Scale Model for Chemicals, Runoff and Erosion from Agricultural Management Systems, W. G. Knisel, ed. Conservation Research Report No. 26. Washington, D.C.: U.S. Department of Agriculture.

Letson, D. 1992. Point/nonpoint source pollution reduction trading: An interpretive survey. Natural Resources Journal 32:219–232.

Lindau, C. W., R. D. DeLaune, W. H. Patrick, Jr., and P. K. Bollich. 1990. Fertilizer effects on dinitrogen, nitrous oxide, and methane emissions from lowland rice. Soil Science Society of America Journal 54:1789–1794.

Lindstrom, M. J., and W. B. Voorhees. 1980. Planting wheel traffic effects on interrow runoff and infiltration. Soil Science Society of America Journal 44:84–88.

Lindstrom, M. J., W. B. Voorhees, and G. W. Randall. 1981. Long-term tillage effects on interrow runoff and infiltration. Soil Science Society of America Journal 45:945–947.

Lins, I. D. G., and F. R. Cox. 1989. Effect of extractant and selected soil properties on predicting the correct phosphorus fertilization of soybeans. Soil Science Society of America Journal 53:813–816.

Logan, T. J., and E. O. McLean. 1973. Nature of phosphorus retention and adsorption with depth in soil columns. Soil Science Society of America Proceedings 37:351–355.

Logan, T. J., T. O. Oloya, and S. M. Yaksich. 1979. Phosphate characteristics and bioavailability of suspended sediments from streams draining into Lake Erie. Journal of Great Lakes Research 5:112–123.

Lopez-Hernandez, D., G. Siegert, and J. V. Rodriguez. 1986. Competitive adsorption of phosphate with malate and oxalate by tropical soils. Soil Science Society of America Journal 50:1460–1462.

Lowdermilk, W. C. 1953. Conquest of the Land Through 7,000 Years. Bulletin 99. Washington, D.C.: U.S. Department of Agriculture, Soil Conservation Service.

Lowery, B., and R. T. Schuler. 1991. Temporal effects of subsoil compaction on soil strength and plant growth. Soil Science Society of America Journal 55:216–223.

Lowrance, R. R. 1992a. Nitrogen outputs from a field-size agricultural watershed. Journal of Environmental Quality 21:602–607.

Lowrance, R. R. 1992b. Groundwater nitrate and denitrification in a coastal plain riparian forest. Journal of Environmental Quality 21:401–405.

Lowrance, R. R., R. A. Leonard, L. E. Asmussen, and R. L. Todd. 1985. Nutrient budgets for agricultural watersheds in the southeastern Coastal Plain. Ecology 66:287–296.

Lowrance, R. R., S. McIntyre, and J. C. Lance. 1988. Erosion and deposition in a coastal plain watershed measured using CS-137. Journal of Soil and Water Conservation 43:195–198.

Lowrance, R. R., R. L. Todd, and L. E. Asmussen. 1984b. Nutrient cycling in an agricultural watershed. I. Phreatic movement. Journal of Environmental Quality 13:22–27.

Lowrance, R., R. Todd, J. Fail, Jr., O. Hendrickson, Jr., R. Leonard, and L. Asmussen. 1984a. Riparian forests as nutrient filters in agricultural watersheds. BioScience 34:374–377.

Luna, J. M., and G. J. House. 1990. Pest management in sustainable agricultural systems. Pp. 157–173 in Sustainable Agricultural Systems, C. A. Edwards, R. Lal, P. Madden, R. H. Miller, and G. House, eds. Ankeny, Iowa: Soil and Water Conservation Society.

Maas, E. V. 1990. Crop salt tolerance. Pp. 262–304 in Agricultural Salinity Assessment and Management K. K. Tanji, ed. ASCE Manuals and Reports on Engineering Practice No. 71. New York: American Society of Civil Engineers.

Mackay, A. D., E. J. Kladivko, S. A. Barber, and D. R. Griffith. 1987. Phosphorus and potassium uptake by corn in conservation tillage systems. Soil Science Society of America Journal 51:970–974.

Madigan, E. 1991. Introduction. Pp. i-v in Agriculture and the Environment: The 1991 Yearbook of Agriculture. Washington, D.C.: U.S. Government Printing Office.

Madison, F., K. Kelling, J. Petersen, T. Daniel, G. Jackson, and L. Massie. 1986. Managing Manure and Waste: Guidelines for Applying Manure to Pasture and Cropland in Wisconsin. Report A3392. Madison: University of Wisconsin-Extension.

Magdoff, F. 1991a. Understanding the Magdoff pre-sidedress nitrate test for corn. Journal of Production Agriculture 4:297–305.

Magdoff, F. 1991b. Managing nitrogen for sustainable corn systems: Problems and possibilities. American Journal of Alternative Agriculture 6:3–8.

Magette, W. L., R. B. Brinsfield, R. E. Palmer, J. D. Wood, T. A. Dillaha, and R. B. Reneau. 1987. Vegetative filter strips for agricultural runoff treatment. Report No. CBP/TRS 2/87. Washington, D.C.: U.S. Environmental Protection Agency.

Magette, W. L., R. A. Weismiller, J. S. Angle, and R. B. Brinsfield. 1989. A nitrate groundwater standard for the 1990 farm bill. Journal of Soil and Water Conservation 44:491–494.

Malik, A. S., B. A. Larson, and M. Ribaudo. 1992. Agricultural Nonpoint Source Pollution and Economic Incentive Policies: Issues in the Reauthorization of the Clean Water Act. Staff Report No. AGES 9229. Washington, D.C.: U.S. Department of Agriculture, Economic Research Service, Resources and Technology Division.

Mallarino, A. P., J. R. Webb, and A. M. Blackmer. 1991. Corn and soybean yields during 11 years of phosphorus and potassium fertilization on a high-testing soil. Journal of Production Agriculture 4:312–317.

Mannering, J. V., D. L. Schertz, and B. A. Julian. 1987. Overview of conservation tillage. Pp. 3–17 in Effects of Conservation Tillage on Groundwater Quality, T. J. Logan, J. M. Davidson, J. L. Baker, and M. R. Overcash, eds. Chelsea, Mich.: Lewis Publishers.

Manrique, L. A., and C. A. Jones. 1991. Bulk density of soils in relation to soil physical and chemical properties. Soil Science Society of America Journal 55:476–481.

Mansell, R. S., P. J. McKenna, E. Flaig, and M. Hall. 1985. Phosphate movement in columns of sandy soil from a wastewater-irrigated site. Soil Science 140:59–68.

Marshall, T. J. 1958. A relation between permeability and size distribution of pores. Journal of Soil Science 9:1–8.

Martin, D. L., E. C. Stegman, and E. Ferers. 1990. Irrigation scheduling principles. Pp. 155–201 in Management of Farm Irrigation Systems, G. J. Hoffman, T. A. Howell, and K. H. Solomon, eds. St. Joseph, Mich.: American Society of Agricultural Engineers.

Martin, R. R., R. S. C. Smart, and K. Tazaki. 1988. Direct observation of phosphate precipitation in the goethite/phosphate system. Soil Science Society of America Journal 52:1492–1500.

Massey, H. F., and M. L. Jackson. 1952. Selective erosion of soil fertility constituents. Soil Science Society of America Proceedings 16:353–356.

McArthur, J. V., Gurtz, M. E., Tate, C. M., and Gilliam, F. S. 1985. The interaction of biological and hydrological phenomena that mediate the qualities of water draining native tallgrass prairie on the Konza Prairie Research Natural Area. Pp. 478–482 in Perspectives on Nonpoint Source Pollution. Report No. EPA-440/5–85–001. Washington, D.C.: U.S. Environmental Protection Agency.

McCall, P. L., M. J. S. Teresz, and S. F. Schwelgien. 1979. Sediment mixing by Lampsilis radiata siliquoidea (Mollusca) from western Lake Erie. Journal of Great Lakes Research 5:105–111.

McCallister, D. L., and T. J. Logan. 1978. Phosphate adsorption-desorption characteristics of soils and bottom sediments in the Maumee River basin of Ohio. Journal of Environmental Quality 7:87–92.

McCallister, D. L., C. A. Shapiro, W. R. Raun, F. N. Anderson, G. W. Rehm, O. P. Engelstad, M. P. Russelle, and R. A. Olson. 1987. Rate of phosphorus and potassium buildup/decline with fertilization for corn and wheat on Nebraska Mollisols. Soil Science Society of America Journal 51:1646–1652.

McCollum, R. E. 1991. Buildup and decline in soil phosphorus: 30-year trends on a Typic Umprabuult. Agronomy Journal 83:77–85.

McDonald, A. 1941. Early American Soil Conservationists. Miscellaneous Publication No. 449. Washington, D.C.: Soil Conservation Service, U.S. Department of Agriculture.

McDowell, L. L., and K. C. McGregor. 1984. Plant nutrient losses in runoff from conservation tillage corn. Soil Tillage Research 4:79–91.

McKeague, J. A. 1981. Phosphorus-enriched soil nodules detected by energy-dispersive x-ray spectrometry. Soil Science Society of America Journal 45:910–912.

McKeague, J. A., C. Wang, and G. C. Tapp. 1982. Estimating saturated hydraulic conductivity from soil morphology. Soil Science Society of America Journal 46:1239–1244.

McKelvey, V. E. 1939. Stream and Valley Sedimentation in the Coon Creek Drainage Basin, Wisconsin. M.A. thesis. University of Wisconsin, Madison.

McLean, E. O. 1982. Soil pH and lime requirement. II. Methods of soil analysis. Agronomy 9:199–224.

McManus, M. L. 1989. Biopesticides: An overview. Pp. 60–68 in Biotechnology and Sustainable Agriculture: Policy Alternatives, J. F. MacDonald, ed. Ithaca, N.Y.: National Agricultural Biotechnology Council.

Meade, R. H. 1982. Sources, sinks, and storage of river sediment in the Atlantic drainage of the United States. Journal of Geology 90:235–252.

Meek, B., L. Graham, and T. Donovan. 1982. Long-term effects of manure on soil nitrogen, phosphorus, potassium, sodium, organic matter, and water infiltration rate. Soil Science Society of America Journal 46:1014–1022.

Meisinger, J. J. 1984. Evaluating plant-available nitrogen in soil-crop systems. Pp. 391–416 in Nitrogen in Crop Production; R. D. Hauck, ed. Madison, Wis.: American Society of Agronomy.

Meisinger, J. J., V. A. Bandel, J. S. Angle, B. E., O'Keefe, and C. M. Reynolds. 1992. Presidedress soil nitrate test evaluation in Maryland. Soil Science Society of America Journal 56:1527–1532.

Meisinger, J. J., V. A. Bandel, G. Stanford, and J. O. Legg. 1985. Nitrogen utilization of corn under minimal tillage and moldboard plow tillage: I. Four-year results using labeled N fertilizer on an Atlantic Coastal Plain soil. Agronomy Journal 77:602–611.

Meisinger, J. J., W. L. Hargrove, R. L. Mikkelsen, J. R. Williams, and V. W. Benson. 1991. Effect of cover crops on groundwater quality. Pp. 57–68 in Cover Crops for Clean Water, W. L. Hargrove, ed. Ankeny, Iowa: Soil and Water Conservation Society.

Meisinger, J. J., and G. W. Randall. 1991. Estimating nitrogen budgets for soil-crop systems. Pp. 85–124 in Managing Nitrogen for Groundwater Quality and Farm Profitability, R. F. Follett, D. R. Keeney, and R. M. Cruse, eds. Madison, Wis.: Soil Science Society of America.

Mendoza, R. E., and N. J. Barrow. 1987. Ability of three soil extractants to reflect the factors that determine the availability of soil phosphate. Soil Science 144:319–329.

Menzel, B. W. 1983. Agricultural management practices and the integrity of instream biological habitat. Pp. 305–329 in Agricultural Management and Water Quality, F. W. Schaller and G. W. Bailey, eds. Ames: Iowa State University Press.

Meyer, J. L., and J. van Schilfgaarde. 1984. Case History—Salton Sea Basin. California Agriculture 38(10):13–16.

Meyer, L. D., and K. G. Renard. 1991. How research improves land management. Pp. 20–26 in Agriculture and the Environment: The 1991 Yearbook of Agriculture. Washington, D. C.: U.S. Government Printing Office.

Midwest Planning Service—Livestock Waste Subcommittee. 1985. Livestock Waste Facilities Handbook. Midwest Planning Service Report MWPS-18, 2nd ed. Ames, Iowa: Iowa State University.

Miller, F. P., and W. E. Larson. 1990. Lower input effects on soil productivity and nutrient cycling. Pp. 549–568 in Sustainable Agricultural Systems, C. A. Edwards, R. Lal, P. Madden, R. H. Miller, and G. House, eds. Ankeny, Iowa: Soil and Water Conservation Society.

Miller, M. R., P. L. Brown, J. J. Donovan, R. N. Bergatino, J. L. Sonderegger, and F. A. Schmidt. 1981. Saline seep development and control in the North American Great Plains—Hydrologic aspects. Agricultural Water Management 4:115–141.

Millington, R. J., and J. P. Quirk. 1961. Permeability of porous solids. Transactions of the Faraday Society 57:1200–1206.

Million, J. B., J. B. Sartain, R. B. Forbes, and N. R. Usherwood. 1989. Effects of K scheduling on soybeans multicropped with sweet corn and cabbage: Crop yields and tissue and soil K. Journal of Production Agriculture 2:120–126.

Mills, W. C., and R. A. Leonard. 1984. Pesticide pollution probabilities. Transactions of the American Society of Agricultural Engineers 27:1704–1710.

Mills, W. C., A. W. Thomas, and G. W. Langdale. 1991. Conservation tillage and season affects on soil erosion risk. Journal of Soil and Water Conservation 46:457–460.

Moore, I. C., F. W. Madison, and R. R. Schneider. 1978. Estimating phosphorus loading from livestock wastes: Some Wisconsin results. Pp. 175–192 in Best Management Practices for Agriculture and Silviculture, R. C. Loehr, D. A. Haith, M. F. Walter, and C. S. Martin, eds. Proceedings of the 1978 Cornell Agricultural Waste Conference. Ann Arbor, Mich.: Ann Arbor Science.

Moore, J., and J. J. Hefner. 1977. Irrigation with saline water in the Pecos Valley of West Texas. Pp. 339–344 in Managing Saline Water for Irrigation. Proceedings of the International Salinity Conference, Texas Technical University, Lubbock, Texas, August 1976.

Morse, D. 1989. Studies of modification of phosphorus concentration in diets, hydrolysis of phytate bound phosphorus, and excretion of phosphorus by dairy cows. Ph.D. dissertation. Gainesville: University of Florida, Dairy Science Department.

Mortimer, C. H. 1941. The exchange of dissolved substances between mud and water in lakes, I. Journal of Ecology 29:280–329.

Mortimer, C. H. 1942. The exchange of dissolved substances between mud and water in lakes, II. Journal of Ecology 30:147–201.

Motavelli, P. P., L. G. Bundy, T. W. Andraski, and A. E. Peterson. 1992. Residual effects of long-term nitrogen fertilization on nitrogen availability to corn. Journal of Production Agriculture 5:363–368.

Mueller, D. H., R. C. Wendt, and T. C. Daniel. 1984. Phosphorus losses as affected by tillage and manure application. Soil Science Society of America Journal 48:901–905.

Mussman H. C. 1991. The President's water quality initiative. Pp. 67–85 in Agriculture and the Environment: The 1991 Yearbook of Agriculture. Washington, D.C.: U.S. Government Printing Office.

Namken, L. N., C. L. Wiegand, and R. G. Brown. 1969. Water use by cotton from low and moderately saline static water tables. Agronomy Journal 61:305–310.

National Association of Conservation Districts, Conservation Tillage Information Center. 1991. National Survey of Conservation Tillage Practices. West Lafayette, Ind.: National Association of Conservation Districts.

National Research Council. 1961. Status and Methods of Research in Economic and Agronomic Aspects of Fertilizer Response and Use. Washington, D.C.: National Academy Press.

National Research Council. 1974. Productive Agriculture and a Quality Environment. Washington, D.C.: National Academy of Sciences.

National Research Council. 1989a. Alternative Agriculture. Washington, D.C.: National Academy Press.

National Research Council. 1989b. Irrigation-Induced Water Quality Problems. Washington, D.C.: National Academy Press.

National Research Council. 1989c. Nutrient Requirements of Dairy Cattle. Sixth Revised Edition. Washington, D.C.: National Academy Press.

National Research Council. 1990. Ground Water Models, Scientific and Regulatory Applications. Washington, D.C.: National Academy Press.

Nearing, M. A., L. J. Lane, E. E. Alberts, and J. M. Laflen. 1990. Prediction technology for soil erosion by water: Status and research needs. Soil Science Society of America Journal 54:1702–1711.

Nelson, D. W. 1982. Gaseous losses of nitrogen other than through denitrification. Pp. 327–363 in Nitrogen in Agricultural Soils. Agronomy Monographs No. 22, F. J. Stevenson, ed. Madison, Wis.: American Society of Agronomy, Madison, Wisconsin.

Nelson, D. W., and T. J. Logan. 1983. Chemical processes and transport of phosphorus. Pp. 65–91 in Agricultural Management and Water Quality, F. W. Schaller and G. W. Bailey, eds. Ames: Iowa State University Press.

Nelson, D. W., E. J. Monke, A. D. Bottcher, and L. E. Sommers. 1978. Sediment and nutrient contributions to the Maumee River from an agricultural watershed. Pp. 491–505 in Best Management Practices for Agriculture and Silviculture, R. C. Loehr, D. A. Haith, M. F. Walter, and C. S. Martin, eds. Proceedings of the 1978 Cornell Agricultural Waste Conference. Ann Arbor, Mich.: Ann Arbor Science.

Nelson, L. A., R. D. Voss, and J. T. Pesek. 1985. Agronomic and statistical evaluation of fertilizer response. Pp. 53–90 in Fertilizer Technology and Use, 3rd Ed., O. P. Engelstad, ed. Madison, Wis.: Agronomy Society of America.

Nielson, J. 1986. Conservation targeting. Success or failure? Journal of Soil and Water Conservation 41:70–76.

Nofziger, D. L., and A. G. Hornsby. 1986. A microcomputer-based management tool for chemical movement in Soil. Applied Agricultural Research 1:50–56.

Nonpoint Source Evaluation Panel. 1990. Report and Recommendations of the Chesapeake Bay Nonpoint Source Program Evaluation Panel. Washington, D.C.: U.S. Environmental Protection Agency.

Novais, R., and E. J. Kamprath. 1978. Phosphorus supplying capacities of previously heavily fertilized soils. Soil Science Society of America Journal 42:931–935.

Nowak, P., and R. Shepard. 1991. The Farm Assessment Technique: Utilizing a Needs Assessment in Water Quality Program Implementation. Madison: University of Wisconsin Extension, Nutrient and Pest Management Program.

Nowak, P. 1992. Why farmers adopt production technology. Journal of Soil and Water Conservation 47:14–16.

Nowak, P. J. 1983. Obstacles to adoption of conservation tillage. Journal of Soil and Water Conservation 38:162–165.

Nowak, P. J. 1985. Farmers attitudes and behaviors in implementing conservation tillage decisions. Pp. 327–340 in A Systems Approach to Conservation Tillage, F. M. D'Itri, ed. Chelsea, Mich.: Lewis Publishers.

Nowak, P., and M. Schnepf. 1987. Implementation of the conservation provisions in the 1985 farm bill: A survey of the county-level U.S. Department of Agriculture agency personnel. Journal of Soil and Water Conservation 42:285–290.

Obenhuber, D. C., and R. Lowrance. 1991. Reduction of nitrate in aquifer microcosms by carbon additions. Journal of Environmental Quality 20:255–258.

Oberle, S. L. and D. R. Keeney. 1990. A case for agricultural systems research. Journal of Environmental Quality 20:4–7.

Obreza, T. A., and F. M. Rhoads. 1988. Irrigated corn response to soil-test indices and fertilizer nitrogen, phosphorus, potassium, and magnesium. Soil Science Society of America Journal 52:701–706.

Olsen, S. R., and F. S. Watanabe. 1957. A method to determine a phosphorus adsorption maximum of soils as measured by the Langmuir isotherm. Soil Science Society of America Proceedings 21:144–148.

Olson, K. R. 1992. Soil physical properties as a measure of cropland productivity. Pp. 41–51 in Proceedings of the Soil Quality Standards Symposium. Report No. W0-WSA-2. Washington, D.C.: U.S. Department of Agriculture, Forest Service.

Olson, R. A. 1985. Nitrogen problems. Pp. 115–138 in Plant Nutrient Use and the Environment. Washington, D.C.: The Fertilizer Institute.

Olson, R. A., A. F. Drier, D. A. Hoover, and H. F. Rhoads. 1962. Factors responsible for poor response of corn and grain sorghum to phosphorus fertilization. I. Soil phosphorus level and climatic factors. Soil Science Society of America Proceedings 26:571–574.

Omernik. 1976. The influence of [watershed] land use on stream nutrient levels. P. 105 in Ecological Research Service EPA Publication No. 600/3–76–014. Washington, D.C.: U.S. Environmental Protection Agency.

O'Neil, W. B. 1983a. Transferable discharge permit trading under varying stream conditions: A simulation of multiperiod permit market performance on the Fox River, Wisconsin. Water Resources Research 19:608–612.

O'Neil, W. B. 1983b. The regulation of water pollution permit trading under conditions of varying streamflow and temperature. Pp. 219–231 in Buying a Better Environment: Cost-Effective Regulation Through Permit Trading, E. F. Joeres and M. H. David, eds. Madison: University of Wisconsin Press.

Oster, J. D., and J. D. Rhoades. 1984. Water management for salinity and sodicity control. Pp. 7-1-7–20 in Irrigation with Reclaimed Municipal Wastewater—A Guideline Manual, G. S. Pettygrove and T. Asano, eds. Sacramento: California State Water Resources Control Board.

Pacific Southwest Inter-Agency Committee. 1968. Factors Affecting Sediment Yield and Measures for the Reduction of Erosion and Sediment Yield. Pacific Southwest Inter-Agency Committee.

Padgitt, S. C. 1989. Farm Practices and Attitudes Toward Groundwater Policies, A Statewide Survey. Report No. IFM 3. Ames: Iowa State University, Cooperative Extension Service.

Page, A. L., A. C. Chang, and D. C. Adriano. 1990. Deficiencies and toxicities of trace elements. Pp. 138–160 in Agricultural Salinity Assessment and Management, K. K. Tanji, ed. ASCE Manuals and Reports on Engineering Practice No. 71. New York: American Society of Civil Engineers.

Parish, D. H. 1971. Soil conditions as they affect plant establishment, root development, and yield. Pp. 277–291 in Compaction of Agricultural Soils. St. Joseph, Mich.: American Society of Agricultural Engineers.

Parkin, T. B., and J. J. Meisinger. 1989. Denitrification below the crop rooting zone as influenced by surface tillage. Journal of Environmental Quality 18:12–16.

Parr, J. F., R. I. Papendick, S. B. Hornick, and R. E. Meyer. 1992. Soil quality: Attributes and relationship to alternative and sustainable agriculture. American Journal of Alternative Agriculture 7:5–11.

Pasternak, D. 1987. Salt tolerance and crop production—a comprehensive approach. Annual Review of Phytopathology 25:271–291.

Patni, N. K., and J. L. B. Culley. 1989. Corn silage yield, shallow groundwater quality and soil properties under different methods and times of manure application. Transactions of the American Society of Agricultural Engineers 32:2123–2129.

Patrick, W. H., Jr., and R. A. Khalid. 1974. Phosphate release and sorption by soils and sediments: Effect of aerobic and anaerobic conditions. Science 186:53–55.

Paul, E. A., and F. E. Clark. 1989. Soil Microbiology and Biochemistry. New York: Academic Press.

Pennell, K. D., A. G. Hornsby, R. E. Jessup, and P. S. C. Rao. 1990. Evaluation of five simulation models for predicting aldicarb and bromide behavior under field conditions. Water Resources Research 26:2679–2693.

Pennsylvania State University, College of Agriculture. 1989. Groundwater and Agriculture in Pennsylvania. Circular 341. College Station: Pennsylvania State University.

Peterjohn, W. T., and D. L. Correll. 1984. Nutrient dynamics in an agricultural watershed. Observations on the role of a riparian forest. Ecology 65:1466–1475.

Peterson, G. A., and W. W. Frye. 1989. Fertilizer nitrogen management. Pp. 183–220 in Nitrogen Management and Ground Water Protection. Developments in Agricultural and Managed-Forest Ecology No. 21., R. F. Follet, ed. Amsterdam: Elsevier.

Peterson, T. A., and M. P. Russelle. 1991. Alfalfa and the nitrogen cycle in the Corn Belt. Journal of Soil and Water Conservation 46:229–235.

Peverly, J. H. 1982. Stream transport of nutrients through a wetland. Journal of Environmental Quality 11:38–43.

Phillips, D. A., and T. M. DeJong. 1984. Dinitrogen fixation in leguminous crop plants. Pp. 121–132 in Nitrogen in Crop Production., R. D. Hauck, ed. Madison, Wis: American Society of Agronomy.

Phillips, J. D. 1989. Nonpoint source pollution control effectiveness of riparian forests along a coastal plain river. Journal of Hydrology 110:221–237.

Phillips, P. A., J. L. B. Culley, F. R. Hore, and N. K. Patni. 1982. Dissolved inorganic nitrogen and phosphate concentrations in discharge from two agricultural catchments in eastern Ontario. Agricultural Water Management 5:29–40.

Piekielek, W. P., and R. H. Fox. 1992. Use of a chlorophyll meter to predict sidedress nitrogen requirements for maize. Agronomy Journal 84:59–65.

Pierce, F. J. 1991. Erosion productivity impact prediction. Pp. 53–83 in Soil Management for Sustainability, R. Lal and F. J. Pierce, eds. Ankeny, Iowa: Soil and Water Conservation Society of America.

Pierce, F. J., R. H. Dowdy, W. E. Larson, and W. A. P. Graham. 1984. Soil productivity in the cornbelt: An assessment of erosion's long-term effects. Journal of Soil and Water Conservation 39:131–138.

Pierce, F. J., and W. E. Larson. 1993. Developing better criteria to evaluate sustainable land management. Pp. 7–14 in Utilization of Soil Survey Information for Sustainable Land Use, J. M. Kimble, ed. Washington, D.C.: U.S. Department of Agriculture, Soil Conservation Service.

Pierce, F. J., W. E. Larson, R. H. Dowdy, and W. A. P. Graham. 1983. Productivity of soils: Assessing long-term changes due to erosion. Journal of Soil and Water Conservation 38:39–44.

Pierce, F. J., and C. W. Rice. 1988. Crop rotation and its impact on efficiency of water and nitrogen use. Pp. 21–42 in Cropping Strategies for Efficient Use of Water and Nitrogen, W. L. Hargrove, ed. ASA Special Publication No. 51. Madison, Wis.:

American Society of Agronomy, Crop Science Society of America, and Soil Science Society of America.

Pierre, W. H., J. Meisinger, and J. R. Birchett. 1970. Cation-anion balance in crops as a factor in determining the effect of nitrogen fertilizers on soil acidity. Agronomy Journal 62:106–112.

Pinkowski, R. H., G. L. Rolfe, and L. E. Arnold. 1985. Effect of feedlot runoff on a southern Illinois forested watershed. Journal of Environmental Quality 14:47–54.

Porter, K. S., and M. W. Stimman. 1988. Protecting Groundwater. A Guide for the Pesticide User. Ithaca, N.Y.: Cornell University and University of California Cooperative Extension.

Postel, S. 1989. Water for Agriculture: Facing the Limits. Worldwatch Paper 93. Washington, D.C.: Worldwatch Institute.

Postel, S. 1990. Saving water for agriculture. Pp. 39–58 in State of the World, L. R. Brown, project director. New York: Norton.

Potash and Phosphate Institute. 1990. Soil test summaries: Phosphorus, potassium and pH. Better Crops with Plant Food 74(2):16–18.

Power, J. F. 1981. Nitrogen in the cultivated ecosystem. Pp. 529–546 in Terrestrial Nitrogen Cycles—Processes, Ecosystem Strategies and Management Impacts, F. E. Clark and T. Rosswall, eds. Ecological Bulletin No. 33. Stockholm: Swedish Natural Science Research Council.

Power, J. F. 1990. Erosion effects on soil chemistry and fertility. Pp. 27–30 in Proceedings of Soil Erosion and Productivity Workshop, W. E. Larson, G. R. Foster, R. R. Allmaras, and C. M. Smith, eds. St. Paul: University of Minnesota.

Power, J. F., and V. O. Biederbeck. 1991. Role of cover crops in integrated crop production systems. Pp. 167–173 in Cover Crops for Clean Water, W. L. Hargrove, ed. Ankeny, Iowa: Soil and Water Conservation Society.

Power, J. F., F. M. Sandoval, and R. E. Ries. 1981. Effect of topsoil and subsoil thickness on soil water content and crop production. Soil Science Society of America Journal 45:124–129.

Power, J. F., and J. S. Schepers. 1989. Nitrate contamination of groundwater in North America. Agriculture, Ecosystems and Environment 26:165–188.

Prabhakaran Nair, K. P., and K. Mengel. 1984. Importance of phosphate buffer power for phosphate uptake by rye. Soil Science Society of America Journal 48:92–95.

Prato, T., H. S. R. Rhew, and M. Brusuen. 1989. Soil erosion and nonpoint source pollution control in an Idaho watershed. Journal of Soil and Water Conservation 44:323–328.

Pratt, P. F. 1984. Nitrogen use and nitrate leaching in irrigated agriculture. Pp. 319–334 in Nitrogen in Crop Production, R. D. Hauck, ed. Madison, Wis.: American Society of Agronomy, Crop Science Society of America and Soil Science Society of America.

Pratt, P. F., and A. E. Laag. 1981. Effect of manure and irrigation on sodium bicarbonate-extractable phosphorus. Soil Science Society of America Journal 45:887–888.

Pratt, P. F., and D. L. Suarez. 1990. Irrigation water quality assessments. Pp. 220–236 in Agricultural Salinity Assessment and Management, K. K. Tanji, ed. ASCE Manuals and Reports on Engineering Practice No. 71. New York: American Society of Civil Engineers.

Putnam, J., J. Williams, and D. Sawyer. 1988. Using the erosion-productivity impact calculator (EPIC) model to estimate the impact of soil erosion for the 1985 RCA appraisal. Journal of Soil and Water Conservation 43:321–326.

Randall, G. W. 1984. Efficiency of fertilizer nitrogen use as related to application methods. Pp. 521–534 in Nitrogen in Crop Production, R. D. Hauck, ed. Madison, Wis.:

American Society of Agronomy, Crop Science Society of America, and Soil Science Society of America.

Rao, P. S. C., J. M. Davidson, and L. C. Hammond. 1976. Estimation of nonreactive and reactive solute front locations in soils. Pp. 235–241 in Proceedings of the Hazard Wastes Research Symposium. Publication No. EPA-600/19–76–015. Washington, D.C.: U.S. Environmental Protection Agency.

Rao, P. S. C., and A. G. Hornsby. 1989. Behavior of Pesticides in Soils and Waters. Soil Science Fact Sheet SL 40 (revised). Gainesville: University of Florida.

Rasmussen, P. E., and H. P. Collins. 1991. Long-term impacts of tillage, fertilizer, and crop residue on soil organic matter in temperate semiarid regions. Advances in Agronomy 45:93–134.

Rasmussen, W. D. 1983. History of soil conservation. Pp. 3–18 in Soil Conservation Policies, Institutions, and Incentives, H. G. Halcrow, E. O. Heady, and M. L. Cotner, eds. Ankeny, Iowa: Soil Conservation Society of America.

Rawls, W. J., et al. 1992. Estimating soil hydraulic properties from soils data. Pp. 329–340 in Indirect Methods for Estimating the Hydraulic Properties of Unsaturated Soils, M. Th. van Genuchten et al., eds. Riverside: University of California.

Ray, H. 1987. Implementing communication in development projects: New directions. In Agricultural Development Systems and Communication. Kingston, R.I.: University of Rhode Island.

Reddy, G. Y., E. O. McLean, G. D. Hoyt, and T. J. Logan. 1978. Effects of soil, cover crop, and nutrient source on amounts and forms of phosphorus movement under simulated rainfall conditions. Journal of Environmental Quality 7:50–54.

Reddy, K. R., R. Khaleel, M. R. Overcash, and P. W. Westerman. 1978. Phosphorus—A potential nonpoint source pollution problem in the land areas receiving long-term application of wastes. Pp. 193–211 in Best Management Practices for Agriculture and Silviculture, R. C. Loehr, D. A. Haith, M. F. Walter, and C. S. Martin, eds. Proceedings of the 1978 Cornell Agricultural Waste Conference. Ann Arbor, Mich.: Ann Arbor Science.

Reganold, J. P., A. S. Palmer, J. C. Lockhart, and A. N. Macgregor. 1993. Soil quality and financial performance of biodynamic and conventional farms in New Zealand. Science 260:344–349.

Rehm, G. W. 1986. Response of irrigated soybeans to rate and placement of fertilizer phosphorus. Soil Science Society of America Journal 50:1227–1230.

Rehm, G. W., R. C. Sorenson, and R. A. Wiese. 1981. Application of phosphorus, potassium, and zinc to corn grown for grain or silage: Early growth and yield. Soil Science Society of America Journal 45:523–528.

Reichelderfer, K. M. 1985. Do USDA Farm Program Participants Contribute to Soil Erosion? Agricultural Economics Report 532. Washington, D.C.: U.S. Department of Agriculture, Economic Research Service.

Reichenberger, L., and J. Russnogle. 1989. Farm by the foot. Farm Journal (mid-March):11–15.

Reilly, W. K. 1991. A new way with wetlands. Journal of Soil and Water Conservation 46:192–194.

Reimund, D. A., and F. Gale. 1992. Structural Change in the U.S. Farm Sector. Agricultural Information Bulletin 647. Washington, D.C.: U.S. Department of Agriculture, Economic Research Service.

Renard, K. G., and G. R. Foster. 1983. Soil conservation: Principles of erosion by water. Pp. 155–176 in Soil Conservation. Dryland Agriculture, Agronomy Monograph No. 23. Madison, Wis.: American Society of Agronomy.

Renard, K. G., G. R. Foster, G. A. Weesies, and J. P. Porter. 1991. RUSLE: The revised universal soil loss equation. Journal of Soil and Water Conservation 46:30–33.

Renard, K. G., W. J. Rawls, and M. M. Fogel. 1982. Currently available models. Pp. 507–522 in Hydrologic Modeling of Small Watersheds. ASAE Monograph No. 5. St. Joseph, Mich.: American Society of Agricultural Engineers.

Rhoton, F. E., N. E. Smeck, and L. P. Wilding. 1979. Preferential clay mineral erosion from watersheds in the Maumee River basin. Journal of Environmental Quality 8:547–550.

Ribaudo, M. 1986. Consideration of offsite impacts in targeting soil conservation programs. Land Economics 62:402–411.

Ribaudo, M., and D. Woo. 1991. Summary of state water quality laws affecting agriculture. Pp. 50–54 in Agricultural Resources: Cropland, Water and Conservation Situation and Outlook. Report No. AR-23. Washington, D.C.: U.S. Department of Agriculture, Economic Research Service, Research and Technology Division.

Ribaudo, M. O., and C. E. Young. 1989. Estimating the water quality benefits from soil erosion control. Water Resources Bulletin. Minneapolis, Minn: American Water Resources Association. 25:71–78.

Richardson, J. W. 1975. Farm programs, pesticide use, and social costs. Southern Journal of Agricultural Economics 5:155–163.

Rigler, F. W. 1968. Further observations inconsistent with the hypothesis that the molybdenum blue method measures orthophosphate in lake water. Limnology and Oceanography 13:7–13.

Rijsberman, F. R., and M. G. Wolman. 1985. Effect of soil erosion on soil productivity: An international comparison. Journal of Soil and Water Conservation 40:349–354.

Ritchie, J. T. 1981. Soil water availability. Plant and Soil 58:327–338.

Robert, P. C., and J. L. Anderson. 1986. Use of computerized soil survey reports in county extension offices. In Proceedings, International Conference on Computers, A. B. Bottcher and F. S. Zazueta, eds. Gainesville: University of Florida, Cooperative Extension Service.

Robert, P. C., R. H. Rust, and W. E. Larson. 1992. Soil Specific Crop Management. Madison, Wis.: American Society of Agronomy, Soil Science Society of America, and Crop Science Society of America.

Roberts, R. S., and D. R. Lighthall. 1991. The political economy of agriculture, ground water quality management, and agricultural research. Water Resources Bulletin 27:437–445.

Rogers, E. M. 1983. Diffusion of Innovations. New York: The Free Press.

Rogers, H. T. 1941. Plant nutrient losses by erosion from a corn, wheat, clover rotation on a Dunmore silt loam. Soil Science Society of America Proceedings 6:263–271.

Romkens, M. J. M., and D. W. Nelson. 1974. Phosphorus relationships in runoff from fertilized soils. Journal of Environmental Quality 3:10–13.

Romkens, M. J. M., D. W. Nelson, and J. V. Mannering. 1973. Nitrogen and phosphorus composition of surface runoff as affected by tillage method. Journal of Environmental Quality 2:292–295.

Rosenthal, A. 1990. State agricultural pollution regulation. Water Environment and Technology 2(8):51–57.

Roth, G. W., D. B. Beegle, and P. J. Bohn. 1992. Field evaluation of a presidedress soil nitrate test and quicktest for corn in Pennsylvania. Journal of Production Agriculture 5:476–481.

Runge, C. F., R. D. Munson, E. Lotterman, and J. Creason. 1990. Agricultural Competitiveness, Farm Fertilizer, and Chemical Use, and Environmental Quality: A Descriptive Analysis. St. Paul: University of Minnesota, Center for International Food and Agriculture Policy.

Russelle, M. P., and W. L. Hargrove. 1989. Cropping systems: Ecology and management. Pp. 277–318 in Nitrogen Management and Ground Water Protection. Developments in Agricultural and Managed-Forest Ecology 21, R. F. Follet, ed. Amsterdam: Elsevier.

Ryden, J. C., J. K. Syers, and R. F. Harris. 1974. Phosphorus in runoff and streams. Advances in Agronomy 25:1–45.

Saffigna, P. G., and D. R. Keeney. 1977. Nitrate and chloride in ground water under irrigated agriculture in central Wisconsin. Ground Water 15:170–177.

San Joaquin Valley Drainage Program. 1990. A Management Plan for Agricultural Subsurface Drainage and Related Problems on the Westside San Joaquin Valley. Final Report. Sacramento, Calif.: San Joaquin Valley Drainage Program.

Sanchez, C. A., and A. M. Blackmer. 1988. Recovery of anhydrous ammonia-derived nitrogen-15 during three years of corn production in Iowa. Agronomy Journal 80:102–108.

Sawhney, B. L., and K. Brown. 1989. Reactions and Movement of Organic Chemicals in Soils. Special Publication No. 22. Madison, Wis.: Soil Science Society of America.

Sawyer, C. N. 1947. Fertilization of lakes by agricultural and urban drainage. New England Water Works Association 61:109–127.

Schepers, J. S., and R. H. Fox. 1989. Estimation of N budgets for crops. Pp. 221–246 in Nitrogen Management and Ground Water Protection. Developments in Agricultural and Managed-Forest Ecology 21, R. F. Follet, ed. Amsterdam: Elsevier.

Schepers, J. S., K. D. Frank, and C. Bourg. 1986. Effect of yield goal and residual soil nitrogen considerations on nitrogen fertilizer recommendations for irrigated maize in Nebraska. Journal of Fertilizer Issues 3:133–139.

Schepers, J. S., M. G. Moravek, E. E. Alberts, and K. D. Frank. 1991. Maize production impacts on groundwater quality. Journal of Environmental Quality 20:12–16.

Schepers, J. S., and A. R. Mosier. 1991. Accounting for nitrogen in nonequilibrium soil-crop systems. Pp. 125–138 in Managing Nitrogen for Groundwater Quality and Farm Profitability, R. F. Follet, D. R. Keeney, and R. M. Cruse, eds. Madison, Wis.: Soil Science Society of America.

Schimel, D. S., D. C. Coleman, and K. A. Horton. 1985. Soil organic matter dynamics in paired rangeland and cropland toposequences in North Dakota. Geoderma 36:201–214.

Schindler, D. W. 1977. Evolution of phosphorus limitation in lakes. Science 195:260–262.

Schlosser, I. J., and J. R. Karr. 1978. Water resources and the land-water interface. Science 201:229–234.

Schuman, G. E., R. G. Spomer, and R. F. Piest. 1973. Phosphorus losses from four agricultural watersheds on Missouri Valley loess. Soil Science Society of America Proceedings 37:424–427.

Schwab, A. P., and S. Kulyingyong. 1989. Changes in phosphate activities and availability indexes with depth after 40 years of fertilization. Soil Science 149:179–186.

Schwer, C. B., and J. C. Clausen. 1989. Vegetative filter treatment of dairy milkhouse wastewater. Journal of Environmental Quality 18:446–451.

Science Advisory Board. 1991. Reducing Risk: Setting Priorities and Strategies for Environmental Protection. SAB-EC-90-121. Washington, D.C.: U.S. Environmental Protection Agency, Science Advisory Board.

Setia, P., and R. Magleby. 1988. Measuring physical and economic impacts of controlling water pollution in a watershed. Lake and Reservoir Management 4:63–71.

Shainberg, I., and M. J. Singer. 1990. Soil response to saline and sodic conditions. Pp.

91–112 in Agricultural Salinity Assessment and Management, K. K. Tanji, ed. ASCE Manuals and Reports on Engineering Practice No. 71. New York: American Society of Civil Engineers.

Sharma, M. L., and D. R. Williamson. 1984. Secondary salinization of water resources in Southern Australia. Pp. 571–582 in Salinity in Watercourses and Reservoirs, R. H. French, ed. Boston: Butterworth.

Sharpley, A. N. 1980. The enrichment of soil phosphorus in runoff sediments. Journal of Environmental Quality 9:521–526.

Sharpley, A. N. 1981. The contribution of phosphorus leached from crop canopy to losses in surface runoff. Journal of Environmental Quality 10:160–165.

Sharpley, A. N. 1985. The selective erosion of plant nutrients in runoff. Soil Science Society of America Journal 49:1527–1534.

Sharpley, A. N., and R. G. Menzel. 1987. The impact of soil and fertilizer phosphorus on the environment. Advances in Agronomy 41:297–324.

Sharpley, A. N., R. G. Menzel, S. J. Smith, E. D. Rhoades, and A. E. Olness. 1981. The sorption of soluble phosphorus by soil material during transport in runoff from cropped and grassed watersheds. Journal of Environmental Quality 10:211–215.

Sharpley, A. N., I. Singh, G. Uehara, and J. Kimble. 1989. Modeling soil and plant phosphorus dynamics in calcareous and highly weathered soils. Soil Science Society of America Journal 53:153–158.

Sharpley, A. N., and S. J. Smith. 1989. Prediction of soluble phosphorus transport in agricultural runoff. Journal of Environmental Quality 18:313–316.

Sharpley, A. N., and S. J. Smith. 1991. Effects of cover crops on surface water quality. Pp. 41–49 in Cover Crops for Clean Water, W. L. Hargrove, ed. Ankeny, Iowa: Soil and Water Conservation Society.

Sharpley, A. N., S. J. Smith, B. A. Stewart, and A. C. Mathers. 1984. Forms of phosphorus in soil receiving cattle feedlot waste. Journal of Environmental Quality 13:211–215.

Sharpley, A. N., W. W. Troeger, and S. J. Smith. 1991. The measurement of bioavailable phosphorus in agricultural runoff. Journal of Environmental Quality 20:235–238.

Sharpley, A. N., and J. R. Williams, eds. 1990. EPIC—Erosion/Productivity Impact Calculator. 1. Model Documentation. Technical Bulletin No. 1768. Washington, D.C.: U.S. Department of Agriculture, Agricultural Research Service.

Shoemaker, R., M. Anderson, and J. Hrubovcak. 1989. From the 1985 farm bill to 1990 and beyond: The resource effects of commodity programs. Pp. 44–47 in Agricultural Resources: Cropland, Water and Conservation Situation and Outlook. Report No. AR-16. Washington, D.C.: U.S. Department of Agriculture, Economic Research Service.

Shortle, J. S., and J. W. Dunn. 1986. The relative efficiency of agricultural source water pollution control policies. American Journal of Agricultural Economics 68:668–677.

Simmons, R. C., P. M. Groffman, and A. J. Gold. 1992. Nitrate dynamics in riparian forests: Groundwater studies. Journal of Environmental Quality 21:656–665.

Sims, G. K. 1990. Biological degradation of soil. Advances in Soil Science 11:289–329.

Skidmore, E. L., and N. P. Woodruff. 1968. Wind Erosion Forces in the United States and Their Use in Predicting Soil Loss. Agriculture Handbook 346. Washington, D.C.: U.S. Department of Agriculture.

Skorupa, J. P., and H. M. Ohlendorf. 1991. Contaminants in drainage water and avian risk thresholds. Pp. 345–385 in The Economics and Management of Water and Drainage in Agriculture, A. Dinar and D. Zilberman, eds. Boston: Kluwer Academic.

Slater, J. M., and D. G. Capone. 1987. Denitrification in aquifer soil and nearshore marine sediments influenced by groundwater nitrate. Applied and Environmental Microbiology 53:1292–1297.

Smith, M. S., W. W. Frye, and J. J. Varco. 1987. Legume winter cover crops. Advances in Soil Science 7:95–139.

Smith, R. A., R. B. Alexander, and M. G. Wolman. 1987. Water-quality trends in the nation's rivers. Science 235:1607–1615.

Smith, R. L., and J. H. Duff. 1988. Denitrification in a sand and gravel aquifer. Applied and Environmental Microbiology 54:171–178.

Smith, S. J., A. N. Sharpley, J. W. Naney, W. A. Berg, and O. R. Jones. 1991. Water quality impacts associated with wheat culture in the Southern Plains. Journal of Environmental Quality 20:244–249.

Soane, B. D., J. W. Dickson, and D. J. Campbell. 1982. Compaction by agricultural vehicles: A review. III. Incidence and control of compaction in crop production. Soil Tillage Research 2:3–36.

Soil Science Society of America. 1984. Glossary of Soil Science Terms. Madison, Wisc.: Soil Science Society of America.

Solis, P., and J. Torrent. 1989a. Phosphate fractions in calcareous Vertisols and Inceptisols of Spain. Soil Science Society of America Journal 53:462–466.

Solis, P., and J. Torrent. 1989b. Phosphate sorption by calcareous Vertisols and Inceptisols of Spain. Soil Science Society of America Journal 53:456–459.

Sonzogni, W. C., S. C. Chapra, D. E. Armstrong, and T. J. Logan. 1982. Bioavailability of phosphorus inputs to lakes. Journal of Environmental Quality 11:555–563.

Spalding, R. F., M. E. Exner, and J. R. Gormly. 1976. A note on an in-situ groundwater sampling procedure. Water Resources Research 12:1318–1321.

Spalding, R. F., J. R. Gormly, B. H. Curtiss, and M. E. Exner. 1978. Nonpoint nitrate contamination of ground water in Merrick County, Nebraska. Ground Water 16:86–95.

Spencer, W. F. 1957. Distribution and availability of phosphates added to a Lakeland fine sand. Soil Science Society of America Proceedings 21:141–149.

Spiker, M., S. Dabrekow, and H. Taylor. 1990. Soils tests and 1989 fertilizer application rates. Pp. 46–49 in Agricultural Resources: Inputs Outlook and Situation. Report No. AR-17. Washington, D.C.: U.S. Department of Agriculture, Economic Research Service, Resources and Technology Division.

Sposito, G. 1989. The Chemistry of Soils. New York: Oxford University Press.

Steehuis, T. S., S. Pacenka, and K. S. Porter. 1987. MOUSE: A management model for evaluating groundwater contamination from diffuse surface sources aided by computer graphics. Applied Agricultural Research 2:277–289.

Steenvoorden, J. H. A. M. 1986. Nutrient leaching losses following application of farm slurry and water quality considerations in the Netherlands. Pp. 168–176 in Efficient Land Use of Sludge and Manure, A. D. Kofoed, J. H. Williams, and P. L'Hermite, eds. London: Elsevier Applied Science Publishers.

Stevenson, F. J. 1982. Origin and distribution of nitrogen in soil. Pp. 1–42 in Nitrogen in Agricultural Soils. Agronomy Monographs No. 22, F. J. Stevenson, ed. Madison, Wis.: Soil Science Society of America.

Stewart, B. A., R. Lal, and S. A. El-Swaify. 1991. Sustaining the resource base of an expanding world agriculture. Pp. 125–144 Soil Management for Sustainability, R. Lal and F. J. Pierce, eds. Ankeny, Iowa: Soil and Water Conservation Society.

Stinner, B. R., D. A. Crossley, Jr., E. P. Odum, and R. L. Todd. 1984. Nutrient budgets and internal cycling of N, P, K, Ca, and Mg in conventional tillage, no-tillage, and old-field ecosystems on the Georgia piedmont. Ecology 65:354–369.

Stoltenberg, N. L., and J. L. White. 1953. Selective loss of plant nutrients by erosion. Soil Science Society of America Proceedings 17:406–410.

Stork, N. E., and P. Eggleton. 1992. Invertebrates as determinants and indicators of soil quality. American Journal of Alternative Agriculture 7:38–47.

Suarez, D. L., and J. D. Rhoades. 1977. Effect of leaching fraction on river salinity. Journal of Irrigation and Drainage Division, American Society of Civil Engineers 103(IR2):245–257.

Sweeten, J. S. 1991. Livestock and Poultry Waste Management: A National Overview. National Workshop on Livestock, Poultry and Aquaculture Waste Management, July 29–31, 1991, Kansas City, Mo.

Syers, J. K., R. F. Harris, and D. E. Armstrong. 1973. Phosphate chemistry in lake sediments. Journal of Environmental Quality 2:1–14.

Szabolcs, I. 1989. Salt-Affected Soils. Boca Raton, Fla.: CRC Press.

Tabatabai, M. A., R. E. Burwell, B. G. Ellis, D. R. Keeney, T. J. Logan, D. W. Nelson, R. A. Olson, G. W. Randall, D. R. Timmons, E. S. Verry, and E. M. White. 1981. Nutrient concentrations and accumulations in precipitation over the north central region. North Central Region Research Publication 282. Ames, Iowa: Iowa State University.

Tanji, K. K. 1991a. Migration and attenuation of DBCP in groundwater systems. Proceedings of a seminar presented to the California DBCP Advisory Committee, Fresno.

Tanji, K. K. 1991b. Principal Accomplishments, 1985–90. Davis: University of California, Division of Agriculture and Natural Resources, UC Salinity/Drainage Task Force.

Tanji, K. K., ed. 1990. Agricultural Salinity Assessment and Management. ASCE Manuals and Reports on Engineering Practice No. 71. New York: American Society of Civil Engineers.

Tanji, K. K., and R. A. Dahlgren. 1990. Efficacy of Evaporation Ponds for Disposal of Saline Drainage Water. Reports to California Department of Water Resources. Davis: University of California.

Tanji, K. K., and F. F. Karajeh. 1991. Agricultural drainage reuse in agroforestry systems. Pp. 53–59 in Proceedings of the 1991 National Conference on Irrigation and Drainage, W. F. Ritter, ed. New York: American Society of Civil Engineers.

Tanji, K. K., and L. Valoppi. 1989. Ground water contamination by trace elements. Agriculture, Ecosystems and the Environment 26:229–274.

Taylor, C. R. 1975. A regional market for rights to use fertilizer as a means of achieving water quality standards. Journal of Environmental Economics and Management (2):7–17.

Taylor, H. H. 1991. Fertilizer application timing. Pp. 30–38 in Agricultural Resources: Inputs and Situation Outlook. Report No. AR-24. Washington, D.C.: U.S. Department of Agriculture, Economic Research Service, Resources and Technology Division.

Taylor, H. M., and E. E. Terrell. 1982. CRC Handbook of Agricultural Productivity, Vol. 1, J. Rechigl, Jr., ed. Boca Raton, Fla.: CRC Press.

Tennessee Valley Authority, National Fertilizer Development Center. 1989. Summary: Soil Nitrate Testing Workshop, Research and Extension Needs in the Humid Regions of the United States. Circular Z-250, TVA/NFDC-89/9. Muscle Shoals, Ala.: Tennessee Valley Authority.

Thomas, G. W. 1967. Problems encountered in soil testing methods. Pp. 37–54 in Soil Testing and Plant Analysis. Part I. Soil Science Society of America Special Publication No. 2. Madison, Wis.: Soil Science Society of America.

Thomas, G. W. 1989. The soil bank account and the farmer's bank account. Journal of Production Agriculture 2:122–124.

Thomas, G. W., and Gilliam, J. W. 1978. Agro-ecosystems in the U.S.A. Pp. 182–243 in Cycling of Mineral Nutrients in Agricultural Ecosystems, M. J. Frissel, ed. New York: Elsevier Scientific.

Thompson, D. B., T. L. Loudon, and J. B. Gerrish. 1979. Animal manure movement in winter runoff for different surface conditions. Pp. 145–157 in Best Management Practices for Agriculture and Silviculture, R. C. Loehr, D. A. Haith, M. F. Walter, and C. S. Martin, eds. Proceedings of the 1978 Cornell Agricultural Waste Conference. Ann Arbor, Mich.: Ann Arbor Science.

Thorne, C. R. 1991. Analysis of channel instability due to catchment land-use change. Pp. 111–122 in Sediment and Stream Water Quality in a Changing Environment—Trends and Explanations. IAHS Publication No. 203.

Thurlow, D. L., and Hiltbold, A. E. 1985. Dinitrogen fixation by soybeans in Alabama. Agronomy Journal 77:432–436.

Thurman, E. M., D. A. Goolsby, M. T. Myer, and D. W. Kolpin. 1991. Herbicides in Surface Waters of the Midwestern United States: The effect of spring flush. Environmental Science and Technology 25:1794–1796.

Tietenberg, T. H. 1985. Emissions Trading: An Exercise in Reforming Pollution Policy. Washington, D.C.: Resources for the Future.

Tim, U. S. 1992. Identification of critical support nonpoint pollution source areas using geographic information systems and water monitoring. Water Resources Bulletin 28: 877–887.

Timmons, D. R., and J. L. Baker. 1991. Recovery of point-injected labeled nitrogen by corn as affected by timing, rate, and tillage. Agronomy Journal 83:850–857.

Tisdale, S. L., and Nelson, W. L. 1966. Soil Fertility and Fertilizers. New York: Macmillan.

Tobey, J. A., and K. A. Reinert. 1991. The effects of domestic agricultural policy reform on environmental quality. Journal of Agricultural Economics Research 43(2):20–28.

Trautman, M. B. 1957. The fishes of Ohio. Columbus, Ohio: Ohio State University Press.

Trautman, M. B. 1977. The Ohio country from 1750 to 1977: A naturalist's view. Biology Note 10. Columbus: Ohio Biological Survey,

Trimble, S. W. 1975. Denudation studies: Can we assume stream steady state? Science 188:1207–1208.

Trimble, S. W. 1983. Changes in sediment storage in the Coon Creek Basin in the Driftless Area, Wisconsin, 1853–1977. American Journal of Science 283:454–474.

Trimble, S. W., and S. W. Lund. 1982. Soil conservation and the reduction of erosion and sedimentation in the Coon Creek Basin, Wisconsin. U.S. Geological Survey Professional Paper 1234. Washington, D.C.: U.S. Government Printing Office.

Trudell, M. R., R. W. Gillham, and J. A. Cherry. 1986. An in-situ study of the occurrence and rate of denitrification in a shallow unconfined sand aquifer. Journal of Hydrology 83:251–268.

Tweeten, L. 1993. Is it Time to Phase Out Commodity Programs? Paper presented at Farm Policy Conference March 5, 1993, Columbus, OH.: Ohio State University.

U.S. Congress, Office of Technology Assessment. 1990. Technologies to improve nutrient and pest management. Pp. 81–167 in Beneath the Bottom Line: Agricultural Approaches to Reduce Agrichemical Contamination of Groundwater. Report OTA-F-418. Washington, D.C.: U.S. Government Printing Office.

U.S. Congress, Office of Technology Assessment. 1992. A New Technological Era for American Agriculture. Report OTA-F-474. Washington, D.C.: U.S. Government Printing Office.

U.S. Department of Agriculture. 1938. Soils and Men; The 1938 Yearbook of Agriculture. Washington, D.C.: U.S. Government Printing Office.

U.S. Department of Agriculture. 1988. Agricultural Statistics. Washington, D.C.: U.S. Government Printing Office.

U.S. Department of Agriculture. 1992. 1993 Budget Summary. Washington, D.C.: U.S. Department of Agriculture.

U.S. Department of Agriculture, Agricultural Stabilization and Conservation Service. 1992. Agricultural Conservation Program: 1991 Fiscal Year Statistical Summary. Washington, D.C.: U.S. Department of Agriculture.

U.S. Department of Agriculture, Economic Research Service. 1990. Conservation and water quality. Pp. 28–41 in Agricultural Resources: Cropland, Water, and Conservation Situation and Outlook Report. Report No. AR-19. Washington, D.C.: U.S. Department of Agriculture.

U.S. Department of Agriculture, Economic Research Service and National Agricultural Statistics Service. 1992. Statistical indicators—Farm income. Pp. 58–61 in Agricultural Outlook. Rockville, Md.: U.S. Department of Agriculture.

U.S. Department of Agriculture, Economic Research Service, Resources and Technology Division. 1989. Conservation and water quality. Pp. 21–35 in Agricultural Resources: Cropland, Water, and Conservation Situation and Outlook. Report No. AR-16. Washington, D.C.: U.S. Department of Agriculture.

U.S. Department of Agriculture, Economic Research Service, Resources and Technology Division. 1991a. Agricultural Resources: Inputs and Situation Outlook. Report No. AR-21. Washington, D.C.: U.S. Department of Agriculture.

U.S. Department of Agriculture, Economic Research Service, Resources and Technology Division. 1991b. Conservation and water quality. Pp. 23–41 in Agricultural Resources: Cropland, Water, and Conservation Situation and Outlook. Report No. AR-23. Washington, D.C.: U.S. Department of Agriculture.

U.S. Department of Agriculture, Soil Conservation Service. 1981. Land resource regions and major land resource areas of the United States. Agricultural Handbook No. 296. Washington, D.C.: U.S. Government Printing Office.

U.S. Department of Agriculture, Soil Conservation Service. 1983. California's Soil Salinity. Davis, Calif.: U.S. Department of Agriculture, Soil Conservation Service.

U.S. Department of Agriculture, Soil Conservation Service. 1989a. The Second RCA Appraisal: Soil, Water, and Related Resources on Nonfederal Land in the United States. Washington, D.C.: U.S. Department of Agriculture.

U.S. Department of Agriculture, Soil Conservation Service. 1989b. Summary Report: 1987 National Resources Inventory. Statistical Bulletin No. 790. Washington, D.C.: U.S. Department of Agriculture.

U.S. Department of Agriculture, Soil Conservation Service. 1989c. Water Quality Indicators Guide: Surface Waters. Report No. SCS-TP-161. Washington, D.C.: U.S. Department of Agriculture.

U.S. Department of Commerce, Bureau of the Census. 1989. 1987 Census of Agriculture. Washington, D.C.: U.S. Government Priniting Office.

U.S. Environmental Protection Agency. 1976. State Program Elements Necessary for Participation in the National Pollutant Discharge Elimination System—Concentrated Animal Feeding Operations. Federal Register 40 CFR 124.82.

U.S. Environmental Protection Agency. 1986a. Pesticides in Groundwater: Background Document. Washington, D.C.: U.S. Environmental Protection Agency.

U.S. Environmental Protection Agency. 1986b. Quality Criteria for Water. Report No. EPA-440/5–86–001. Washington, D.C.: U.S. Environmental Protection Agency.

U.S. Environmental Protection Agency. 1988. Pesticides in Ground Water Data Base: 1988 Interim Report. Washington, D.C.: U.S. Environmental Protection Agency, Office of

Pesticide Programs, Environmental Fate and Effects Division, Environmental Fate and Ground Water Branch.

U.S. Environmental Protection Agency. 1990a. National Water Quality Inventory, 1988 Report to Congress. Report No. EPA-440-4-90-003. Washington D.C.: U.S. Environmental Protection Agency, Office of Water.

U.S. Environmental Protection Agency. 1990b. National Pesticide Survey: Phase I Report. Report No. PB91–125765. Springfield, Va.: U.S. Department of Commerce, National Technical Information Service.

U.S. Environmental Protection Agency. 1991. Economic Incentives: Options for Environmental Protection. Policy Planning and Evaluation PM-220. Washington, D.C.: U.S. Environmental Protection Agency.

U.S. Environmental Protection Agency. 1992. Securing Our Legacy: An EPA Progress Report 1989–1991. Report No. 175 R-92–001. Washington, D.C.: U.S. Environmental Protection Agency.

U.S. Environmental Protection Agency, Science Advisory Board. 1990. Reducing Risk: Setting Priorities and Strategies for Environmental Protection. Report No. SAB-EC-90–021. Washington, D.C.: U.S. Environmental Protection Agency.

U.S. General Accounting Office. 1977. To Protect Tomorrow's Food Supply, Soil Conservation Needs Priority Attention. Washington, D.C.: Comptroller General of the United States.

U.S. General Accounting Office. 1992. Pollutant Trading Could Reduce Compliance Costs If Uncertainties Are Resolved. Report No. GAO/RCD-92–153. Washington, D.C.: U.S. General Accounting Office.

U.S. General Accounting Office. 1993. Conservation Reserve Program: Cost-Effectiveness Is Uncertain. Report No. GAO/RCED-93–132. Washington, D.C.: U.S. General Accounting Office.

University of California, Study Group on Biological Approaches to Pest Management. 1992. Beyond Pesticides: Biological Approaches to Pest Management in California. Oakland: University of California, ANR Publications.

University of Wisconsin-Extension and Wisconsin Department of Agriculture, Trade and Consumer Protection. 1989. Nutrient and Pesticide Best Management Practices for Wisconsin Farms, R. W. Schmidt, ed. Technical Bulletin ARM-1. Madison: University of Wisconsin-Extension.

Van Diepen, C. A., H. van Keulen, J. Wolf, and J. A. A. Berkhout. 1991. Land evaluation: From intuition to quantification. Advances in Soil Science 15:139–204.

Van Dyne, D. L., and C. B. Gilbertson. 1978. Estimating U.S. Livestock and Poultry Manure Production. Report ESCS-12. Washington, D.C.: U.S. Department of Agriculture. Economics, Statistics and Cooperative Service.

Van Horn, H. H. 1991. Achieving environmental balance of nutrient flow through animal production systems. The Professional Animal Scientist 7:22–33.

Van-Loon, C. I., L. A. H. Smet, and F. R. Boone. 1985. The effect of a plowpan in marine loam soils on potato growth. 2. Potato plant responses. Potato Research. Wageningen, The Netherlands: The European Association for Potato Research 28:315–330.

Van Reimsdijk, W. H., T. M. Lexmond, C. G. Enfield, and S. E. A. T. M. Van Der Zee. 1987. Phosphorus and heavy metals: Accumulation and Consequences. Pp. 213–227 in Animal Manure on Grassland and Fodder Crops. Fertilizer or Waste? H. G. Van Der Meer, R. J. Unwin, T. A. Van Dijk, and G. C. Enik, eds. The Hague, The Netherlands: Martinus Nijhoff.

van Schilfgaarde, J. 1990. Irrigated Agriculture: Is it Sustainable? Pp. 432–449 in Agricultural Salinity Assessment and Management, K. K. Tanji, ed. ASCE Manuals and

Reports on Engineering Practice No. 71. New York: American Society of Civil Engineers.

Varco, J. J., W. W. Frye, M. S. Smith, and C. T. MacKown. 1989. Tillage effects on N recovery by corn from a nitrogen-15 labeled legume cover crop. Soil Science Society of America Journal 53:822–827.

Varvel, G. E., and T. A. Peterson. 1990. Residual soil nitrogen as affected by continuous, two-year, and four-year crop rotation systems. Agronomy Journal 82:958–962.

Viets, F. G. 1975. The environmental impact of fertilizers. CRC Critiques and Reviews in Environmental Control 5:423–453.

Violante, A., C. Colombo, and A. Buondonno. 1991. Competitive adsorption of phosphate and oxalate by aluminum oxides. Soil Science Society of America Journal 55:65–70.

Visser, S., and D. Parkinson. 1992. Soil biological criteria as indicators of soil quality: Soil microorganisms. American Journal of Alternative Agriculture 7:33–37.

Voorhees, W. B. 1983. Relative effectiveness of tillage and natural forces in alleviating wheel-induced soil compaction. Soil Science Society of America Journal 47:129–133.

Voorhees, W. B. 1987. Assessment of soil susceptibility to compaction using soil and climatic data bases. Soil Tillage Research 10:29–38.

Voorhees, W. B., J. F. Johnson, G. W. Randall, and W. W. Nelson. 1989. Corn growth and yield as affected by surface and subsoil compaction. Agronomy Journal. Madison, Wis.: American Society of Agronomy. 81:294–303.

Voorhees, W. B., J. F. Johnson, G. W. Randall, and W. W. Nelson. 1990. Corn growth and yield as affected by surface and subsoil compaction. Soil Science Society of America Journal 82:973–979.

Vroomen, H. 1989. Fertilizer Use and Price Statistics: 1960–88. Statistical Bulletin 780. Washington, D.C.: U.S. Department of Agriculture, Economic Research Service, Resources and Technology Division.

Wagenet, R. J., and J. L. Hutson. 1989. LEACHM: A Finite Difference Model for Simulating Water, Salt and Pesticide Movement in the Plant Root Zone, Version 2.0. Continuum, Vol. 2. Ithaca: Cornell University, N.Y. State Resources Institute.

Wagenet, R. J., and P. S. C. Rao. 1990. Modeling Pesticide Fate in Soils. Pp. 351–399 in Pesticides in the Soil Environment: Processes, Impacts and Modeling, H. H. Cheng, ed. Soil Science Society of America Book Series No. 2. Madison, Wis.: Soil Science Society of America.

Wagger, M. G., and D. B. Mengel. 1988. The role of nonleguminous cover crops in the efficient use of water and nitrogen. Pp. 115–128 in Cropping Strategies for Efficient Use of Water and Nitrogen, W. L. Hargrove, ed. ASA Special Publication No. 51. Madison, Wis.: American Society of Agronomy, Crop Science Society of America, and Soil Science Society of America.

Wahl, R. 1989. Western Water Rights. Washington, D.C.: Resources for the Future.

Walker, D. J., and D. L. Young. 1986. Assessing soil erosion productivity damage. Pp. 21–62 in Soil Conservation: Assessing the National Resources Inventory, Vol. II. Washington, D.C.: National Academy Press.

Wall, G. J., and L. P. Wilding. 1976. Mineralogy and related parameters of fluvial suspended sediments in northwestern Ohio. Journal of Environmental Quality 5:168–173.

Wallace, H. A. 1938. Soils and Men (Foreword). Yearbook of Agriculture, 1938. Washington, D.C.: U.S. Government Printing Office.

Walter, M. F., T. L. Richard, P. D. Robillard, and R. Muck. 1987. Manure management with conservation tillage. Pp. 253–270 in Effects of Conservation Tillage on Groundwater Quality: Nitrates and Pesticides, T. J. Logan, J. M. Davidson, J. L. Baker, and M. R. Overcash, eds. Chelsea, Mich.: Lewis Publishers.

Watson, A. K. 1991. The classical approach with plant pathogens. Pp. 3–23 in Microbial Control of Weeds, D. O. TeBeest, ed. London: Chapman and Hall.

Wauchope, R. D. 1978. The pesticide content of surface water draining from agricultural fields: A review. Journal of Environmental Quality 7:459–472.

Wauchope, R. D., T. M. Butler, A. G. Hornsby, P. W. M. Augustijn-Beckers, and J. P. Burt. 1992. The SCS/ARS/CES pesticide properties database for environmental decision-making. Reviews of Environmental Contamination and Toxicology 123:1–164.

Wauchope, R. D., and R. A. Leonard. 1980. Maximum pesticide concentrations in agricultural runoff: A semi-empirical prediction formula. Journal of Environmental Quality 9:665–672.

Webb, J. R. 1982. Rotation-fertility experiment. Pp. 16–18 in Annual Progress Report Northwest Research Center. Ames, Iowa: Iowa State University.

Webster, C. P., and K. W. T. Goulding. 1989. Influence of soil carbon content on denitrification from fallow land during autumn. Journal of Science and Food Agriculture 49:131–142.

Welsch, D. J. 1991. Riparian forest buffers: Function and design for protection and enhancement of water resources. Publication NA-AR-07–91. Radnor, Pa.: U.S. Department of Agriculture, U.S. Forest Service.

Wendt, R. C., and R. B. Corey. 1980. Phosphorus variations in surface runoff from agricultural lands as a function of land use. Journal of Environmental Quality 9:130–136.

Wescott, M. P., V. R. Stewart, and R. E. Lund. 1991. Critical petiole nitrate levels in potato. Agronomy Journal 83:844–850.

Wichelns, D. 1991. Increasing block-rate prices for irrigation water molecule drain water reduction. Pp. 275–294 in The Economics and Management of Water and Drainage in Agriculture, A. Dinar and D. Zilberman, eds. Boston: Kluwer Academic Publishers.

Wilding, L. P. 1988. Improving our understanding of the composition of the soil-landscape. Pp. 13–39 in Proceedings of an International Interactive Workshop on Soil Resources: Their Inventory, Analysis and Interpretation for Use in the 1990s. St. Paul: University of Minnesota.

Wildung, R. E., R. L. Schmidt, and A. R. Gahler. 1974. The phosphorus status of eutrophic lake sediments as related to changes in limnological conditions—Total, inorganic, and organic phosphorus. Journal of Environmental Quality 3:133–138.

Williams, J. D. H., H. Shear, and R. L. Thomas. 1980. Availability to *Scenedesmus quadricauda* of different forms of phosphorus in sedimentary materials from the Great Lakes. Limnology and Oceanography 25:1–11.

Williams, J. R., A. D. Nicks, and J. G. Arnold. 1985. Simulator for water resources in rural basins. Journal of Hydrology Engineers, American Society of Civil Engineers 111:970–986.

Williams, R. D., and A. D. Nicks. 1988. Using CREAMS to simulate filter strip effectiveness in erosion control. Journal of Soil and Water Conservation 43:108–112.

Wischmeier, W. H., and D. D. Smith. 1978. Predicting Rainfall Erosion Losses—A Guide to Conservation Planning. Agriculture Handbook 537. Washington, D.C.: U.S. Department of Agriculture, Science and Education Administration.

Woodward, D. E., Merkel, W. H., and C. D. Clarke. 1991. Ephemeral gully erosion model. Pp. PS-9-PS-17 in Proceedings of the 5th Federal Interagency Sedimentation Conference.

Woolhiser, D. A., R. E. Smith, and D. C. Goodrich. 1990. KINEROS, A Kinematic Runoff and Erosion Model: Documentation and User Manual. Report ARS-77. Washington, D.C.: U.S. Department of Agriculture, Agricultural Research Service.

Yalin, Y. S. 1963. An expression for bedload transportation. Journal of Hydraulics Division, Proceedings American Society of Civil Engineers 89:221–250.

Yang, J. E., and J. S. Jacobsen. 1990. Soil inorganic phosphorus fractions and their uptake relationships in calcareous soils. Soil Science Society of America Journal 54:1666–1669.

Yerokun, O. A., and D. R. Christenson. 1990. Relating high soil test phosphorus concentrations to plant phosphorus uptake. Soil Science Society of America Journal 54:796–799.

Yost, R. S., E. J. Kamprath, G. C. Naderman, and E. Lobato. 1981. Residual effects of phosphorus applications on a high phosphorus adsorbing Oxisol of central Brazil. Soil Science Society of America Journal 45:540–543.

Young, A. 1991. Soil monitoring: A basic task for soil survey organizations. Soil Use and Management 7:126–130.

Young, C. E., B. M. Crowder, J. S. Shortle, and J. R. Alwang. 1985. Nutrient management on dairy farms in southeastern Pennsylvania. Journal of Soil and Water Conservation 40:443–445.

Young, D. L. 1984. Modeling agricultural productivity impacts of soil erosion and future technology. Pp. 60–85 in Future Agricultural Technology and Resource Conservation. Ames: Iowa State University Press.

Young, R. A., T. Huntrods, and W. Anderson. 1980. Effectiveness of vegetated buffer strips in controlling pollution from feedlot runoff. Journal of Environmental Quality 9:483–487.

Young, R. A., and C. K. Mutchler. 1976. Pollution potential of manure spread on frozen ground. Journal of Environmental Quality 5:174–179.

Young, R. A., A. E. Olness, C. K. Mitchler, and W. C. Moldenhauer. 1985. Erosion and Soil Productivity. St. Joseph, Mich.: American Society of Agricultural Engineers.

Young, R. A., C. A. Onstad, D. D. Bosch, and W. P. Anderson. 1987. AGNPS, Agricultural Non-Point-Source Pollution Model: A Watershed Analysis Tool. Conservation Research Report No. 35. Washington, D.C.: U.S. Department of Agriculture, Agricultural Research Service.

Young, R. A., and W. B. Voorhees. 1982. Soil erosion and runoff from planting to canopy development as influenced by tractor wheel traffic. Transactions of the American Society of Agricultural Engineers 25:708–712.

Zalom, F. G., R. E. Ford, R. E. Frisbie, C. R. Edwards, and J. P. Tette. 1992. Integrated Pest Management: Addressing the Economic and Environmental Issues of Contemporary Agriculture. St. Paul, Minn.: APS Press.

# Glossary

**acidification**   a chemical process by which the soil or water environment becomes more acidic; occurs naturally over time in humid regions and is also caused by the application of acid-forming, nitrogenous fertilizers. Soils become acidic when bases ($Ca^{+2}$, $Mg^{+2}$, $K^+$, $Na^+$) are leached out and replaced by $H^+$, resulting in pH levels less than 7.0. Highly acidic soils may limit plant growth due to insufficient calcium and magnesium, toxic levels of exchangeable aluminum (in soils with pH <5.5), decreasing availability of nutrients, and changing decomposition rates of soil organic matter and organic residues.

**advection**   a method of nutrient transport through water flows.

**aerobic**   characterized by the presence of oxygen; pertains to organisms (such as some bacteria) that require oxygen to maintain life processes (such as composting or biological treatment processes). Carbon dioxide is a product of aerobic processes.

**allelopathy**   the suppression of growth of one species of plant or microorganism by another due to the release of toxic substances by the organism.

**alluvial deposits**   materials such as clay, silt, sand, gravel, and mud that have been eroded, transported, and deposited by running water: alluvion; alluvium.

**anaerobic**   characterized by the absence of oxygen; pertains to organisms (such as some bacteria) that do not require oxygen to maintain life processes (such as digestion of sewage sludge or manure and decomposition). Denitrification occurs under anaerobic conditions.

**carbonation**   a process of chemical weathering of minerals that contain soda, lime, potash, or basic oxides contributing to an increase in the dissolved mineral load in the soil solution and in waters; the conversion to a carbonate—a salt or ester compound—as a reaction to some corrosive agent or dissolved $CO_2$.

**denitrification**   the bacterial reduction of nitrate or nitrite to gaseous molecular nitrogen ($N_2$), and nitrogen oxides ($NO_xs$), such as $NO_2$, $N_2O$, and $NO$; the process by which nitrogen is returned from the soil environment to the atmosphere; occurs under anaerobic conditions and results in loss of available nitrogen from surface soils of wet riparian forests and vegetative filter strips.

**dissolution**   a chemical weathering mechanism that involves the dissolving or breaking up of a material such as salt or minerals in water. The deposition of salts in soil solution and water affects properties such as electrical conductivity (EC) and total dissolved solids (TDS).

**ephemeral erosion**   erosion of the soil by running water that creates small channels, called rills, that cannot be completely filled in by tillage. Ephemeral rills will collect water during heavy rains and increase the severity of erosion.

**erosion**   the loosening, transportation, and wearing away of the land surface by running water, wind, ice, or other geological agents; the single most important process of soil degradation resulting in loss of soil productivity and increased water pollution from sediment and agricultural chemicals.

**eutrophication**   the process by which a body of water becomes rich in nutrients, characterized by high concentrations of phosphorus (P) and nitrogen (N), frequently shallow depths, and seasonal oxygen deficiency in the deeper areas; occurs naturally and by human activity, usually in the form of industrial or municipal wastewater or agricultural runoff. Eutrophication causes algal blooms, fish kills, and other water quality problems.

**evapotranspiration**   the loss of pure water from the soil as a result of both soil surface evaporation (the process by which liquid is changed to a vapor or gas) and plant transpiration (the photosynthetic and physiological process by which plants release water vapor into the air).

**fertility (soil)**   the quality of a soil that enables it to provide nutrients in adequate amounts and proper balance for plant growth when other growth factors such as light, moisture, and temperature are favorable. Two main methods for ensuring fertility are biological nutrient cycling and the addition of chemical fertilizers.

**greenhouse effect** warming of the earth's surface partly caused by concentration of gases such as water vapor ($H_2O$), carbon dioxide ($CO_2$), methane ($CH_4$), and chlorofluorocarbons (CFCs) in the earth's atmosphere. The concentration of gases acts as a cover—absorbing longer, infrared waves and trapping heat in the atmosphere. Soil management can have a significant impact on greenhouse gas concentrations inasmuch as soils can be sources or sinks of greenhouse gases.

**gully erosion** severe erosion of the soil by running water that creates deep channels: gully; gullies.

**halophytes** a plant or microorganism adapted to high-saline soil environments. Examples include greasewood (*Sarcobatus*), and saltgrass (*Distichlis*). Halophytes can be irrigated with reused drainage water that is higher in salinities.

**humus** the fraction of organic matter in the soil that is relatively resistant to further breakdown and decomposition; contains complex and more stable (less reactive or labile) organic matter including decaying plant and animal tissue.

**hydrolysis** a process that dissolves minerals and breaks down chemical compounds by reaction with hydrogen and oxygen; a method of pesticide degradation and a primary source of salt deposits.

**labile carbon** that fraction of organic matter in soils that is most readily decomposable by soil microorganisms. The amount of labile carbon in the soil has an important effect on soil biological activities such as mineralization of nutrients, generation of soil structure, and specific enzyme activities.

**laterization** the desiccation and hardening of exposed plinthitic materials.

**laterite** refers to a zonal group of red soils, rare in the United States, developed in hot, humid climates characterized by intense weathering, chemical change, and residues of aluminum and iron oxides.

**ligand** the molecule, ion, or group bound to the central atom in a chelate (cyclical structure with a central metallic ion) or coordination compound forming a complex. Trace elements such as copper (Cu) and molybdenate ($MoO_4^{-2}$) that tend to form complexes with ligands have greater mobilities than those that are not complexed.

**loess**   an essentially unconsolidated, unstratified, calcareous silt transported and deposited by wind; usually homogeneous, permeable, and buff to grey in color.

**organic carbon**   the total amount of carbon held in the organic matter in the soil; chemical compounds based on carbon chains or rings that also contain hydrogen with or without oxygen, nitrogen, or other elements. Cultivation causes marked reductions in total organic carbon content in the soil; but it can be replaced by crop residues, manures, or other sources of organic matter added to soils.

**osmotic potential**   the work per unit quantity of pure water that has to be done to prevent the transport or flow of a solvent across a membrane that separates (1) solutions of differing concentrations or (2) pure solvent from solute (as in pure water and salt water): osmotic pressure; osmotic gradient. Salt water in the root zone decreases the osmotic potential of the soil solution relative to the osmotic potential of the root, thereby reducing the amount of fresh water available to plants.

**oxidation-reduction**   a simultaneous chemical reaction wherein an oxidation reactant (reducing agent) gains a positive charge (loses an electron), while a reduction reactant (oxidizing agent) loses a positive charge (gains an electron); a process of chemical weathering that is a source of salt deposits in water: redox reactions.

**pedon**   the smallest volume of soil capable of representing all the horizons of a soil profile, including soil horizon shapes and relations. The three-dimensional profile is usually hexagonal horizontally ranging from 1 to 10 $m^2$ with depths to the lower limit of the genetic soil horizons.

**pedotransfer functions**   functions that relate different soil attributes and properties to one another (for example, predicting changes in soil's organic matter on the basis of the amount of crop residue added).

**pH**   a measurement scale indicating acidity and alkalinity in which values less than 7 are acidic, the value of 7 is neutral, and values from 7 to 14 are basic (alkaline). pH of a soil environment affects productivity by influencing solubilities of heavy metals and minerals, mobilities of anionic trace elements, and activities of microorganisms.

**photolysis**   chemical decomposition or dissociation by the action of radiant energy (i.e., light); a method of chemical degradation of pesticides with a short (on the order of hours and days) reaction rate.

**plinthitic materials**   earthy material of clay and quartz high in iron oxides, aluminum hydroxides, and some silica; poor in humus, highly weathered, and usually red in color.

**runoff**   rainfall excess that is not absorbed by the soil.

**rill erosion**   erosion of soil by running water carving out visible channels that are small enough to be filled in by tillage.

**salinity**   the concentration of dissolved salt in water traditionally referring to major anions and cations (Na, Ca, Mg, K, Cl, $SO_4$, $HCO_3$, $CO_3$, $NO_3$) found in irrigation water, but now also including toxic trace elements. Salinity in soils has deleterious effects on physical soil condition and is frequently accompanied by waterlogging, which results in poor aeration of the root zone. Salinity reduces crop yields, affects germination, seedling and vegetative growth, and has adverse effects on water quality. Salinity is measured using electroconductivity ($\mu$S/cm) or total dissolved solids (mg/L).

**salinization**   the process by which salts accumulate in soil. Chemical weathering of minerals in soils and rocks—dissolution, hydrolysis, carbonation, acidification, and oxidation-reduction—is the primary cause of dissolved mineral load (salinity). Secondary causes of salinity include evaporation, release and dissolution of fossil salts, atmospheric deposition, seawater intrusion, human-induced irrigation, and salt seeps.

**seep**   an area, generally small, where a fluid (water, soil, gas, saline water) contained in the ground percolates slowly to the surface and forms a pool. Saline seeps, where salty groundwater moves to the surface, are a source of salinization of soils.

**sheet erosion**   erosion of the soil by running water that removes soil in thin uniform sheets.

**slaking**   the breakdown of soil aggregates as a result of exposure to air or water. Slaking can destroy pore interstices, reducing the hydraulic conductivity of a soil, leading to reduced permeability, poor crop establishment, inadequate water intake rates, and increased runoff and erosion.

**sodicity**   of, relating to, or containing sodium (Na). Soils are classified as sodic when Na is the prevalent cation in saline soils. Sodic soils may limit plant growth by having toxic concentrations of exchangeable Na that keep soil dispersed and maintain poor soil structure.

The sodium absorption rate (SAR) is a comparison of the concentration of sodium ions ($Na^+$) to that of calcium ions ($Ca^{+2}$) and magnesium ions ($Mg^{+2}$) and is considered with the electroconductivity measurement of salinity when assessing the potential effects of water quality on soil water penetration.

**soil horizon**   an approximately horizontal layer of soil differing from adjacent layers in physical, chemical, or biological properties such as color, structure, texture, consistence, kinds and numbers of organisms present, or degree of acidity or alkalinity.

**solum**   the upper horizons of the soil profile in which the natural processes of soil formation take place; true soil. The solum is where most plant roots grow.

**tailwater**   irrigation water that reaches the lower end of a field. Tailwater ratio is an indication of irrigation water application systems; tailwater return systems increase efficiency of surface water irrigation systems.

**turgor**   the normal state of turgidity and tension in plant cells caused by water pressure when plant cells are full of water. This pressure keeps stems upright and leaves expanded to receive sunlight.

**turgor pressure**   the pressure developed as the result of fluid in a turgid (swollen) plant cell caused by the osmotic diffusion toward the inside of a cell. Saline soil environments can affect the necessary plant cell turgor pressure by disrupting the osmotic balance.

**vadose zone**   the unsaturated zone of the soil above the permanent groundwater level. The vadose zone is of concern in considering the potential contamination, transport, degradation, and mobilization of nutrients, pesticides, salts, and trace elements.

**water-holding capacity**   the capability of soils to store and release available water to plants. Estimation of water-holding capacities are made using regression models from parameters such as particle size distributions, organic matter content, and bulk density.

# About the Authors

SANDRA S. BATIE (*Chair*) is the Elton R. Smith Professor of Food and Agricultural Policy at Michigan State University. She received her Ph.D. in agricultural economics from Oregon State University. Her research includes work in natural resource and agricultural policy, soil conservation, water quality, research methodology, and rural development. She was professor of agricultural economics at Virginia Polytechnic and State University during most of the preparation of this report.

J. WENDELL GILLIAM is a professor of soil science at North Carolina State University. He earned his Ph.D. from Mississippi State University. His research work includes studies of the contribution of fertilizers to pollution, effect of large-scale agricultural development on water resources, nonpoint source pollution, pollutant removal effectiveness and the hydraulics of wetland filter areas, effect of erosion on soil productivity, the influence of riparian vegetation on water quality, and the impact of land drainage on estuarine habitat.

PETER MARK GROFFMAN is an associate scientist at the Institute of Ecosystem Studies, Millbrook, New York. His Ph.D. in ecology he earned from the University of Georgia. His research interests include terrestrial microbial ecology, dynamics of microbial processes at the landscape level, and transformation of environmental pollutants.

GEORGE R. HALLBERG is chief of environmental geology for the Iowa Department of Natural Resources. He earned his Ph.D. in geology from the University of Iowa. He has been actively involved in research of agricultural and nonpoint source pollution, soil genesis, hydrogeology, and groundwater quality. He directs a variety of environmental programs related to agriculture including the Big Spring Basin Demonstration Project and the Model Farms Demonstration Project.

NEIL D. HAMILTON is the Ellis and Nelle Levitt Distinguished Professor of Law, and the director of the Agricultural Law Center at Drake University Law School. He earned his B.S. from Iowa State University and his J.D. from the University of Iowa College of Law. He has been involved in research covering the legal aspects of agricultural finance, land tenure and land use, property rights and conservation and environmental law. His area of primary interest is agricultural law and agricultural law as it applies to agricultural and environmental policy and land use.

WILLIAM E. LARSON received his Ph.D. in 1949 from Iowa State University. Now retired, he was the head of the Department of Soil Sciences at the University of Minnesota. During his career he was involved in research of soil structure and mechanics, water infiltration, nutrient interrelations in plants, crop response to soil moisture levels and soil temperature, tillage requirements of crops, utilization of sewage wastes on land, and effects of erosion on productivity and fertility.

LINDA K. LEE is a visiting associate professor teaching agricultural economics at the University of Connecticut. Her Ph.D. is in economics from Iowa State University. Research areas in which she has worked include natural resources and environmental economics, agricultural impacts on ground- and surface water, soil conservation, and land use.

PETER J. NOWAK is an associate professor at the University of Wisconsin-Madison. He received his Ph.D. in rural sociology from the University of Minnesota. His research work involves the application of sociological models to agricultural problems, transfer of technology to the rural community, and the organizational and informational obstacles to adoption of conservation practices.

KENNETH G. RENARD is a hydraulic engineer at the Southwestern Watershed Research Center, Agricultural Research Service, U.S. Department of Agriculture, Tucson, Arizona. His Ph.D. is in hydrology from the University of Arizona. Focal points of his research have been watershed hydrology relating land practices to water yields and peak rates of discharge, and sediment transport phenomenon in ephemeral stream beds specifically as these phenomena relate to hydraulic engineering.

RICHARD E. ROMINGER was part owner and operator of A. H. Rominger and Sons, a family farm in Winters, California prior to being sworn in as Deputy Secretary, U.S. Department of Agriculture, on May 12, 1993. He received his B.S. in plant science from the University of California at Davis. From 1977 to 1982 he was Director of the California Department of Food and Agriculture and has been on the board of American Farmland Trust since 1986.

B. A. STEWART is laboratory director and research soil scientist at the Conservation and Production Research Laboratory, Agricultural Research Service, U.S. Department of Agriculture, Bushland, Texas. He received his Ph.D. in soils from Colorado State University. Research in which he has been involved includes soil fertility, nitrogen management nutrient cycling, effects of agricultural practices on environment, soil and water management, water use efficiency, and erosion control. His work in soil chemistry, nitrogen management, and water management has been of primary importance.

KENNETH K. TANJI is a professor of hydrologic science in the Department of Land, Air, and Water Resources, University of California, Davis. He previously served as chair of the Department of Land, Air, and Water Resources, director of the Kearney Foundation of Soil Science, assistant director of the agricultural experiment station, director of the University of California salinity/drainage task force, and coordinator of the USDA water quality program at the University of California. He earned his M.S. in soil science from the University of California. He teaches courses in chemical hydrology and modeling. His research program includes the chemistry of salt-affected soils and waters, salinity control and management, trace element chemistry, and hydrochemical modeling.

JAN VAN SCHILFGAARDE is associate deputy administrator for natural resources and systems of the Agricultural Research Service, U.S. Department of Agriculture. His Ph.D. in agricultural engineering and soil physics was earned from Iowa State University. His research has focused primarily on management of drainage water in crop production. He directs national research programs including soil management and erosion, water quality and management, and watershed engineering.

He was elected a member of the National Academy of Engineering in 1989.

R. J. WAGENET is Professor and Chair of the Department of Soil, Crop, and Atmospheric Sciences, Cornell University. He received his Ph.D. in soil science from the University of California at Davis. His areas of research work include simulation modeling of soil water and solutes, description of transient nitrogen and pesticide fluxes under field conditions, and the utilization and improvement of salt-affected soils and saline soils.

DOUGLAS L. YOUNG is a professor and agricultural economist at Washington State University. He earned his Ph.D. in agricultural economics from Oregon State University. Research in which he has been involved includes risk management, economics of soil conservation, economies of size in agriculture, economics of low-input agriculture, agricultural and environmental policy, agricultural development, and farming systems.

## About the Cartoonist

J. N. "DING" DARLING was an eminent political cartoonist whose career spanned the 1930s through the 1950s. Through cartoons, he brought the idea of conservation to the forefront of public awareness. He became chief of the organization that became the U.S. Fish and Wildlife Service, founded the National Wildlife Federation, and originated the Federal Duck Stamp Program, which has generated millions of dollars for wildlife refuges. After his death in 1962, the J. N. "Ding" Darling Foundation was created to continue his efforts in conservation education.

# INDEX

# Index

*(Page numbers in bold refer to figures)*

## O

# R

**Recent Publications of the Board on Agriculture**

*Policy and Resources*

Pesticides in the Diets of Infants and Children (1993), 408 pp., ISBN 0-309-04875-3.

Managing Global Genetic Resources: Livestock (1993), 294 pp., ISBN 0-309-04394-8.

Sustainable Agriculture and the Environment in the Humid Tropics (1993), 720 pp., ISBN 0-309-04749-8.

Agriculture and the Undergraduate: Proceedings (1992), 296 pp., ISBN 0-309-04682-3.

Water Transfers in the West: Efficiency, Equity, and the Environment (1992), 320 pp., ISBN 0-309-04528-2.

Managing Global Genetic Resources: Forest Trees (1991), 244 pp., ISBN 0-309-04034-5.

Managing Global Genetic Resources: The U.S. National Plant Germplasm System (1991), 198 pp., ISBN 0-309-04390-5.

Sustainable Agriculture Research and Education in the Field: A Proceedings (1991), 448 pp., ISBN 0-309-04578-9.

Toward Sustainability: A Plan for Collaborative Research on Agriculture and Natural Resource Management (1991), 164 pp., ISBN 0-309-04540-1.

Investing in Research: A Proposal to Strengthen the Agricultural, Food, and Environmental System (1989), 156 pp., ISBN 0-309-04127-9.

Alternative Agriculture (1989), 464 pp., ISBN 0-309-03985-1.

Understanding Agriculture: New Directions for Education (1988), 80 pp., ISBN 0-309-03936-3.

Designing Foods: Animal Product Options in the Marketplace (1988), 394 pp., ISBN 0-309-03798-0; ISBN 0-309-03795-6 (pbk).

Agricultural Biotechnology: Strategies for National Competitiveness (1987), 224 pp., ISBN 0-309-03745-X.

Regulating Pesticides in Food: The Delaney Paradox (1987), 288 pp., ISBN 0-309-03746-8.

Pesticide Resistance: Strategies and Tactics for Management (1986), 480 pp., ISBN 0-309-03627-5.

Pesticides and Groundwater Quality: Issues and Problems in Four States (1986), 136 pp., ISBN 0-309-03676-3.

Soil Conservation: Assessing the National Resources Inventory, Volume 1 (1986), 134 pp., ISBN 0-309-03649-9; Volume 2 (1986), 314 pp., ISBN 0-309-03675-5.

New Directions for Biosciences Research in Agriculture: High-Reward Opportunities (1985), 122 pp., ISBN 0-309-03542-2.

Genetic Engineering of Plants: Agricultural Research Opportunities and Policy Concerns (1984), 96 pp., ISBN 0-309-03434-5.

*Nutrient Requirements of Domestic Animals Series and Related Titles*

Nutrient Requirements of Horses, Fifth Revised Edition (1989), 128 pp., ISBN 0-309-03989-4; diskette included.

Nutrient Requirements of Dairy Cattle, Sixth Revised Edition, Update 1989 (1989), 168 pp., ISBN 0-309-03826-X; diskette included.

Nutrient Requirements of Swine, Ninth Revised Edition (1988), 96 pp., ISBN 0-309-03779-4.

Vitamin Tolerance of Animals (1987), 105 pp., ISBN 0-309-03728-X.

Predicting Feed Intake of Food-Producing Animals (1986), 95 pp., ISBN 0-309-03695-X.

Nutrient Requirements of Cats, Revised Edition (1986), 87 pp., ISBN 0-309-03682-8.

Nutrient Requirements of Dogs, Revised Edition (1985), 79 pp., ISBN 0-309-03496-5.

Nutrient Requirements of Sheep, Sixth Revised Edition (1985), 106 pp., ISBN 0-309-03596-1.

Nutrient Requirements of Beef Cattle, Sixth Revised Edition (1984), 90 pp., ISBN 0-309-03447-7.

Nutrient Requirements of Poultry, Eighth Revised Edition (1984), 71 pp., ISBN 0-309-03486-8.

Further information, additional titles (prior to 1984), and prices are available from the National Academy Press, 2101 Constitution Avenue, NW, Washington, DC 20418, 202/334-3313 (information only); 800/624-6242 (orders only); 202/334-2451 (fax).